Mechanisms and Machine Science

Volume 119

This book series establishes a well-defined forum for monographs, edited Books, and proceedings on mechanical engineering with particular emphasis on MMS (Mechanism and Machine Science). The final goal is the publication of research that shows the development of mechanical engineering and particularly MMS in all technical aspects, even in very recent assessments. Published works share an approach by which technical details and formulation are discussed, and discuss modern formalisms with the aim to circulate research and technical achievements for use in professional, research, academic, and teaching activities.

This technical approach is an essential characteristic of the series. By discussing technical details and formulations in terms of modern formalisms, the possibility is created not only to show technical developments but also to explain achievements for technical teaching and research activity today and for the future.

The book series is intended to collect technical views on developments of the broad field of MMS in a unique frame that can be seen in its totality as an Encyclopaedia of MMS but with the additional purpose of archiving and teaching MMS achievements. Therefore, the book series will be of use not only for researchers and teachers in Mechanical Engineering but also for professionals and students for their formation and future work.

The series is promoted under the auspices of International Federation for the Promotion of Mechanism and Machine Science (IFToMM).

Prospective authors and editors can contact Mr. Pierpaolo Riva (publishing editor, Springer) at: pierpaolo.riva@springer.com

Indexed by SCOPUS and Google Scholar.

Honghua Dai
Editor

Computational and Experimental Simulations in Engineering

Proceedings of ICCES 2022

Springer

Editor
Honghua Dai
School of Astronautics
Northwestern Polytechnical University
Xi'an, China

ISSN 2211-0984 ISSN 2211-0992 (electronic)
Mechanisms and Machine Science
ISBN 978-3-031-02096-4 ISBN 978-3-031-02097-1 (eBook)
https://doi.org/10.1007/978-3-031-02097-1

This Springer imprint is published by the registered company Springer Nature Switzerland AG
The registered company address is: Gewerbestrasse 11, 6330 Cham, Switzerland

Preface

This book gathers the latest advances, innovations, and applications in the field of computational engineering, as presented by leading international researchers and engineers at the 28th International Conference on Computational & Experimental Engineering and Sciences (ICCES). ICCES covers all aspects of applied sciences and engineering: theoretical, analytical, computational, and experimental studies and solutions to problems in the physical, chemical, biological, mechanical, electrical, and mathematical sciences. As such, the book discusses highly diverse topics, including Data-Driven Computational Modeling; Biomedical Engineering & Biomechanics; Sound & Vibration; Computational & Experimental Materials & Design; Engineering and Experimental Sciences; Modern Computational Methods; Modern Developments in Mechanics of Materials & Structures; multi-scale & multi-physics fluid engineering; structural integrity & longevity; materials design & simulation. The contributions, which were selected by means of a rigorous international peer-review process, highlight numerous exciting ideas that will spur novel research directions and foster multidisciplinary collaborations.

Xi'an, China

Honghua Dai

Contents

Role of Gluex in the Ion Exchange Mechanism of CLC^F F^-/H^+ Antiporter

Akihiro Nakamura, Takashi Tokumasu, and Takuya Mabuchi

Abstract In recent years, the construction of artificial cells using molecular robots have attracted much attention. In order to achieve selective transport of multiple ion species in artificial ion channels, it is essential to understand the mechanism of ion transport in antiporters and symporters of natural membrane proteins. The CLC^F F^-/H^+ Antiporter (CLC^F) has been attracting attention for its ability to specifically transport F^- as an antiporter. The CLC^F exchanges intracellular F^- with extracellular H^+ and releases F^- from the cell when bacteria's intracellular F^- concentration reaches the toxic concentration of 10–100 μM. On the other hand, it has been reported that certain drugs inhibit bacterial growth by regulating membrane transport proteins and decreasing ion transport function. Regulation of CLC^F is also expected to inhibit bacterial growth, andit is necessary to understand the ion exchange mechanism for the regulation of CLC^F. The structure of CLC^F is similar to that of CLC-ec1, which exchanges Cl^- and H^+. The CLC-ec1utilizes Gluex for ion exchange, and CLC^F also possesses Gluex and is thought to contribute to ion exchange. However, the role of Gluex in the ion exchange mechanism is still unclear. In this study, we performed MD simulations to investigate the role of Gluex in the ion exchange mechanism of the CLC^F protein by analyzing the relationship between Gluex and surrounding Gluex structures in the different states of Gluex.

Keywords Membrane transport protein · Ion exchange mechanism · Molecular dynamics simulation

A. Nakamura
Graduate School of Engineering, Tohoku University, Miyagi 980-8579, Japan

A. Nakamura · T. Tokumasu · T. Mabuchi (✉)
Institute of Fluid Science, Tohoku University, Miyagi 980-8577, Japan
e-mail: mabuchi@tohoku.ac.jp

T. Mabuchi
Frontier Research Institute for Interdisciplinary Sciences, Miyagi 980-8578, Japan

H. Dai (ed.), *Computational and Experimental Simulations in Engineering*,
Mechanisms and Machine Science 119,
https://doi.org/10.1007/978-3-031-02097-1_1

1 Introduction

Artificial cells are artificial systems that mimic cells with the characteristics of living cells, and are being developed for practical use. In creating artificial cells, molecular robots which use biomolecules and biochemical processes, are known to be very powerful tools [1]. Molecular robots can be designed at the molecular level, allowing for delicate control, and are attracting attention for applications such as drug delivery that mimic biological membranes [2]. In particular, the construction of artificial membrane nanopores with selective ion permeability for the control of ion absorption and efflux, which is necessary for maintaining cellular homeostasis, is highly important from the viewpoint of a bottom-up understanding of biological phenomena [3, 4]. However, it is difficult to control proton transport, which has a different transport mechanism from other ions, or to control multiple ion transport in a single nanopore at this stage. Since the control of these ion transport mechanisms requires atomic-level analysis, molecular simulations are effective. In particular, proton-specific transport phenomena including the Grotthuss mechanism [5] have been analyzed using ab initio MD simulation [6] and reactive MD simulation [7–12]. In order to realize proton and multiple ion transport in artificial transmembrane nanopores, it is important to understand the ion transport mechanisms in natural membrane proteins that have these specific transport mechanisms. As an example, CLC^F is known as an antiporter that specifically transports H^+ and F^- [13–15]. Many bacteria have been shown to release F^- from the cell when the intracellular F^- concentration reaches toxic concentrations of 10–100 μM, using CLCF as a membrane transport protein in the cell wall to exchange intracellular F^- for extracellular H^+ [16, 17]. CLC^F belongs to the CLC superfamily. CLC-ec1, which belongs to the same family, mainly transports Cl^- into the cell by exchanging it with intracellular H^+ [18, 19]. CLC-ec1 uses glutamate (Gluex) as a key residue in ion exchange. Gluex normally blocks the Chloride pathway, but it regulates ion exchange by a switching mechanism that opens the Chloride pathway temporarily when Gluex is protonated [20–23]. Although CLC^F also has Gluex, the peripheral structure of Gluex in CLC^F is different from that of CLC-ec1, and the mechanism for exchanging F^- and H^+ is also thought to be different from that of CLC-ec1 [15, 24]. Miller et al. proposed the Gluex windmill mechanism for F^- and H^+ ion exchange in CLC^F based on experimental structure and ion flux measurements [15]. However, with respect to the ion exchange mechanism, the effect of the protonation of Gluex on Gluex and its surrounding structure is not clear. In this study, in order to clarify the role of Gluex in the ion exchange mechanism of the CLC^F protein, we performed MD simulations to analyze Gluex and its surrounding structures with and without protonation, and investigated the effect of Gluex on the opening and closing of the fluoride pathway and proton transport (Fig. 1).

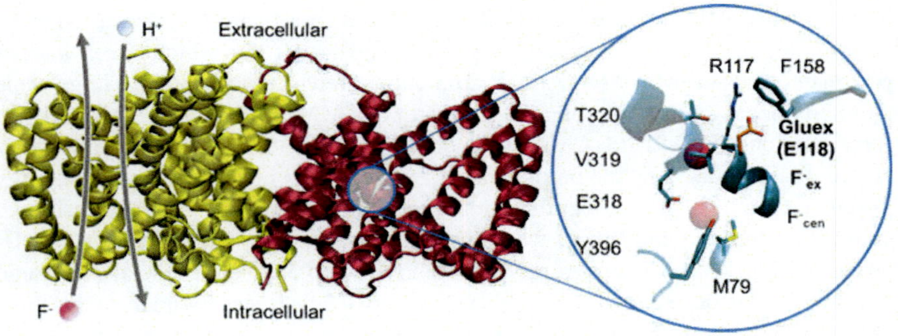

Fig. 1 Relationship between Gluex and F^- in CLC^F (PDBID: 6D0J)

2 Simulation Details

The system consists of CLC^F (PDB: 6D0J) that has two symmetrical subunits shown in Fig. 2, embedded into a 1-palmitoyl-2-oleoyl-sn-glycero-3phosphoethanolamine (POPE) bilayer using CHARMM-GUI, and solvated with water, 150 mM NaCl, resulting in a 130 Å × 90 Å × 84 Å box with ~100,000 atoms (CLC^F, POPE:272, Water molecule:15,264, F^-:2, Na^+:46, Cl^-:52) [25–27]. In this system, the Gluex is not protonated, and there is one F^- coordinated to each subunit. In addition,

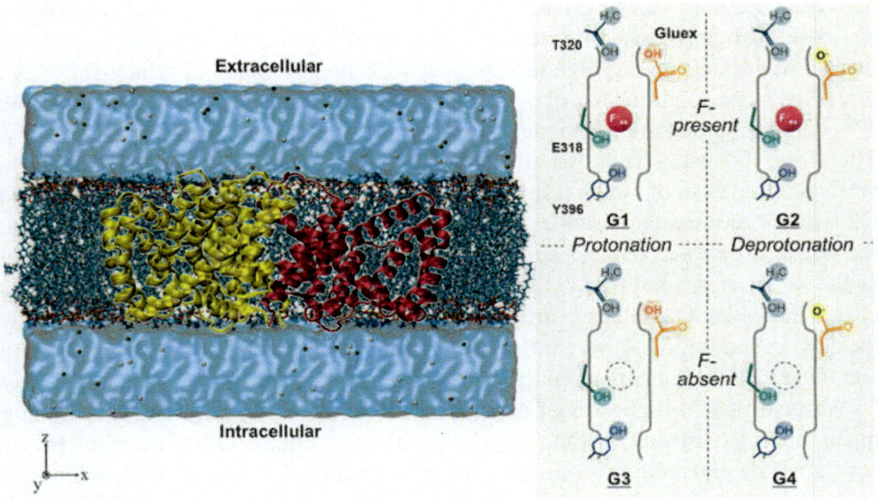

Fig. 2 The left is the calculation system. Protein (new cartoons), POPE lipid bilayer (licorice), Na^+ and Cl^- ions (gray and white spheres), and water molecules (quick surface). The right is four models G1-G4 that conditions of Gluex in MD simulations. G1:Protonation / $F^-{}_{ex}$ present, G2:Deprotonation/$F^-{}_{ex}$ present, G3:Protonation/$F^-{}_{ex}$ absent, G4:Deprotonation/$F^-{}_{ex}$ absent. Gluex (orange sticks) and $F^-{}_{ex}$ (magenta spheres)

in order to explore the effect of the Gluex on the coordination structure with the surrounding residues of Gluex, we constructed and a system in which Gluex was not protonated and absence of F^-. All the simulations were performed with a large-scale atomic/molecular massively parallel simulator (LAMMPS) and the CHARMM36m force field and CHARMM–CMAP was employed [28–30]. The electrostatic interaction was calculated by the particle mesh Ewald (PME) method [31]. After energy minimization 6 stages of equilibration (pre-equilibrium) considering positional constraints (protein and lipids) were conducted by following the CHARMM-GUI protocol [25]. Then, all constraints were removed, and NPT ensemble at 310 K and 1.0 atm was performed for 400 ns with a time step of 2 fs. It was found that additional 100 ns of equilibration is needed in addition to the pre-equilibrium from the results of the root mean square deviation (RMSD) of *Ca* atoms in CLC^F using MDanalysis [32, 33].

3 Results and Discussion

3.1 *Orientation of Gluex*

In order to investigate the role of Gluex in ion transport in CLC^F, the orientation of Gluex is important to understand the status of Gluex in each ion transport. Therefore, the orientation of Gluex in each condition was investigated based on the distance between OE2 o the carboxy group in Gluex and the surrounding residues of Gluex. The results of the vertical orientation of Gluex and Y396 are shown in Fig. 3a. Each dashed line shows the average distance between the backbone of Gluex and Y396 under the same conditions as in practice. Each dashed line shows a value of 11 Å $\pm$ 1 Å, confirming that the backbone of Gluex is not structurally changed significantly. Therefore, the results from the Gluex backbone were used as the reference for the vertical orientation of Gluex. Figure 3a shows that the values in conditions G1 and G2 with F^- are relatively stable, and the distance from the backbone indicates that the Gluex is oriented upward and outward from the cell. In the absence of F^- and in the protonated condition G3, the distance between Gluex and the backbone is closer, indicating that Gluex is oriented downward and can access the solution inside the cell. In condition G4, where the Gluex is not protonated without F^-, there is no significant difference compared to the backbone and the values are unstable.

We investigated the effect of protonation of Gluex on the opening and closing of the fluoride pathway. In the initial protonation of Gluex, Gluex acquires H^+ by accessing the extracellular solution, so we focused on G1 and G2 where Gluex is facing upward toward the extracellular solution. The target residue was T320, which has a hydroxy group that facilitates formation of the fluoride pathway and hydrogen bonding with Gluex. The results are shown in Fig. 3b. It can be confirmed that G1 and G2 have a distance of 8 Å and 5 Å, respectively, from Fig. 3b. The distance from the measurement point (C_β) of T320 to OE2 of the carboxy group of the same residue is

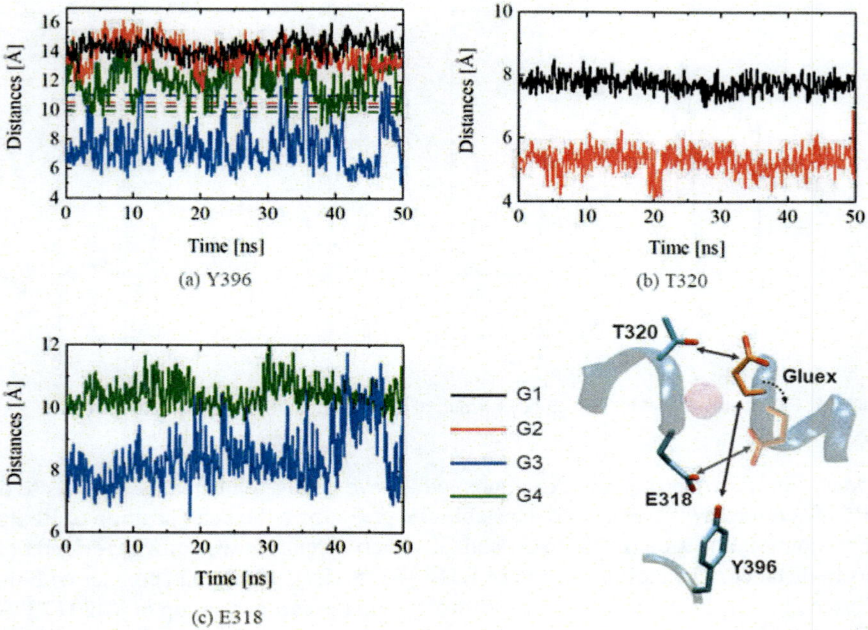

Fig. 3 Distances between Gluex and each residue

about 1.5 Å. Therefore, the distances between the carboxy groups of Gluex and T320 are 6.5 Å for G1 and 3.5 Å for G2. The distance of 3.5 Å between each carboxy group in G2 corresponds to the fluoride diameter, which is not sufficient for F^- to penetrate. Therefore, it is considered that the fluoride pathway is closed in deprotonated Gluex, while it is open in protonated Gluex. We will also investigate the effect of protonation of Gluex on the intracellular access of Gluex. For G3 and G4, which show orientations other than upward, we calculate the distance from Gluex to E318, which exists in the intracellular direction. Figure 3c shows that the distance between protonated and deprotonated G3 and G4 is farther from E318 in the deprotonated condition. It can be confirmed that the deprotonated condition G4 without F^- is coordinated between the protonated condition G3 without F- and the deprotonated condition G2 with F^-, together with the same condition in Fig. 3a. These results indicate that Gluex may have a rotational mechanism that is the windmill mechanism proposed by Miller et al.

3.2 Position of F^- in the Fluoride Pathway

The orientation of Gluex suggests that Gluex with F^- opens and closes the fluoride pathway under the influence of protonation. This suggests that T320 can hydrogen

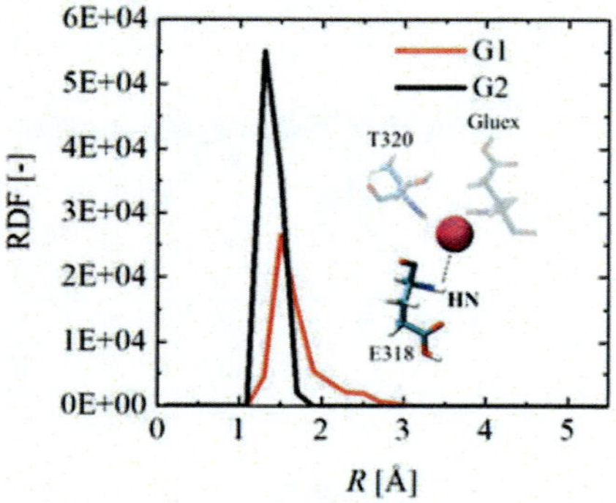

(a) E318

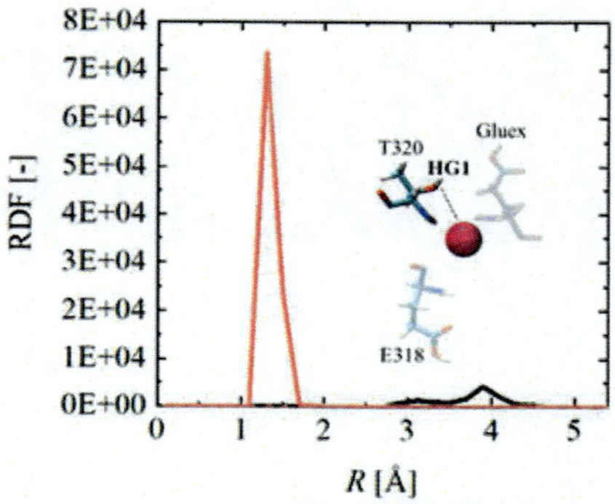

(b) T320

Fig. 4 RDF between the F^- and carboxy group of each residue

bond with F^- by protonation of Gluex, which we believe changes the stable site of F^-. Therefore, we investigated the stable site of F^- from the peak position of Radial Distribution Function (RDF) in G1 and G2, where Gluex showed upward orientation as in distance. The carboxy group of E318 (E318-HN), which is in contact with the intracellular solvent in the Fluoride pathway, and the carboxy group of T320 (T320-HG1), which is expected to be hydrogen-bonded by protonation, were selected for analysis. Figure 4a shows that the peaks of protonated G1 and deprotonated G2 are located around the distance $R = 1.5$ Å, where F^- can form hydrogen bonds. At the same time, G2 has a peak in a wide region. Figure 4b shows that the peak of G2 is at a distance of $R = 1.5$ Å, while the peak of G1 is almost invisible. It is clear that F^- is coordinated to E318-HN in the G1 condition and to T320-HG1 in the G2 condition. In order to understand the pore state of the fluoride pathway in each condition, the pore diameter was analyzed by HOLE, and the results are shown in Fig. 5 as a representative visualization image [34]. It was confirmed that the deprotonated Gluex blocked the fluoride pathway and the protonation opened the fluoride pathway, allowing F^- to migrate to the outside of the cell, from Fig. 5. These results suggest that Gluex regulates the opening and closing of the Fluoride pathway upon protonation, thereby enabling the transport of F^- to the extracellular side.

3.3 Hydrogen Bonding at Gluex with Water in Solvents

The effect of the up and down orientations of protonated Gluex on proton transport was investigated for hydrogen bonds between gluex and water molecules around Gluex at G1 and G3, the conditions under which Gluex was protonated. The results are shown in Fig. 6, where Z = 0 Å is *Ca* in Gluex, and the extracellular direction is H^+ and the intracellular direction is F^-. The position of OE2 in Gluex (Z ≈

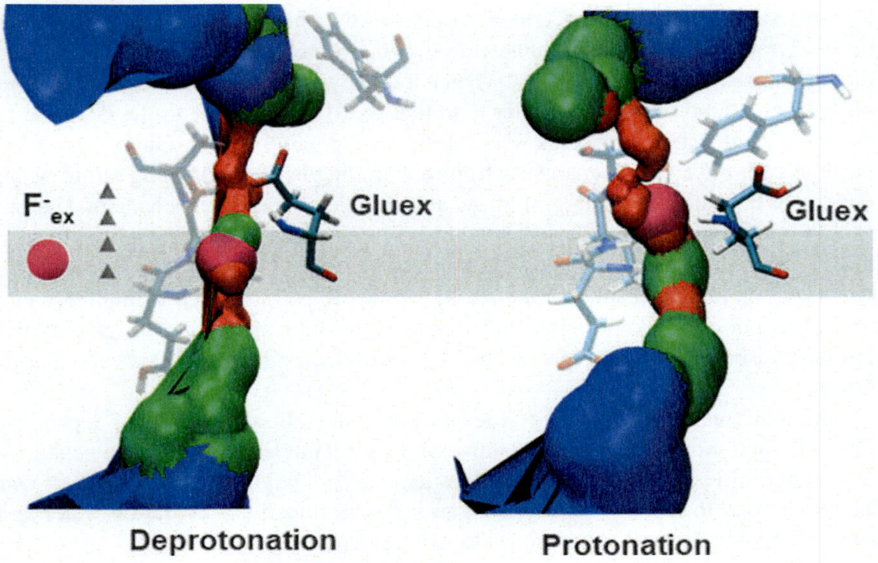

Fig. 5 Comparison of pore radius profiles of the crystal structure (black) and those of MD simulations (averaged over the whole trajectory) with different anions bound (colored traces)

Fig. 6 Number of hydration around the protonates Gluex

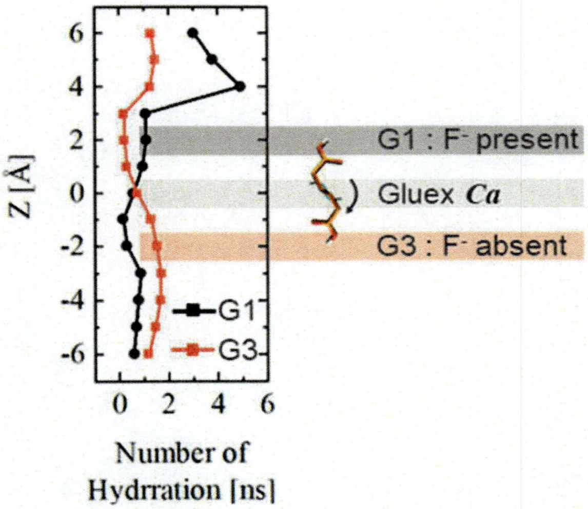

± 2 Å) under each condition is shown on each sheet, because Gluex coordinates upward and downward depending on the presence or absence of F^-. In state G1, where Gluex is upward, there is a greater number of hydrogen bonds at a distance of +4 Å from the extracellular direction, suggesting that the protonated Gluex may access the extracellular solvent and transfer H^+ when it is upward. The number of hydrogen bonds in G3, where Gluex is downward, is higher than in G1 around 4 Å in the intracellular direction, suggesting that the protonated Gluex may access the intracellular solvent and transfer H^+ when it is downward. The reason why the number of hydrogen bonds in G3 is smaller than that in G1 may be due to the decrease in the number of water molecules near the Gluex by Y396. When Gluex passes H^+ into the cell, different mechanisms from our findingds of transferring H^+ to water molecules have also been suggested. As an example of another mechanism, Carloni et al. proposed that Gluex temporarily holds F^-cen with H^+ and the carboxyl group of E318, and then transports H^+ in the form of hydrogen fluoride (HF) bound to F^-. However, the predominant mechanism by which H^+ is transferred from Gluex into the cell is not clear. For a better understanding of the extracellular to intracellular proton transport by Gluex, further investigation of the predominant mechanism is needed.

The above results indicate that Gluex may have the following functions depending on the F^- and protonation state, as shown in Fig. 7. Gluex has a rotating mechanism like a windmill and opens and closes the fluoride pathway upon protonation. It was also suggested that protonated Gluex can access intracellular hydration water and pass H^+ through several mechanisms to help transport H^+ into the cell.

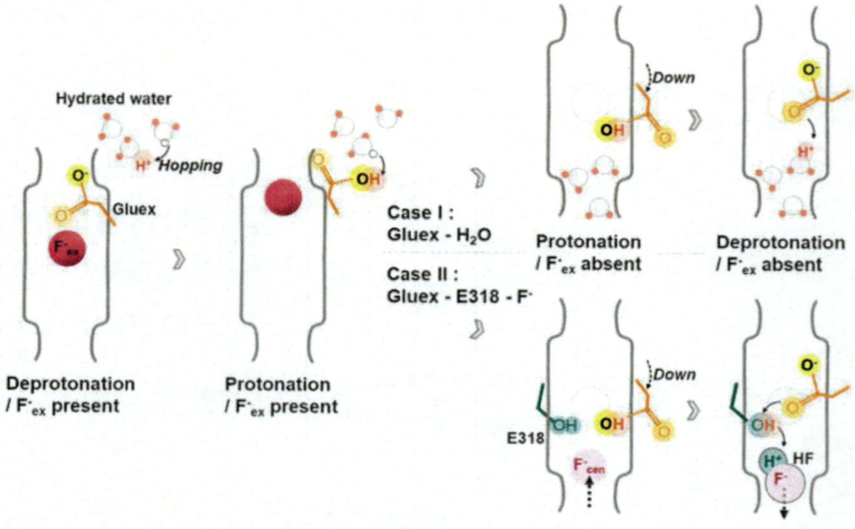

Fig. 7 Relationship between Gluex windmill model and Fluoride

4 Conclusions

MD simulations were performed to investigate the role of Gluex in ion exchange in CLC^F. Our findings suggested that Gluex is responsible for opening and closing the fluoride pathway and proton transport by utilizing the rotation mechanism, supporting the Gluex windmill that is proposed based on experimental measurements. However, it is still unclear how Gluex passes H^+ into the cell, which is associated with the Grotthuss mehanism. Therefore, the mechanism needs to be further investigated. In the future, we plan to investigate the details of proton transport mechanism using reactive MD methods such as the Reax FF.

References

1. Shoji, K., Kawano, R.: Recent advances in liposome-based molecular robots. Micromachines. **11**, 1–20 (2020). https://doi.org/10.3390/MI11090788
2. Chao, J., Liu, H., Su, S., Wang, L., Huang, W., Fan, C.: Structural DNA nanotechnology for intelligent drug delivery. Small **10**, 4626–4635 (2014). https://doi.org/10.1002/smll.201401309
3. Zheng, S.P., Huang, L.B., Sun, Z., Barboiu, M.: Self-assembled artificial ion-channels toward natural selection of functions. Angewandte Chemie—Int. Edition. **60**, 566–597 (2021). https://doi.org/10.1002/anie.201915287
4. Zhang, Z., Huang, X., Qian, Y., Chen, W., Wen, L., Jiang, L.: Engineering smart nanofluidic systems for artificial ion channels and ion pumps: from single-pore to multichannel membranes. Adv. Mater. **32**, 1–15 (2020). https://doi.org/10.1002/adma.201904351
5. Agmon, N.: The Grotthuss mechanism. Chem. Phys. Lett. **244**, 456–462 (1995). https://doi.org/10.1016/0009-2614(95)00905-J
6. Tuckerman, M.E., Chandra, A., Marx, D.: A statistical mechanical theory of proton transport kinetics in hydrogen-bonded networks based on population correlation functions with applications to acids and bases. J. Chem. Phys **133** (2010). https://doi.org/10.1063/1.3474625
7. Ma, X., Li, C., Martinson, A.B.F., Voth, G.A.: Water-assisted proton transport in confined nanochannels. J. Phys. Chem. C **124**, 16186–16201 (2020). https://doi.org/10.1021/acs.jpcc.0c04493
8. Mabuchi, T., Fukushima, A., Tokumasu, T.: A modified two-state empirical valence bond model for proton transport in aqueous solutions. J. Chem. Phys. **143** (2015). https://doi.org/10.1063/1.4926394
9. Mabuchi, T., Tokumasu, T.: Effects of water nanochannel diameter on proton transport in proton-exchange membranes. J. Polym. Sci., Part B: Polym. Phys. **57**, 867–878 (2019). https://doi.org/10.1002/polb.24842
10. Mabuchi, T., Tokumasu, T.: Relationship between proton transport and morphology of perfluorosulfonic acid membranes: a reactive molecular dynamics approach. J. Phys. Chem. B **122**, 5922–5932 (2018). https://doi.org/10.1021/acs.jpcb.8b02318
11. Chen, H., Wu, Y., Voth, G.A.: Proton transport behavior through the influenza a M2 channel: insights from molecular simulation, (n.d.). https://doi.org/10.1529/biophysj.107.105742
12. Lee, S., Swanson, J.M.J., Voth, G.A.: Multiscale simulations reveal key aspects of the proton transport mechanism in the ClC-ec1 antiporter. Biophys. J. **110**, 1334–1345 (2016). https://doi.org/10.1016/j.bpj.2016.02.014
13. Stockbridge, R.B., Kolmakova-Partensky, L., Shane, T., Koide, A., Koide, S., Miller, C., Newstead, S.: Crystal structures of a double-barrelled fluoride ion channel. Nature **525**, 548–551 (2015). https://doi.org/10.1038/nature14981

14. Brammer, A.E., Stockbridge, R.B., Miller, C.: F-/Cl- selectivity in CLCF-type F-/H+ antiporters. J. Gen. Physiol. **144**, 129–136 (2014). https://doi.org/10.1085/jgp.201411225
15. Last, N.B., Stockbridge, R.B., Wilson, A.E., Shane, T., Kolmakova-Partensky, L., Koide, A., Koide, S., Miller, C.: A clc-type f - /h + antiporter in ion-swapped conformations. Nature Struct. Mol. Biol. **25** (2018). https://doi.org/10.1038/s41594-018-0082-0
16. Stockbridge, R.B., Lim, H.H., Otten, R., Williams, C., Shane, T., Weinberg, Z., Miller, C.: Fluoride resistance and transport by riboswitch-controlled CLC antiporters. Proc. Natl. Acad. Sci. U.S.A. **109**, 15289–15294 (2012). https://doi.org/10.1073/pnas.1210896109
17. Baker, J.L., Sudarsan, N., Weinberg, Z., Roth, A., Stockbridge, R.B., Breaker, R.R.: Widespread genetic switches and toxicity resistance proteins for fluoride. Science **335**, 233–235 (2012). https://doi.org/10.1126/science.1215063
18. Maduke, M., Miller, C., Mindell, J.A.: A Decade of CLC chloride channels: structure, mechanism, and many unsettled questions. Annu. Rev. Biophys. Biomol. Struct. **29**, 411–438 (2000). https://doi.org/10.1146/annurev.biophys.29.1.411
19. Dutzler, R.: The ClC family of chloride channels and transporters. Curr. Opin. Struct. Biol. **16**, 439–446 (2006). https://doi.org/10.1016/j.sbi.2006.06.002
20. Jentsch, T.J., Pusch, M.: CLC chloride channels and transporters: structure, function, physiology, and disease. Physiol. Rev. **98**, 1493–1590 (2018). https://doi.org/10.1152/physrev.00047.2017
21. Feng, L., Campbell, E.B., MacKinnon, R.: Molecular mechanism of proton transport in CLC Cl-/H+ exchange transporters. Proc. Natl. Acad. Sci. U.S.A. **109**, 11699–11704 (2012). https://doi.org/10.1073/pnas.1205764109
22. Chavan, T.S., Cheng, R.C., Jiang, T., Mathews, I.I., Stein, R.A., Koehl, A., McHaourab, H.S., Tajkhorshid, E., Maduke, M.: A CLC-EC1 mutant reveals global conformational change and suggests a unifying mechanism for the CLC CL– /H+ transport cycle. Elife **9**, 1–30 (2020). https://doi.org/10.7554/eLife.53479
23. Miller, C.: ClC chloride channels viewed through a transporter lens. Nature **440**, 484–489 (2006). https://doi.org/10.1038/nature04713
24. Dutzler, R., Campbell, E.B., Cadene, M., Chait, B.T., MacKinnon, R.: X-ray structure of a CIC chloride channel at 3.0 Å reveals the molecular basis of anion selectivity. Nature **415**, 287–294 (2002). https://doi.org/10.1038/415287a
25. Jo, S., Kim, T., Iyer, V.G., Im, W.: CHARMM-GUI: a web-based graphical user interface for CHARMM. J. Comput. Chem. **29**, 1859–1865 (2008). https://doi.org/10.1002/jcc.20945
26. Jo, S., Lim, J.B., Klauda, J.B., Im, W.: CHARMM-GUI membrane builder for mixed bilayers and its application to yeast membranes. Biophys. J . **97**, 50–58 (2009). https://doi.org/10.1016/j.bpj.2009.04.013
27. Lee, J., Cheng, X., Swails, J.M., Yeom, M.S., Eastman, P.K., Lemkul, J.A., Wei, S., Buckner, J., Jeong, J.C., Qi, Y., Jo, S., Pande, V.S., Case, D.A., Brooks, C.L., MacKerell, A.D., Klauda, J.B., Im, W.: CHARMM-GUI input generator for NAMD, GROMACS, AMBER, OpenMM, and CHARMM/OpenMM simulations using the CHARMM36 additive force field. J. Chem. Theory Comput. **12**, 405–413 (2016). https://doi.org/10.1021/acs.jctc.5b00935
28. Plimpton, S.: Fast parallel algorithms for short-range molecular dynamics. J. Comput. Phys. **117**, 1–19 (1995). https://doi.org/10.1006/jcph.1995.1039
29. Klauda, J.B., Venable, R.M., Freites, J.A., O'Connor, J.W., Tobias, D.J., Mondragon-Ramirez, C., Vorobyov, I., MacKerell, A.D., Pastor, R.W.: Update of the CHARMM all-atom additive force field for lipids: validation on six lipid types. J. Phys. Chem. B **114**, 7830–7843 (2010). https://doi.org/10.1021/jp101759q
30. Sarmoria, C., Miller, D.R.: Spanning-tree models for A(f) homopolymerizations with intramolecular reactions. Comput. Theor. Polym. Sci. **11**, 113–127 (2001). https://doi.org/10.1016/S1089-3156(99)00088-4
31. Chughtai, B., Jarvis, T., Kaplan, S.: Testosterone and benign prostatic hyperplasia. Asian J. Androl. **17**, 212 (2015). https://doi.org/10.4103/1008-682X.140966
32. Theobald, D.L.: Rapid calculation of RMSDs using a quaternion-based characteristic polynomial. Acta Crystallogr. A **61**, 478–480 (2005). https://doi.org/10.1107/S0108767305015266

33. Liu, P., Agrafiotis, D.K., Theobald, D.L.: Fast determination of the optimal rotational matrix for macromolecular superpositions. J. Comput. Chem. **31**, 1561–1563 (2010). https://doi.org/10.1002/jcc.21439
34. Smart, O.S., Neduvelil, J.G., Wang, X., Wallace, B.A., Sansom, M.S.P.: HOLE: a program for the analysis of the pore dimensions of ion channel structural models. J. Mol. Graph. **14**, 354–360 (1996). https://doi.org/10.1016/S0263-7855(97)00009-X

Influence of the Train Speed on the Long Term Performance of the Subgrade of the Ballasted and Ballastless Tracks

Ana Ramos, António Gomes Correia, and Rui Calçada

Abstract The long term performance of the railway tracks is a relevant topic and a major concern of the railway Infrastructure Managers due to the technical and economical aspects related to the degradation of the track. Thus, there are several factors that have a significant influence on the short and long term performance of the ballasted and ballastless tracks. One of these factors is the train speed (which also includes the critical speed) since the stresses and strains are significantly amplified. This work tries to compare the short but mostly the long term performance of the ballasted and the ballastless track (*Rheda system*) and also an optimised version of the *Rheda system*, where the support layers (HBL and FPL) are omitted. This comparison includes the influence of the train speed (and critical speed) and it is based on the stress levels and cumulative permanent deformation induced by the passage of the train. The results were obtained using a hybrid methodology between the 2.5D FEM-PML (to model the short term behaviour) and the implementation of an empirical permanent deformation model (to analyse the long term performance).

Keywords Ballasted track · Ballastless track · Train speed · Critical speed long-term performance

1 Introduction

The short and long term performance of the railway structures in general and the subgrade, in particular, have an important influence on the track maintenance operations and respective costs [1, 2]. In order to reduce the maintenance procedures and increase the robustness and efficiency of the railway structures, it is important to fully understand the short-term but also the long-term behaviour of the subgrade

A. Ramos (✉) · A. G. Correia
Department of Civil Engineering, University of Minho, ISISE, Guimarães, Portugal
e-mail: id6629@alunos.uminho.pt

R. Calçada
CONSTRUCT – LESE, Faculty of Engineering, University of Porto, Porto, Portugal

H. Dai (ed.), *Computational and Experimental Simulations in Engineering*, Mechanisms and Machine Science 119,
https://doi.org/10.1007/978-3-031-02097-1_2

when submitted to the cyclic loads such as the passage of the train. The short-term behaviour is characterised by the resilient modulus (M_r) and the long-term behaviour is characterised by the permanent deformation (ε_p). Using these two concepts, the subgrade can be characterised through the laboratory tests such as the cyclic triaxial tests. Where the samples are submitted to cyclic loads [3–13]. However, it is important to have an integrated approach where the whole performance of the subgrade is analysed in the track environment considering the passage of the vehicle.

This work aims to compare the performance of the conventional ballasted track, *Rheda* system ballastless track, and an optimised ballastless track only constituted by the concrete slab. This optimised ballastless track allows understanding the importance of the support layers in the response of the track. The obtained conclusions can be used as guidelines for possible optimisations. Furthermore, the influence of the train speed in the short but mostly in the long-term performance of the subgrade is analysed in detail. This study also includes the analysis of the influence of the critical speed since the strong amplification of the response can significantly increase the track degradation [14–17].

The analysis is focused on the subgrade layer. The stresses induced by the passage of the train are obtained using the 2.5D FEM-PML approach. This methodology is articulated with the implementation of an empirical permanent deformation model that uses these stress results and the number of loading cycles as the main inputs to determine and compare the permanent deformation and respective cumulative permanent deformation.

2 Railway Track Modelling: Train-Track-Ground System

Sub-structured models can be used to model the train-track systems and the interaction between both. This approach allows simplifying the modelling since both structures are modelled separately keeping the compatibility between both and respecting the equilibrium restrictions [18].

In this work, the track-ground system is modelled using the 2.5D FEM-PML approach. With this methodology, it is possible to reduce the computational effort and consider the 3D nature of the problem [18–20]. The methodology used in this work is described in more detail in the work developed by [18].

Due to the 3D and transient characters of the problem, *Perfectly Matched Layers* (PML's) were applied on the boundaries. This special treatment allows avoiding spurious reflections. This methodology was previously implemented with satisfactory results [21]. In this method, an external layer of the interest domain is implemented. This "special" layer absorbs the energy of the waves that impinge the boundaries. This methodology is described in more detail in the work developed by [21].

Thus, the track-ground structure is modelled by the 2.5D-FEM-PML approach and the train is modelled considering a multi-body formulation. The models are coupled numerically following a compliance formulation. The interaction problem implies the compatibility of displacements and load equilibrium at the contact points between the rolling stock and the track. The wheel-rail contact is simulated applying the concept of the linearized *Hertzian* stiffness (this only considers the dead load transmitted by the wheelset and only the vertical movement of the train is taken into account). More details about this methodology and formulation can be found in [22].

The 2.5D FEM-PML methodology allows obtaining the stress levels induced by the passage of the train that are posteriorly used as inputs to simulate the long-term behaviour of the subgrade of the railway structures.

3 Long Term Behaviour: Subgrade Modelling

The selected empirical permanent deformation model to study the long-term performance of the subgrade is based on the work developed by [23]. This model combines the effect of the elastic stress state in the soil with the number of load cycles. Posteriorly, [24] updated the model, suggesting some modifications as the introduction of influence of the initial stress state:

$$\varepsilon_1^p(N) = \varepsilon_1^{p0}\left[1 - e^{-BN}\right]\left(\frac{\sqrt{{p_{am}}^2 + {q_{am}}^2}}{p_a}\right)^a \cdot \frac{1}{m\left(1 + \frac{p_{ini}}{p_{am}}\right) + \frac{s}{p_{am}} - \frac{(q_{ini} + q_{am})}{p_{am}}} \quad (1)$$

where p_{am} and q_{am} are the amplitude of the mean stress and deviator stress induced by the train loadings, m and s are defined by the *yielding* criterion $q = s + mp$; and p_{ini} and q_{ini} are the mean and deviator stresses considering the initial stress state.

The model presents several advantages since it considers the amplitude of the applied load, the proximity of the stress path to the failure line and integrates the influence of the initial stress state. Comparing to other empirical models [25, 26], this model is more complex because it depends on more parameters but. at the same time, it is very easy to implement.

The material parameters used in this study were obtained in [11] considering $\varepsilon_1^{p0} = 0.00093$, $B = 0.2$ and $a = 0.65$. According to the unified soil classification, this material is classified as non-plastic silty sand.

Thus, the model developed by [24] was implemented to study the long-term performance of the railway structures through the determination of the cumulative permanent deformation. The model is applied to the subgrade material (including the FPL—frost protection layer in the case of the ballastless track) and the analysis does not include the ballast. The cumulative permanent deformation (δ) is the sum of the product of the permanent deformation (applying expression 1) of each element of the selected alignment, considering $N = 1{,}000{,}000$ cycles and the thickness of the element:

$$\delta = \sum_{i=1}^{n} \varepsilon_p^i \times h_i \quad (2)$$

where i represent the number of elements of the alignment, h is the thickness of the element (in m) and ε_p is the permanent deformation (dimensionless) and it is obtained through expression 1. The cumulative permanent deformation is maximum at the bottom of the model since corresponds to the sum of the permanent deformation of all elements. However, the elements that most contribute are close to the surface [27].

4 Case Study

In this analysis, the long-term performance of the subgrade is analysed and compared considering three different railway structures: the ballasted track, ballastless track (*Rheda* system), and a special optimised ballastless track. The numerical models of each structure are presented in Fig. 1.

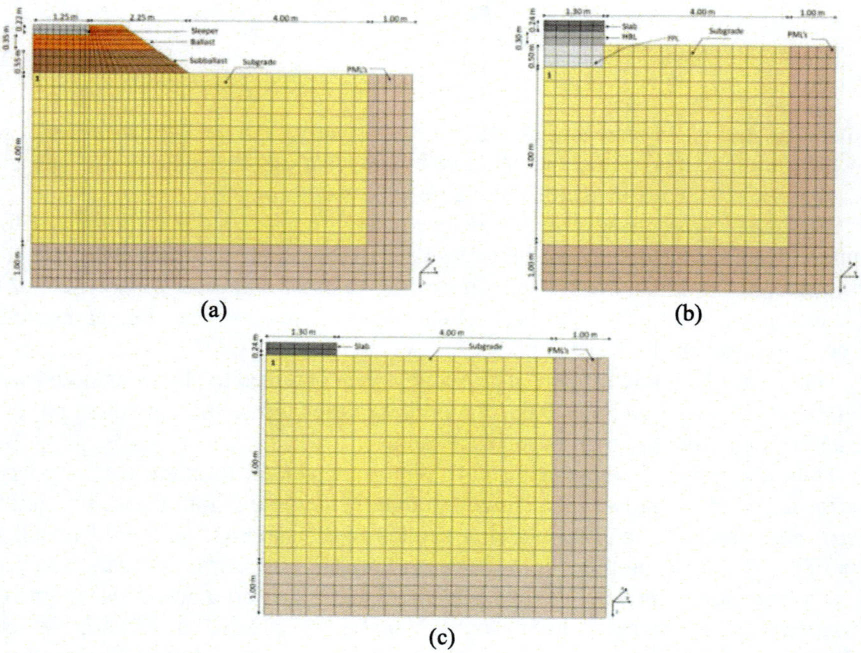

Fig. 1 Numerical models: **a** ballasted track; **b** ballastless track (Rheda system); **c** ballastless track with only a concrete slab (Adapted from [30])

Table 1 Characteristics of the materials—ballasted track

Elements	M_r (MPa)	υ	ξ	ρ (kg/m^3)
Sleepers (ballasted track)	30,000	0.20	0.010	1833.3
Ballast	97	0.12	0.061	1591.0
Sub-ballast	212	0.30	0.054	1913.0
Subgrade[a]	120	0.30	0.030	2040.0

M_r = Resilient modulus, υ = Poisson's ratio, ξ damping and ρ density
[a]The subgrade is characterised by a $\phi' = 30°$ and $c' = 0$ kPa

Table 2 Characteristics of the materials—ballastless tracks

Elements	M_r (MPa)	υ	ξ	ρ (kg/m^3)
Concrete slab	34,000	0.20	0.030	2500
HBL	10,000	0.20	0.030	2500
FPL[a]	120	0.20	0.030	2500
Subgrade[a]	120	0.30	0.030	2040.0

M_r = Resilient modulus, υ = Poisson's ratio, ξ damping and ρ density
[a]The subgrade and FPL are characterised by a $\phi' = 30°$ and $c' = 0$ kPa

The comparison of the performance is carried out under 6 different train speeds: 80, 144, 200, 300, 360 and 500 km/h. The properties of the materials are described in Tables 1 and 2.

Regarding the numerical modelling, the railway structures were modelled by finite elements with 8 nodes. The ballasted track is composed of the rails, railpads, sleepers, ballast, sub-ballast and subgrade. The ballastless track is composed of the rails, railpads, concrete slab, a support layer (hydraulically bound layer—HBL), and also the frost protection layer (FPL). The optimised ballastless track is similar to the *Rheda* system but, in this case, the support layers (HBL and FPL) were omitted. The rails correspond to the UIC60 model and the railpads were modelled with stiffness and damping of 600 kN/mm and 22.5 kNs/m, and 40 kN/mm and 8 kNs/m in the case of the ballasted track and ballastless track, respectively. The remaining materials were modelled as linear elastic. This assumption is valid since we are dealing with small strains. The sleeper was modelled as a continuous and orthotropic element. This means that, in the longitudinal direction, the stiffness of the ballast is adopted [18].

The initial stresses were obtained considering an isotropic stress state ($K_O = 1$) and, as depicted in Fig. 1, the modelling takes into account the symmetric conditions of the problem.

In the ballastless track (*Rheda* system), to simplify the modelling and the analysis of the problem, the FPL was included in the subgrade layer since both materials share the same stiffness. Indeed, they are very similar, and the main difference is related to the density. This simplification able us to compare and discuss the numerical results

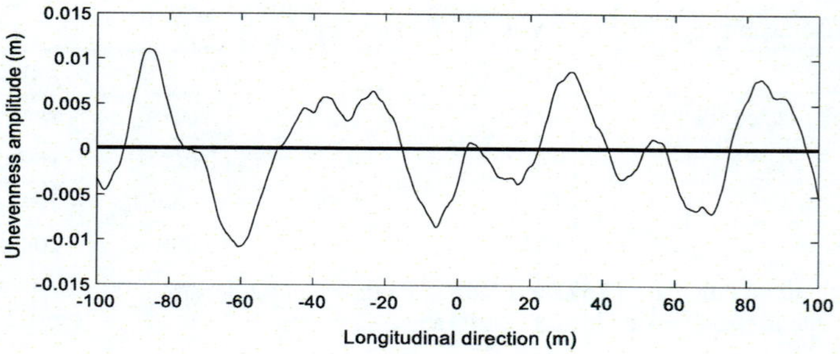

Fig. 2 Unevenness profile

of the finite elements from the different railway structures at the same depth, which means that they would have an analogous initial stress state [27].

The adopted vehicle is the *Alfa Pendular* train. This train presents a symmetry plane and is composed of 6 car bodies. Regarding the loading, the average axle load is close to 135 kN. In this case, only the mass of the wheelsets was considered in the model. This assumption is aligned with the studies performed previously by [28, 29] in the scope of the ground vibrations.

4.1 Generation of the irregularity's Profile

In order to simulate and consider the dynamic mechanism, an unevenness profile was artificially generated. This profile was defined by a sinusoidal function described by a number of harmonics. In this process, the *PSD* (Power Spectral Density) function defined by the *FRA* (*Federal Railroad Administration*) was used. More details about this methodology can be found in [27].

The generated profile is depicted in Fig. 2, it is composed of 120 frequencies and it was defined considering a geometric progression. This profile was determined following the recommendations described in [31].

This study is focused on element 1 (Fig. 1) located at x = 0 m (Fig. 2).

4.2 Short-Term Performance

Firstly, to understand the amplification phenomenon of the stresses (as well as the permanent deformations) for higher speeds, two train speeds closer to the critical speed were evaluated (360 and 500 km/h). The critical speed of each system was determined through the dispersion curves, which is an approximation method but very expeditiously and efficient [32]. Indeed, [14] show that, when the ground is homogeneous, the characteristics of the track can be neglected since the critical speed is almost the same, regardless of the type of structure. In the case of this study, the ballasted track and ballastless track are characterised by a critical speed of 515 km/h and 540 km/h, respectively. The critical speeds were determined based on the results depicted in Fig. 3. More details about this methodology can be found in [32]. Dispersion curves associated with more rigid tracks are curves associated with lower frequencies than tracks with lower rigidity. Thus, it is tempting to think then that the higher the stiffness of the track, the higher the critical speed of the system. But this may not be true, since the ground dispersion curve will also influence the point of intersection, and it is known that the critical speed of the systems is usually conditioned by the characteristics of the ground. The results depicted in Fig. 3 show that, since the ground is homogeneous, to determine the critical speed of the structure, it was only necessary to obtain the slope of the red line because the inverse of the slope is equal to the critical speed.

From experimental results (mostly in the ballasted track) documented in the bibliography ([33] based on the work developed by [34]), it is possible to state that when the train's speed exceeds 75% of the critical speed, the amplitude of the dynamic response of the track increases rapidly. Indeed, 75% of critical speed can be assumed

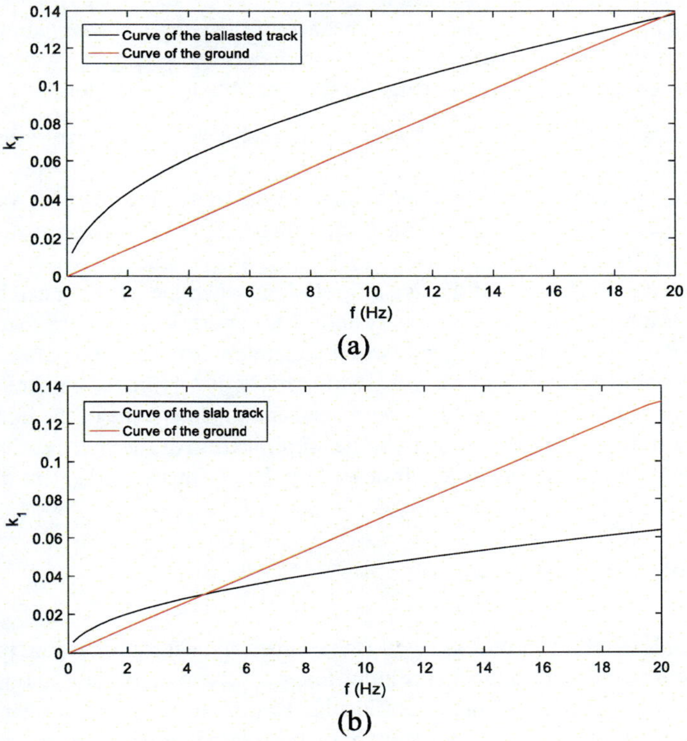

Fig. 3 Dispersion curves: **a** ballasted track; **b** ballastless tracks

as the practical speed limit of the ballasted railway tracks. This means that the value of 360 km/h is perfectly acceptable as a practical speed limit [27]. The superior value of the critical speed of the ballastless track can be justified by the stiffness of the track.

Therefore, to assess the influence of the nearing of the critical speed, a higher train's speed than 360 km/h was considered: 500 km/h. This value is higher than 360 km/h and lower than 515 or 540 km/h. Indeed, it is possible to conclude that this speed is unrealistic, but its results are presented to show the ability of the method to capture the behaviour of the structure when submitted to this range of speeds.

In order to compare the short-term performance, the stress paths were determined for each railway structure (Fig. 4). Considering the previous information about the critical speed, the trains' speed equal to 500 km/h was only considered for the ballastless track (*Rheda* system). In the other railway structures, the stress paths go beyond the *yielding criterion*, which is not a realistic situation since it is expected to have stresses' distribution. Thus, for the ballasted track, and considering this particular stiffness of the subgrade and FPL ($M_r = 120$ MPa), the train´s speed of 500 km/h is too much close to the critical speed (515 km/h), which leads to significant amplification of the stresses. This situation does not occur in the ballastless track (where the critical speed of the system is 540 km/h). In the case of the ballastless track only with concrete slab, the results show that the absence of the support layers can influence the stress results when the speeds are closer to the critical speed.

Analysing the obtained results (Fig. 4), it is possible to conclude:

- The train's speed is one of the factors that most influence the stress levels and stress paths on the subgrade;
- For reduced train speeds, the amplification of the dynamic stresses is very small or residual; With the increase of the train's speed, there is an amplification of the stress levels;
- Significant amplification of the dynamic stresses when the train's speed is getting closer to the critical speed of the system (360 and 500 km/h). In fact, for the ballasted and ballastless track (only with a concrete slab), the stress path of element 1 goes beyond the *yielding criterion.* In the case of the optimised ballastless track, this is due to the amplitude of the stress level and the low value of the initial mean stresses. Indeed, from v = 360 km/h, the amplification of the stress paths is more prominent in the ballasted track, followed by the optimised ballastless track.

4.3 Long-Term Performance

The amplitude of the stress paths as well as the values of the initial stress are important variables and are used as inputs in the empirical permanent deformation model. This means that the stress results presented in Fig. 4 have an impact on the cumulative permanent deformation and, consequently, on the degradation of the railway tracks.

The results of the cumulative permanent deformation are depicted in Fig. 5. Regarding the cumulative permanent deformations, the ballastless track (only with a

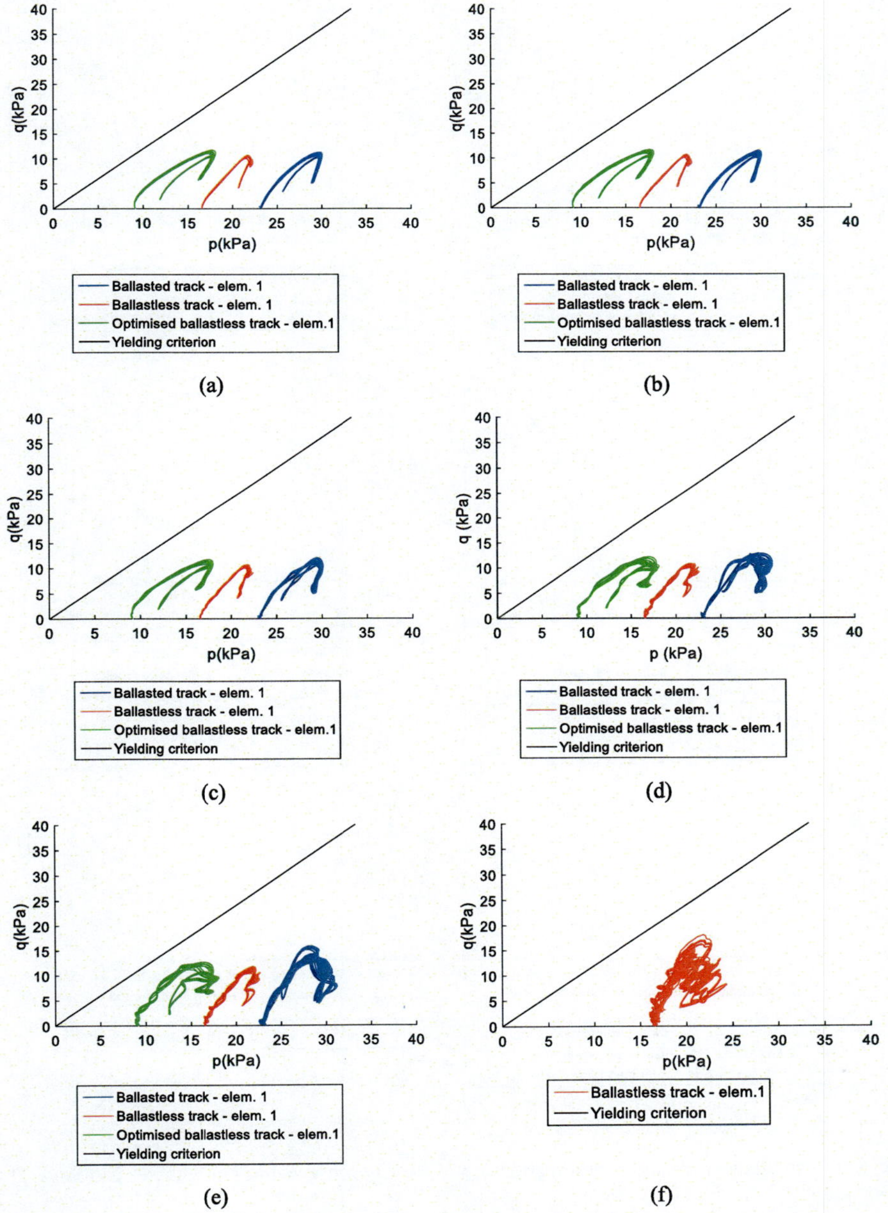

Fig. 4 Stress path: **a** v = 80 km/h; **b** v = 144 km/h; **c** v = 200 km/h; **d** v = 300 km/h; **e** v = 360 km/h, **f** v = 500 km/h

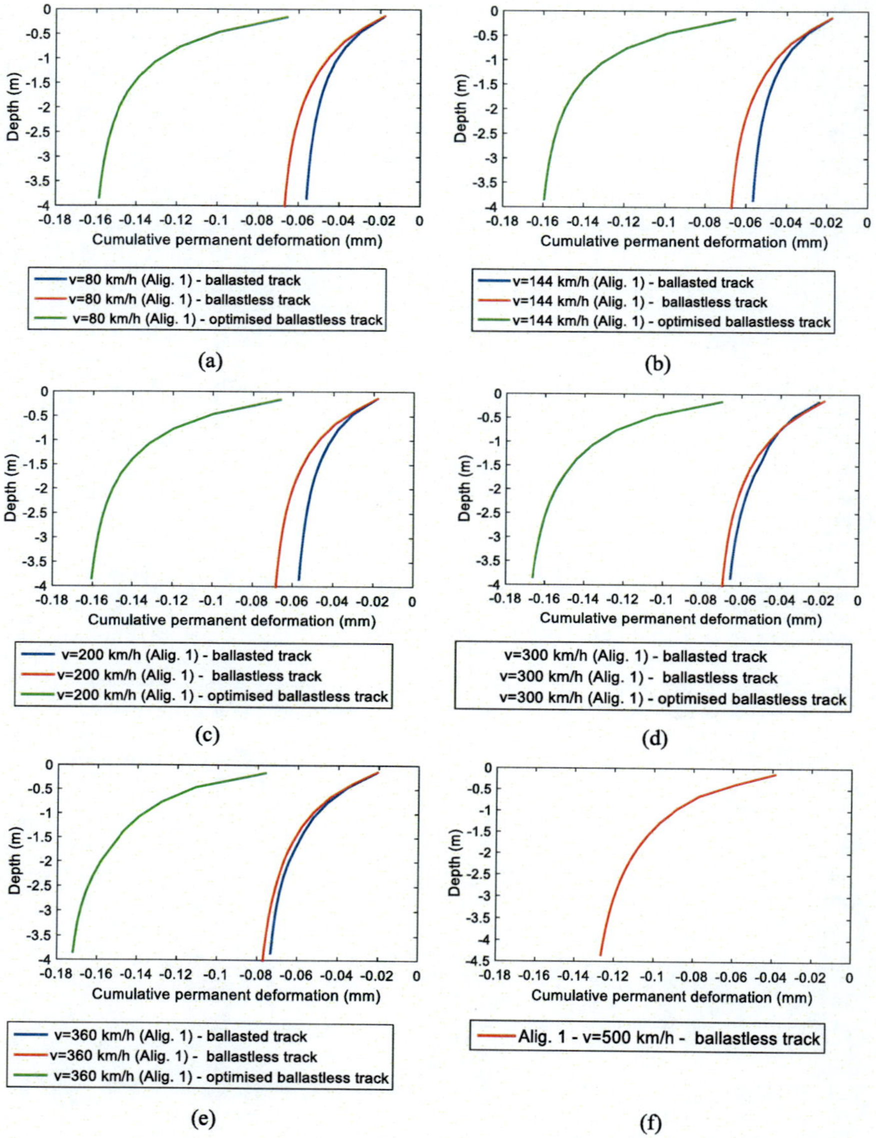

Fig. 5 Cumulative permanent deformation: **a** v = 80 km/h; **b** v = 144 km/h; **c** v = 200 km/h; **d** 300 km/h; **e** 360 km/h; f 500 km/h

concrete slab) shows higher values. In Fig. 5f) are depicted the values for the ballastless track (*Rheda system*) but considering a train's speed of 500 km/h to understand the magnitude of these displacements. The results were obtained in the alignment of element 1 depicted in Fig. 1.

Figure 5 also shows that there is a small difference between the ballasted track and the ballastless track in terms of cumulative permanent deformation. Indeed, this value would be higher if the permanent deformation of the ballast was considered (the influence of the ballast was neglected in this study). Thus, it is important to highlight that this is a comparison at the subgrade level, which justified the similarity of the performance of the subgrade.

The ballastless track with only the concrete slab shows higher cumulative permanent deformation and the difference is significant, which evidences the importance of the support layers (HBL and FPL). This difference is due to reduced initial mean stress, which leads to higher values of cumulative permanent deformation since the stress path is closer to the *yielding criterion* (Fig. 4).

The results also show that the cumulative permanent deformation is almost constant until v = 200 km/h. From this train speed, there is an increase in the cumulative permanent deformation in each structure due to the amplification of stresses. This amplification is higher when the train speed is closer to critical speed.

5 Conclusions

This work aims to compare the performance of the ballasted and ballastless tracks in terms of stresses and cumulative permanent deformations.

The stress levels are analysed by a numerical tool (2.5D FEM-PML approach), the permanent deformation is evaluated through the implementation of an empirical permanent deformation model and the results are presented in terms of cumulative permanent deformations.

Regarding the ballastless track, two structures are modelled to evaluate the influence of the support layers in the response of the subgrade. This analysis may be helpful in future optimizations of the ballastless track. The obtained results are a product of a simplification in the ballastless track (*Rheda* system) since for calculation purposes, the FPL was integrated into the subgrade.

The obtained results show that the train's speed has a significant influence on the response of the subgrade, which also includes the influence of the approaching to the critical speed. When the train's speed is reduced (i.e., $v = 80$ km/h until $v = 200$ km/h), the dynamic responses are not significantly amplified. When the train's speed increase and it is getting closer to the critical speed ($v = 360$ km/h to $v = 500$ km/h), the amplification's effects growth.

This study shows that the presence of irregularities, the moving character of the loads allied to the train speed can be crucial parameters in the design and future performance of railway structures. Furthermore, both ballasted and ballastless tracks present similar responses regarding long-term behaviour. Moreover, this analysis shows that the ballastless track with only a concrete slab can be an option instead of the "ordinary" ballastless track—*Rheda* system—(despite the higher absolute values of the permanent deformation because of the low values of the initial mean stresses) since the amplification of the stresses is similar to the remaining structures.

References

1. Selig, E.T., Waters, J.M.: Track Geotechnology and Substructure Management. Thomas Telford Services Ltd., London (1994)
2. Nielsen, J.C.O., Li, X.: Railway track geometry degradation due to differential settlement of ballast/subgrade—numerical prediction by an iterative procedure. J. Sound Vib. **412**, 441–456 (2018)
3. Li, D., Selig, E.T.: Cumulative plastic deformation for fine-grained subgrade soils. J. Geotech. Geoenvironmental Eng. **122**(12), 1006–1013 (1996)
4. Puppala, A.J., Mohammad, L.N., Allen, A.: Permanent deformation characterization of subgrade soils from RLT test. J. Mater. Civ. Eng. **11**(4), 274–282 (1999)
5. Correia, A.G., Biarez, J.: Stiffness properties of materials to use in pavement and rail track design. In: Geotechnical Engineering for Transportation Infrastructure. Proceedings of the 12th European Conference on Soil Mechanics and Geotechnical Engineering. Amsterdam, Netherlands (1999)
6. Rahim, A.M., George, K.P.: Models to estimate subgrade resilient modulus for pavement design. Int. J. Pavement Eng. **6**(2), 89–96 (2005)
7. Gomes Correia, A.: Innovations in design and construction of granular pavements and railways. In: Advances in Transportation Geotechnics—Proceedings of the 1st International Conference on Transportation Geotechnics. CRC Press, Taylors & Francis Group, Nottingham, UK (2008)
8. Puppala, A.J., Saride, S., Chomtid, S.: Experimental and modeling studies of permanent strains of subgrade soils. J. Geotech. Geoenvironmental Eng. **135**(10), 1379–1389 (2009)
9. Ng, C.W.W., et al.: Resilient modulus of unsaturated subgrade soil: experimental and theoretical investigations. Can. Geotech. J. **50**(2), 223–232 (2013)
10. Gomes Correia, A., Cunha, J.: Analysis of nonlinear soil modelling in the subgrade and rail track responses under HST. Transp. Geotech. **1**(4), 147–156 (2014)
11. Salour, F., Erlingsson, S.: Permanent deformation characteristics of silty sand subgrades from multistage RLT tests. Int. J. Pavement Eng. **18**(3), 236–246 (2015)
12. Ling, X., et al.: Permanent deformation characteristics of coarse grained subgrade soils under train-induced repeated load. Adv. Mater. Sci. Eng. **2017**, 15 (2017)
13. Rahman, M.M., Gassman, S.L.: Permanent deformation characteristics of coarse grained subgrade soils using repeated load triaxial tests. In: Meehan, C.L., et al. (eds.) Geo-Congress 2019, Philadelphia, Pennsylvania, pp. 599–609
14. Alves Costa, P., et al.: Critical speed of railway tracks. Detailed and simplified approaches. Transp. Geotech. **2**, 30–46 (2015)
15. Mezher, S.B., et al.: Railway critical velocity—analytical prediction and analysis. Transp. Geotech. **6**, 84–96 (2016)
16. Tang, Y., Xiao, S., Yang, Q.: Numerical study of dynamic stress developed in the high speed rail foundation under train loads. Soil Dyn. Earthq. Eng. **123**, 36–47 (2019)
17. Hu, J., et al.: Investigation into the critical speed of ballastless track. Transp. Geotech. **18**, 142–148 (2019)
18. Alves Costa, P., et al.: Influence of soil non-linearity on the dynamic response of high-speed railway tracks. Soil Dyn. Earthq. Eng. **30**(4), 221–235 (2010)
19. Yang, Y., Hung, H.: A 2.5D finite/infinite element approach for modelling visco-elastic body subjected to moving loads. Int. J. Numer. Methods Eng. **51**(11), 1317–1336 (2001)

20. Sheng, X., Jones, C.J.C., Thompson, D.J.: Prediction of ground vibration from trains using the wavenumber finite and boundary element methods. J. Sound Vib. **293**(3–5), 575–586 (2006)
21. Lopes, P., et al.: Numerical modeling of vibrations induced by railway traffic in tunnels: from the source to the nearby buildings. Soil Dyn. Earthq. Eng. **61–62**, 269–285 (2014)
22. Alves Costa, P., Calçada, R., Silva Cardoso, A.: Track–ground vibrations induced by railway traffic: in-situ measurements and validation of a 2.5D FEM-BEM model. Soil Dyn. Earthq. Eng. **32**(1), 111–128 (2012)
23. Gidel, G., et al.: A new approach for investigating the permanent deformation behaviour of unbound granular material using the repeated load triaxial apparatus. Bulletin des Laboratoires des Pont et Chaussées **233**, 5–21 (2001)
24. Chen, R., et al.: Cumulative settlement of track subgrade in high-speed railway under varying water levels. Int. J. Rail Transp. **2**(4), 205–220 (2014)
25. Korkiala-Tanttu, L.: A new material model for permanent deformations in pavements. In: Proceedings of the Seventh Conference on Bearing Capacity of Roads and Airfields. Trondheim, Norway (2005)
26. Rahman, M.S., Erlingsson, S.: A model for predicting permanent deformation of unbound granular materials. Road Mater. Pavement Des. **16**(3), 653–673 (2015)
27. Ramos, A., et al.: Stress and permanent deformation amplification factors in subgrade induced by dynamic mechanisms in track structures. Int. J. Rail Transp. 1–33 (2021)
28. Alves Costa, P., Calçada, R., Silva Cardoso, A.: Influence of train dynamic modelling strategy on the prediction of track-ground vibrations induced by railway traffic. In: Proc. Inst. Mech. Eng. Part F J. Rail Rapid Transit **226**(4), 434–450 (2012)
29. Colaço, A., Costa, P.A., Connolly, D.P.: The influence of train properties on railway ground vibrations. Struct. Infrastruct. Eng. **12**(5), 517–534 (2016)
30. Ramos, A.L., et al.: Influence of permanent deformations of substructure on ballasted and ballastless tracks performance. In: Proceedings of 7th Transport Research Arena TRA 2018, 16–19 Apr 2018. Zenodo, Vienna, Austria
31. EN13848-5: Railway applications—track—track geometry quality—part 5: geometric quality levels. In EN13848. European Committee for Standardization (CEN), Brussels, Belgium (2008)
32. Alves Costa, P., et al.: Railway critical speed assessment: a simple experimental-analytical approach. Soil Dyn. Earthq. Eng. **134** (2020)
33. Sayeed, M.A., Shahin, M.A.: Investigation into impact of train speed for behavior of ballasted railway track foundations. In: Advances in Transportation Geotechnics 3. The 3rd International Conference on Transportation Geotechnics (ICTG 2016). Guimarães, Portugal (2016)
34. Madshus, C., Kaynia, A.M.: High-speed railway lines on soft ground: dynamic behaviour at critical train speed. J. Sound Vib. **231**(3), 689–701 (2000)

Qualitative Study of a Class of Exponential Difference Equations with Periodic Coefficient

Anqi Chen, Jing Liu, Penghui Shen, and Kaisu Wu

Abstract This paper mainly studies the exponential difference equation system

$$\mathbf{u}_n = a + Bv_{n\ 1}e^{-v_{n\ 1}}$$
$$v_{\mathrm{n}} = a + bv_{n\ 1}e^{-v_{n\ 1}} \quad \mathrm{n} = 0, 1, 2, \ldots,$$

the bounded persistence of positive solutions, the existence and uniqueness of positive equilibrium points, and the global asymptotic stability. All a, b, B are fixed positive real numbers.

1 Introduction

Difference equations occupies a large proportion in mathematical applications, and are also widely used in physics, chemistry, biology, environment, economy and other fields. In population dynamics, differential equation modeling methods are often used to describe and predict the development process and trends of species.

In [1] Grove discussed the uniqueness of the positive equilibrium point of the exponential difference equation $x_{n+1} = ax_n + bx_{n-1}e^{-x_n}$ and the global asymptotic behavior of the solution. This equation considers the growth of a perennial turf, the number of grasses in the previous year plus this year. The amount of newly grown grass is the total amount of grass on the lawn at the end of this year.

Papaschinopoulos, G. study the exponential difference equation system in [2]

$$\begin{cases} x_{n+1} = a + by_{n-1}e^{-x_n} \\ y_{n+1} = c + dx_{n-1}e^{-y_n} \end{cases}$$

A. Chen (✉) · J. Liu · K. Wu
College of Mathematics and Science, Beijing University of Chemical Technology, Beijing, China
e-mail: wuks@mail.buct.edu.cn

P. Shen
College of Information Science and Technology, Beijing University of Chemical Technology, Beijing, China

H. Dai (ed.), *Computational and Experimental Simulations in Engineering*, Mechanisms and Machine Science 119,
https://doi.org/10.1007/978-3-031-02097-1_3

The existence and uniqueness of the positive equilibrium point, the global progression of the solution is equal, this difference equation describes the competitive relationship between the two populations, the solution of this biological mode (x_n, y_n) n = 0, 1, 2, ... corresponding to time n [3]. The number of the two populations.

Inspired by the above documents, this paper expands the coefficient b of the second-order exponential difference equation $x_{n+1} = a + bx_n e^{-x_{n-1}}$ to a two-period periodic sequence b, B, and obtains a new exponential difference equation system [4]

$$\begin{cases} x_{2n+1} = a + bx_{2n}e^{-x_{2n-1}} \\ x_{2n} = b + Bx_{2n-1}e^{-x_{2n-2}} \quad \text{n} = 0, 1, 2, \ldots \end{cases}$$

The a, b, B are fixed positive real numbers, the x_{2n}, x_{2n+1} is the population in the first half and the second half of the year, and the a is the external migration, which is stable in the long run. Due to the particularity of this species, the growth rate shows periodic characteristics in different periods. In general b $\neq$ B, this paper mainly studies the dynamic properties such as bounded persistence of positive solutions, existence and uniqueness of positive equilibrium points and global asymptotic stability.

In order to solve the difficulties caused by the different recurrences of odd and even terms, this article carries out the following transformation:

$$\begin{cases} x_{2n} = u_n \\ x_{2n+1} = v_n \end{cases}$$

The exponential difference equation system studied in this paper is obtained

$$\mathbf{u}_n = a + Bv_{n\ 1}e^{-v_{n\ 1}}$$
$$v_{\text{n}} = a + bv_{n\ 1}e^{-v_{n\ 1}}$$

n = 0, 1, 2, ... All a, b, B are fixed positive real numbers.

The initial value $u_{-1} = x_{-2}, v_{-1} = x_{-1}$ is a positive real number.

2 Positive Solution Boundedness and Persistence

2.1 Theorem

If a, b, B are positive real numbers, and there are

$$0 < \text{s} = \text{bB}e^{-2a} < 1$$

Then all the solutions of the differential equation system are positive solutions, and their solutions are bounded and persistent.

Proof (1) Boundedness proof

Since $u_{-1} = x_{-2}$, $v_{-1} = x_{-1}$ the initial values u_{-1}, v_{-1} of the difference equation system are all positive real numbers, and because the coefficients in the difference equation system are all positive real numbers (u_n, v_n), any solution of the system is all positive real numbers. In this case, let be any set of solutions (u_n, v_n) of the difference equation system, then this paper can be obtained from the equation

$$u_n \geq a, v_n \geq a, n = 0, 1, 2, \ldots$$

Further, for any n $\geq$ 0, there are:

$$\begin{aligned} u_n &= a + Bv_{n-1}e^{-u_{n-1}} \leq a + Bv_{n-1}e^{-a} \\ v_n &= a + bu_n e^{-v_{n-1}} = a + b\left(a + Bv_{n-1}e^{-u_{n-1}}\right)e^{-v_{n-1}} \\ &= a\left(1 + be^{-a}\right) + bBv_{n-1}e^{-u_{n-1}-v_{n-1}} \leq a\left(1 + be^{-a}\right) + bBv_{n-1}e^{-2a} \end{aligned}$$

Make s = bBe^{-2a}, in this formula

$$v_n \leq a\left(1 + be^{-a}\right) + sv_{n-1}$$

Consider the linear difference equation:

$$k_n = a\left(1 + be^{-a}\right) + sk_{n-1}$$

Make, $k_{-1} = v_{-1}$, can be obtained from the above equation

$$k_0 = a\left(1 + be^{-a}\right) + sk_{-1} = a\left(1 + be^{-a}\right) + sv_{-1} > 0$$

Thus available $k_n > 0, n = 0, 1, 2, \ldots$.

According to the form and structure of the linear difference equation:

$$\begin{aligned} k_0 &= a\left(1 + be^{-a}\right) + sk_{-1} \\ k_1 &= a\left(1 + be^{-a}\right) + sk_0 \\ k_2 &= a\left(1 + be^{-a}\right) + sk_1 \\ k_3 &= a\left(1 + be^{-a}\right) + sk_2 \end{aligned}$$

Therefore

$$\begin{aligned} k_1 - k_0 &= s(k_0 - k_{-1}) \\ k_2 - k_1 &= s(k_1 - k_0) \\ k_3 - k_2 &= s(k_2 - k_1) \end{aligned}$$

$$\vdots$$

$$k_n - k_{n-1} = s(k_{n-1} - k_{-n-2})$$

So the solution of the linear difference equation can be obtained $k_n = s^n k_0 + \frac{a(1+be^{-a})(1-s^n)}{1-s} = s^{n+1}k_{-1} + \frac{a(1+be^{-a})(1-s^{n+1})}{1-s}$.
Since s < 1, k_n is bounded, that is

$$k_n = s^{n+1}k_{-1} + \frac{a\left(1 + be^{-a}\right)\left(1 - s^{n+1}\right)}{1 - s} \leq k_{-1} + \frac{a\left(1 + be^{-a}\right)}{1 - s}$$

$v_n \leq k_n, n = 0, 1, 2, \ldots$, available

$$v_n \leq v_{-1} + \frac{a\left(1 + be^{-a}\right)}{1 - s}$$

u_n is bounded.

Because $u_n = a + Bv_{n-1}e^{-u_{n-1}} \leq a + Bv_{n-1}e^{-a}$, so u_n is bounded.

At this point, this article has been proved u_n, v_n is bounded.

(2) Proof of persistence

Persistence definition: if there is a positive real number K (or L) satisfies

$$\text{supp}\ u_n \subset [K, \infty), n = 0, 1, 2, \ldots$$

either

$$\text{supp}\ v_n \subset (0, L], n = 0, 1, 2, \ldots$$

$\{u_n\}$, $\{v_n\}$ the sequence is persistent.

On the basis of the first question, it is proved in this paper that all are u_n, v_n bounded. Let's suppose that there exists m, M(m > 0, M > 0), such that for any n = 0, $1, 2\ldots$, that is $u_n \in [a, m]$, $v_n \in [a, M]$, it meets the definition of system persistence, so the difference equation system is persistent [5].

3 The Existence and Uniqueness of Positive Equilibrium Point

First of all, the existence of invariant intervals is studied in this paper.

Theorem *Make a, b, B a positive real number and* $0 < s = bBe^{-2a} < 1$. *Then the following conclusion holds: consider the following interval:*

$$I_1 = \left[a, \frac{a + aBe^{-a}}{1 - s}\right], I_2 = \left[a, \frac{a + abe^{-a}}{1 - s}\right],$$
$$I_3 = \left[a, \frac{a + aBe^{-a} + \varepsilon}{1 - s}\right], I_4 = \left[a, \frac{a + abe^{-a} + \varepsilon}{1 - s}\right],$$

ε is any positive real number.

If (u_n, v_n) is any positive solution of the system and satisfies $u_{-1} \in I_1, v_{-1} \in I_2$, there will be $u_n \in I_1, v_n \in I_2, n = 0, 1, 2, \ldots$

If (u_n, v_n) is any positive solution of the system and satisfies $m \in N$, bring $u_n \in I_3, v_n \in I_4, n \geq m$.

Proof (1) (u_n, v_n) is any positive solution to the system, $u_{-1} \in \mathrm{I}_1, v_{-1} \in \mathrm{I}_2$,

$$a \leq u_0 = a + Bv_{-1}e^{-u_{-1}} \leq a + B\frac{a + abe^{-a}}{1 - s}e^{-a} = \frac{a + aBe^{-a}}{1 - s}$$
$$a \leq v_0 = a + bu_0e^{-v_{-1}} \leq a + b\frac{a + aBe^{-a}}{1 - s}e^{-a} = \frac{a + abe^{-a}}{1 - s}$$

As a result, for any positive integer n,

$$a \leq u_n = a + Bv_{n-1}e^{-u_{n-1}} \leq a + B\frac{a + abe^{-a}}{1 - s}e^{-a} = \frac{a + aBe^{-a}}{1 - s}$$
$$a \leq v_n = a + bu_ne^{-v_{n-1}} \leq a + b\frac{a + aBe^{-a}}{1 - s}e^{-a} = \frac{a + abe^{-a}}{1 - s}$$

So I_1, I_2 is the invariant interval of u_n, v_n.

(u_n, v_n) is any positive solution to the system, $\lim_{n\to\infty} sup\, u_n = \mathrm{M} < \infty$, $\lim_{n\to\infty} sup v_n = \mathrm{L} < \infty$,

$\mathrm{M} \leq \frac{a+aBe^{-a}}{1-s}, \mathrm{L} \leq \frac{a+abe^{-a}}{1-s}$. Therefore the nature of the upper bound must exist m $\in$ N, n $\geq$ m, $u_n \in \mathrm{I}_3$, $v_n \in \mathrm{I}_4$. □

Then we discuss the existence of positive equilibrium points. If (u, v) is the equilibrium point of the system.

$$\begin{cases} u = a + Bve^{-u} \\ v = a + bue^{-v} \end{cases}$$

$\mathrm{u} = \frac{v-a}{be^{-v}}, \mathrm{v} = \frac{u-a}{Be^{-u}}$.

$\mathrm{u} = \frac{v-a}{be^{-v}} = \frac{\frac{u-a}{Be^{-u}} - a}{b} e^{\frac{u-a}{Be^{-u}}} = \frac{(u-a)e^u - aB}{bB} e^{\frac{u-a}{Be^{-u}}}$.

Make $\mathrm{F(u)} = \frac{(u-a)e^u - aB}{bB} e^{\frac{u-a}{Be^{-u}}} - \mathrm{u} = \frac{G(u)}{bB}, u \geq \mathrm{a}$. $\mathrm{G(u)} = [(\mathrm{u} - \mathrm{a})\mathrm{e}^{\mathrm{u}} - \mathrm{aB}]e^{\frac{u-a}{Be^{-u}}} - \mathrm{bBu}$.

Take in $\mathrm{G(a)} = -\mathrm{aB} - \mathrm{abB} < 0$

$\mathrm{G(u)} > (\mathrm{u} - \mathrm{a})\mathrm{e}^{\mathrm{u}} - \mathrm{aB} - \mathrm{bBu}$

$> (\mathrm{u} - \mathrm{a})(1 + \mathrm{u}) - \mathrm{aB} - \mathrm{bBu}$

$= u^2 + (1 - a - bB)u - a(1 + B)$.

Make $H(u) = u^2 + (1 - a - bB)u - a(1 + B)$, Access $\lim_{n\to\infty} H(u) = +\infty$, Therefore $\lim_{n\to\infty} G(u) = +\infty$.

F (u) has zero point on $[a, +\infty)$ and the equations have solutions, so there is a positive equilibrium point in the difference equation system.

Finally, the uniqueness of the positive equilibrium point is proved. $G'(u) = [e^u + (u - a)e^u]e^{\frac{u-a}{Be^{-u}}} + [(u - a)e^u - aB]e^{\frac{u-a}{Be^{-u}}} \frac{(u+1-a)e^u}{B} - bB$.

$= [e^u + (u - a)e^u]e^{\frac{u-a}{Be^{-u}}} + [(u - a)e^u - aB]e^{\frac{u-a}{Be^{-u}}} \frac{(u+1-a)e^u}{B} - \frac{(u-a)e^u - aB}{u} e^{\frac{u-a}{Be^{-u}}}$.

To prove $G'(u) > 0$, just $u^2 - au + a > 0$ that is $\Delta = a^2 - 4a < 0$, $a < 4$.

4 Local Progressive Stability

Defined as: $\rho_1 = \frac{a+aBe^{-a}}{1-s}$, $\rho_2 = \frac{a+abe^{-a}}{1-s}$.

Lemma *Let the x, y be real and satisfy* $|x + y| + |xy| < 1$ *and* $|x| < 1$.

Proof Only proof $|x| < 1$, by symmetry can be proved by $|y| < 1$.

Use the countervailing method. Assumption $|x| \geq 1$, $x^2 - (x + y)x + xy = 0$.

$$\begin{aligned} \text{Be } x^2 &= |(x + y)x - xy| \\ &\leq |(x + y)x| + |xy| \\ &\leq |(x + y)x| + |xy||x| \\ &= (|(x + y)| + |xy|)|x| \\ &< |x| \end{aligned}$$

This contradicts the hypothesis, thus $|x| < 1$, $|y| < 1$.

Theorem $0 < s = bBe^{-2a} < 1$, *if* $s(1 + \rho_1\rho_2) + \rho_1 + \rho_2 < 1 + 2a$, *then the only positive equilibrium point of the difference equation system* $(\overline{u}, \overline{v})$ *is locally progressive and stable.*

It is proved that the equivalent deformation of the difference equation system is first obtained:

$$\begin{cases} u_n = a + Bv_{n-1}e^{-u_{n-1}} \\ v_n = a + b(a + Bv_{n-1}e^{-u_{n-1}})e^{-v_{n-1}} \end{cases}$$

Further translation: $\begin{cases} u_n = a + Bv_{n-1}e^{-u_{n-1}} \\ v_n = a + abe^{-v_{n-1}} + bBv_{n-1}e^{-u_{n-1}-v_{n-1}} \end{cases}$.

At the equilibrium point $(\overline{u}, \overline{v})$ the linearization process is obtained

$$u_n = (-B\overline{v} \cdot e^{-\overline{u}})u_{n-1} + (B \cdot e^{-\overline{u}})v_{n-1}$$
$$v_n = (-bB \cdot e^{-\overline{u-\overline{v}}}u_{n-1} + [bB(1 - \overline{v}) \cdot e^{-\overline{u}-\overline{v}} - abe^{-\overline{v}}]v_{n-1}.$$

This is equivalent to the following matrix form difference equation system: $W_n = T\ W_{n-1}$.
其中 $W_n = \begin{bmatrix} u_n \\ v_n \end{bmatrix}, T = \begin{bmatrix} \alpha\ \beta \\ \gamma\ \delta \end{bmatrix}$.
Besides $\alpha = -B\overline{v}e^{-\overline{u}}, \beta = B\, e^{-\overline{u}}, \gamma = -bB\overline{v}e^{-\overline{u-\overline{v}}}, \delta = bB(1-\overline{v})\, e^{-\overline{u}-\overline{v}} - abe^{-\overline{v}}$.
The characteristic equation of the system is $\lambda^2 - (\alpha+\delta)\lambda + \alpha\delta - \beta\gamma = 0$.
$|\alpha + \delta| = |-B\overline{v}e^{-\overline{u}} + bB(1-\overline{v})\, e^{-\overline{u}-\overline{v}} - abe^{-\overline{v}}|$
$= bB\, e^{-\overline{u-\overline{v}}}| -\frac{\overline{v}}{b}e^{\overline{v}} - \frac{a}{B}e^{\overline{u}} - \overline{v+1}|$.
Because $(\overline{u}, \overline{v})$ is the equilibrium point of the differential equation system, then

$$\overline{u} = a + B\overline{v}e^{-\overline{u}}$$
$$\overline{v} = a + b\overline{u}e^{-\overline{v}}$$

Simplified $\overline{u} = \frac{\overline{v}-a}{be^{-\overline{v}}}, \overline{v} = \frac{\overline{u}-a}{Be^{-\overline{u}}}$.
Further, $|\alpha + \delta| = bB\, e^{-\overline{u-\overline{v}}}| - \frac{\overline{uv}}{\overline{v}-a} - \frac{a\overline{v}}{\overline{u}-a} - \overline{v} + 1| = bB\, e^{-\overline{u-\overline{v}}}|1 - \overline{v}(1 + \frac{\overline{u}}{\overline{v}-a} + \frac{a}{\overline{u}-a})|$.

The known conditions $\begin{matrix} \overline{u} - a = B\overline{v}e^{-\overline{u}} \\ \overline{v} - a = b\overline{u}e^{-\overline{v}} \end{matrix}$.
United $|\alpha + \delta| = bB\, e^{-\overline{u-\overline{v}}}|1 - \overline{v}\frac{\overline{uv}-2a\overline{u}+\overline{u}^2}{Bb\overline{uv}e^{-\overline{u}-\overline{v}}}| = bB\, e^{-\overline{u-\overline{v}}}|1 - \frac{\overline{v}-2a+\overline{u}}{Bbe^{-\overline{u}-\overline{v}}}|$
On the other hand $|\alpha\delta - \beta\gamma|=|abB\ \overline{v}e^{-\overline{u-\overline{v}}} + b\ B^2\overline{v}^2e^{-\overline{2u-\overline{v}}}|$
$= bB\, e^{-\overline{u-\overline{v}}}|a\ \overline{v} + B\ \overline{v}^2e^{-\overline{u}}|$
$= bB\, e^{-\overline{u-\overline{v}}}|a\ \overline{v} + (\overline{u} - a)\overline{v}|$
$= bB\, e^{-\overline{u-\overline{v}}}\overline{uv}$.
thereby$|\alpha + \delta| + |\alpha\delta - \beta\gamma| = bB\, e^{-\overline{u-\overline{v}}}|1 - \frac{\overline{v}-2a+\overline{u}}{Bbe^{-\overline{u}-\overline{v}}}| + bB\, e^{-\overline{u-\overline{v}}}\overline{uv}$.
And I know $\overline{u} \in I_1, \overline{v} \in I_2$, there are:
$|\alpha + \delta| + |\alpha\delta - \beta\gamma| \le bB\, e^{-\overline{u-\overline{v}}} + \overline{v} - 2a + \overline{u} + bB\, e^{-\overline{u-\overline{v}}}\overline{uv}$.
$\le s + \rho_1 + \rho_2 + s\ \rho_1\rho_2 - 2a$.
$= s(1 + \rho_1\rho_2) + \rho_1 + \rho_2 - 2a < 1$.
Let the eigenvalue of the characteristic equation be λ_1, λ_2, so $\lambda_1 + \lambda_2 = \alpha + \delta, \lambda_1\lambda_2 = \alpha\delta - \beta\gamma$.

All the roots of the characteristic equation are less than 1, which is known by the linear stability theorem. Under the conditions of the theorem, the only positive equilibrium point $(\overline{u}, \overline{v})$ of the difference equation system is locally asymptotically stable [6].

5 Saddles and Repulsions

Theorem *If* $4s\rho_1\rho_2 + 2(\rho_1 + \rho_2 - 2a)bBe^{-2a} < b^2B^2e^{-2\rho_1-2\rho_2}$.

The equilibrium point is unstable $s\rho_1\rho_2 + \rho_1 + \rho_2 - bBe^{-\rho_1-\rho_2} < 2a - 1$ *and we call the equilibrium point saddle point.*

Proof Because

$$4|A| \leq 4s\rho_1\rho_2$$

and

$$\begin{aligned}|tr(A)|^2 &= b^2B^2e^{-2\overline{u}-2\overline{v}} - 2(\overline{u}+\overline{v}-2a)bBe^{-\overline{u}-\overline{v}} + (\overline{u}+\overline{v}-2a)^2 \\ &\geq b^2B^2e^{-2\rho_1-2\rho_2} - 2(\rho_1+\rho_2-2a)bBe^{-2a}\end{aligned}$$

So if $4s\rho_1\rho_2 + 2(\rho_1 + \rho_2 - 2a)bBe^{-2a} < b^2B^2e^{-2\rho_1-2\rho_2}$ holds, then

$$|tr(A)|^2 > 4|A|$$

In addition, as $|1 + |A|| \leq 1 + s\rho_1\rho_2$

$$\begin{aligned}|tr(A)| = |a + d| &= bBe^{-\overline{u}-\overline{v}}\left|1 - \frac{\overline{u}+\overline{v}-2a}{bBe^{-\overline{u}-\overline{v}}}\right| \\ &= \left|bBe^{-\overline{u}-\overline{v}} - (\overline{u}+\overline{v}-2a)\right| \\ &\geq bBe^{-\overline{u}-\overline{v}} - (\overline{u}+\overline{v}-2a) \\ &\geq bBe^{-\rho_1-\rho_2} - (\rho_1+\rho_2-2a)\end{aligned}$$

So if $s\rho_1\rho_2 + \rho_1 + \rho_2 - bBe^{-\rho_1-\rho_2} < 2a - 1$ holds, then

$$|tr(A)| > 1 + |A|$$

Therefore, the equilibrium point is a saddle point.

Theorem *If a* $a^2bBe^{-\rho_1-\rho_2} > 1$ *and* $s + \rho_1 + \rho_2 - a^2bBe^{-\rho_1-\rho_2} < 1 + 2a$ *come into existence, If the equilibrium point of the difference equation is unstable, we call the equilibrium point a repulsive.*

Proof Because

$$|A| = bBe^{-\overline{u}-\overline{v}}\overline{uv} \geq a^2bBe^{-\rho_1-\rho_2}$$

Therefore, if $a^2bBe^{-\rho_1-\rho_2} > 1$ is established, then $||A|| > 1$.
Also because

$$|\mathrm{tr(A)}| = |\mathrm{a} + \mathrm{d}| = \mathrm{bB}e^{-\overline{u}-\overline{v}}\left|1 - \frac{\overline{u} + \overline{v} - 2a}{\mathrm{bB}e^{-\overline{u}-\overline{v}}}\right|$$

$$\leq bBe^{-\overline{u}-\overline{v}} + \overline{u} + \overline{v} - 2a \leq s + \rho_1 + \rho_2 - 2a$$

moreover

$$1 + |\mathrm{A}| = 1 + \mathrm{bB}e^{-\overline{u}-\overline{v}}\overline{uv} \geq 1 + a^2bBe^{-\rho_1-\rho_2}$$

So when $\mathrm{s} + \rho_1 + \rho_2 - a^2bBe^{-\rho_1-\rho_2} < 1 + 2a$ there are:

$$|\mathrm{tr(A)}| < 1 + |\mathrm{A}|$$

Therefore, the equilibrium point is repulsive.

6 Global Stability

Lemma *$f(u, v)$, $g(u, v)$ are binary continuous functions on the $R^+ \times R^+$, and $R^+ = 0, \infty, a_1, b_1, a_2, b_2$ is a positive real number*: [7, 8]

$$f\colon [a_1, b_1] \times [a_2, b_2] \to [a_1, b_1]$$

$$g\colon [a_1, b_1] \times [a_2, b_2] \to [a_2, b_2]$$

Considering the difference equation system $u_n = f(u_{n-1}, v_{n-1})$, $v_n = g(u_n, v_{n-1})$, $n = 0, 1, 1, \ldots$.

The initial value $u_{-1} \in [a_2, b_2]$, $v_{-1} \in [a_2, b_2]$, and the following three conditions are established:

the fixed u, f (u, v) is the v non-subtractive function, and the g (u, v) is the non-increasing function
the fixed v, f (u, v) is a u non-increased function, and the g (u, v) is a u non-subtractive function.

If the p, P, q, Q is a positive real number and all satisfy

$$P = f(p, Q), p = f(P, q), Q = g(P, q), q = g(p, Q)$$

and $p \leq P, q \leq Q$.

Can be launched $p = P, q = Q$.

Then the differential equation system has a unique equilibrium point, and when n $\to \infty$, each positive solution (u_n, v_n) of the system converges to this unique positive equilibrium point $(\overline{u}, \overline{v})$.

Theorem: if the a, b, B is a positive real number, the following conclusion holds:

Record $\mu_1 = \mathrm{B}e^{-a}$, $\mu_2 = \mathrm{b}e^{-a}$, if $\max\{\mu_1, \mu_2\} < 1$ and $(1 + a)s + a\,\mu_1 < 1$, $(1 + a)\,s + a\mu_2 < 1$,

$\lambda = \frac{s(1-s)^2}{[1-(1+a)s-a\mu_1][1-(1+a)s-a\mu_2]} < 1.$

Then the system has a unique positive equilibrium point $(\overline{u}, \overline{v})$, and any positive solution of the system converges to this unique positive equilibrium point when the n tends to ∞.

Proof let f: $R^+ \times R^+ \to R^+$, g: $R^+ \times R^+ \to R^+$, 并且f(u, v) = a + Bve^{-u}, g(u, v) = a + bue^{-v}.

If $u \in I_3$, $v \in I_4$, $a \leq f(u, v) \leq a + B\,\frac{a+abe^{-a}+\varepsilon}{1-s}e^{-a} = \frac{a+a\mu_1+\varepsilon\mu_1}{1-s} < \frac{a+a\mu_1+\varepsilon}{1-s}$.

$a \leq g(u, v) \leq a + b\,\frac{a+aBe^{-a}+\varepsilon}{1-s}e^{-a} = \frac{a+a\mu_2+\varepsilon\mu_2}{1-s} < \frac{a+a\mu_2+\varepsilon}{1-s}$.

As a result, the binary function f, g satisfied: f: $I_3 \times I_4 \to I_3$, g: $I_3 \times I_4 \to I_4$.

Now the positive real number p, P $\in I_3$, positive real number q, Q $\in I_4$, 且$p \leq P$, $q \leq Q$, bring

$P = a + BQ\,e^{-p}$, $p = a + Bq\,e^{-P}$.

$Q = a + bP\,e^{-q}$, $q = a + bp\,e^{-Q}$.

$p = a + B(a + bp\,e^{-Q})e^{-P} = a + aB\,e^{-P} + bBp\,e^{-P-Q}$.

$q = a + b\,(a + Bq\,e^{-P})e^{-Q} = a + ab\,e^{-Q} + bBq\,e^{-P-Q}$.

Further obtained $p = \frac{a+aBe^{-P}}{1-bBe^{-P-Q}}$, $q = \frac{a+abe^{-Q}}{1-bBe^{-P-Q}}$.

Because $P \geq a$, $Q \geq a$, ministers $p \leq \frac{a+aBe^{-a}}{1-s} = \frac{a+a\mu_1}{1-s}$.

$q \leq \frac{a+abe^{-a}}{1-s} = \frac{a+a\mu_2}{1-s}$.

On the one hand, by the Lagrange mean value theorem, there is $\eta \in [p, P]$ usable $e^P - e^p = e^{\eta}(P - p)$.

$P - p = B(Q\,e^{-p} - q\,e^{-P})$.

$= B\,e^{-p}(Q - q) + Bq\,e^{-P-p}(e^{-p} - e^{-P})$.

$= B\,e^{-p}(Q - q) + Bq\,e^{-P-p+\eta}(P - p)$.

$\leq \mu_1(Q - q) + q\,\mu_1(P - p)$.

Further available $P - p \leq \mu_1(Q - q) + \frac{\mu_1(a+a\mu_2)}{1-s}\,(P - p)$.

Also because $s = \mu_1\mu_2$, $(P - p)\frac{1-s-a\mu_1-as}{1-s} \leq \mu_1(Q - q)$,

$P - p \leq \frac{\mu_1(1-s)}{1-s-a\mu_1-as}(Q - q)$.

In the same way $Q - q \leq \frac{\mu_2(1-s)}{1-s-a\mu_2-as}(P - p)$.

The simultaneous $P - p \leq \lambda(P - p)$ can be deduced $Q = q$, the same principle [9]. According to Lemma, the equilibrium point $(\overline{u}, \overline{v})$ of the difference equation system is unique, and when $n \to \infty$, every positive solution of (u_n, v_n) the system converges to this unique positive equilibrium point $(\overline{u}, \overline{v})$.

References

1. Grove, E.A., Ladas, G., Prokup, N.R., Levins, R.: On the global behavior of solutions of a biological model. Commun. Appl. Nonlinear Anal. **7**(2) (2000)

2. Papaschinopoulos, G., Schinas, C.J.: On the dynamics of two exponential type systems of difference equations. Comput. Math. Appl. **64**(7), 2326–2334 (2012)
3. Papaschinopoulos, G., Radin, M., Schinas, C.J.: Study of the asymptotic behavior of the solutions of three systems of difference equations of exponential form. Appl. Math. Comput. **218**(9), 5310–5318 (2011)
4. El-Metwally, E., Grove, E.A., Ladas, G., Levins, R., Radin, M.: On the difference equation, xn $+ 1 = \alpha + \beta$xn $-$ 1e $-$ xn. Nonlinear Anal. **47**, 4623–4634 (2001)
5. Grove, E.A., Ladas, G.: Periodicities in Nonlinear Difference Equations. CRC Press (2004)
6. Camouzis, E., Ladas, G.: Dynamics of Third-Order Rational Difference Equations with Open Problems and Conjectures. CRC Press (2007)
7. Khan, A.Q.: Global dynamics of two systems of exponential difference equations by Lyapunov function. Adv. Differ. Equ. **2014**(1), 1–21 (2014)
8. Zayed, E.M.E., El-Moneam, M.A.: On the global attractivity of two nonlinear difference equations. J. Math. Sci. **177**(3) (2011)
9. Kocic, V.L., Ladas, G.: Global Behavior of Nonlinear Difference Equations of Higher Order with Applications. Springer, Dordrecht (1993)

Rapid Pattern Recognition of Electric Submersible Pump Ammeter Card Based on Artificial Neural Network

Biao Wang, Guoqing Han, Xin Lu, and Shuai Tan

Abstract Ammeter card diagnosis is a typical diagnostic method for ESP working conditions. The traditional ammeter card pattern recognition method needs to be completed manually, which contains certain technical barriers and induces subjective errors. As a machine learning algorithm, artificial neural network (ANN) can make up for these errors. For the purpose of realizing fast, accurate and objective pattern recognition. We use the relationships between the current characteristics of the collected ammeter card data after pre-processing and the actual working conditions, establishes the ANN models. Using the established ANN models to diagnose the working condition, which has the advantages that the traditional ammeter card manual identification is incomparable. In this paper, the ANN model is established and then used for the pattern recognition by the method mentioned above. By extracting the ammeter card data not involved in the training stage, the diagnostic model of the working condition is verified, and reached high accuracy, which proved the feasibility and reliability of the diagnosis of the working condition using the ANN model in quickly identifying the ammeter cards.

Keywords Ammeter card · Electrical submersible pump (ESP) · Working condition diagnosis · Pattern recognition · Artificial neural network (ANN) · Machine learning

1 Introduction

Electric submersible pump (ESP) is a common lifting tool for offshore platforms. The main working parts of the electric submersible pump unit are underground. The problems are more invisible compared with the ground swabbing equipment such as pumping units. In order to understand the working condition of the electric submersible pump clearly, various diagnostic techniques for the working condition

B. Wang (✉) · G. Han · X. Lu · S. Tan
Key Laboratory of Petroleum Engineering, China University of Petroleum – Beijing, Beijing 102249, China
e-mail: wang.clement@outlook.com

H. Dai (ed.), *Computational and Experimental Simulations in Engineering*, Mechanisms and Machine Science 119,
https://doi.org/10.1007/978-3-031-02097-1_4

of the electric submersible pump had been put forward. As a common diagnosis method of ESPs, current card, recording the changes of current input into the motor of ESPs, has been widely used in the field. Traditional current card diagnosis technology mainly relies on skilled engineers, and infers its possible working condition by manually recognizing the shape of current card, influenced by subjective factors. By training the artificial neural network (ANN) model to learn the characteristics of the current card, the status of current card can be judged in real time, so that the working condition of the ESPs can be grasped at any time. Chen et al. (2004) put forward a method of extracting feature value of current card based on pattern recognition, which was used to assist manual judgment, so that the quantitative index was added [1]. Yu et al. (2005) considered the neural network method to help to judge on the basis of feature engineering, and began to use the machine learning method to identify the current working condition [2]. Gan et al. (2011) further developed the algorithm and with more adaptable Back Propagation algorithm to supplement the machine learning method [3]. Han et al. (2015) supplemented the types of working conditions with BP neural network, which further developed the working condition identification [4]. In this paper, the extraction method of current eigenvalue is further refined on the basis of previous studies, which makes the description of current card eigenvalue more accurate. At the same time, the structure of BP neural network is compared and optimized, and the network model with lower judgment error rate is selected, so as to improve the applicability in intelligent oil fields and the judgment effect.

2 Relationship Between the Current Card and Working Condition

The current value and current fluctuation degree are different in different working conditions of different ESP wells. The fluctuation of current in polar coordinate system is recorded on the current card, which can intuitively understand the current recorded by the current card under different working conditions within the time period recorded by the current card. That is, the current card can diagnose the working condition of ESP by identifying the different characteristics of motor current [4]. Normally, each current card records the current fluctuation in the past 24 h [5].

2.1 Analysis of the Cause of Production Blocking of ESP Wells

In the cases where the problems encountered in ESP wells impacting the production, the problems of reservoir liquid supply capacity and downhole equipment failures can be reflected in the current card of ESP, and there will be different current values and current fluctuations under different working conditions, so the working condition of ESP can be reflected by the current card [6].

2.2 *The Basis of Current Card Classification*

Generally, the working conditions of the ESPs are classified according to the reasons that lead to the change of the working current of the pumps. From the current and its fluctuation shown in the current card, the pump shutdown condition, current fluctuation time, fluctuation frequency and recurrence of fluctuation can be visually observed. Through these conditions, the actual problems reflected in the current card can be analyzed. The detailed classification basis of current cards is shown in Fig. 1.

It can be seen from Fig. 1 that according to which the current cards are classified that the ESPs working under different conditions are different in current values and current fluctuations. Taking the overload shutdown condition as an example, this condition is usually caused by the motor being overloaded for a long period, which exceeds the capacity of the motor, resulting in automatic protective shutdown of the motor, and the current of the electric pump will not recover without artificial restart. In the current card, it should be shown that the current does not fluctuate or fluctuate negligibly within a normal range, and then at a certain moment, the current shut down to zero and does not fluctuate greatly. When diagnosing the current card, we should first observe whether the pump stops, that is, whether the current reaches the zero value. If this state has been reached, analyze whether there is frequent circulation of current, that is, the current card shows frequent and large fluctuations between zero

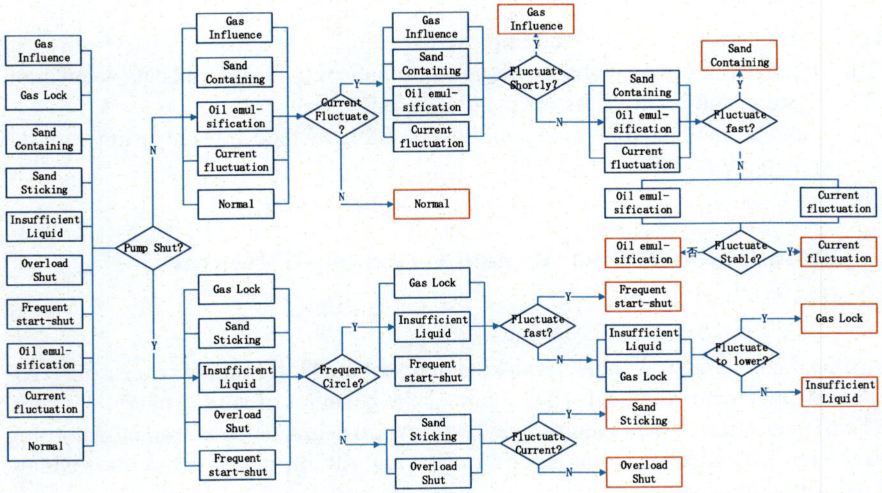

Fig. 1 The basis of classification of Ammeter cards

value and non-zero value. Finally, it is analyzed whether there is current fluctuation in a relatively small range before the shutdown. If there is no current fluctuation, it can be inferred that the electric pump is overloaded and shut down.

2.3 *Shortcomings of Traditional Current Card Diagnosis*

Traditional current card diagnosis is a classification method of ESPs proposed by API in 1982. When there is a problem in the oil well, the operator manually interprets the current card with personal experience to recognize the working condition of the ESP. The judgment flow determines that this manual identification method has its inherent defects of needing the manual identification of experienced field operators. On the one hand, this identification method needs to be noticed by the operators when the fault obviously affects the current range in the current card, which may be difficult to be noticed at the early stage of the failure; On the other hand, the judgment of failure types will vary from operator to operator, which introduces inevitable human error [3].

2.4 *Optimization Target of Current Card Diagnosis Method*

The current card diagnosis method of ESP itself is a typical diagnosis method of ESP working condition, but its shortcomings are also obvious. Based on the analysis of its shortcomings, the optimization objectives of the working condition diagnosis methods are put forward as follow:

(1) Eliminate human error and improve accuracy;
(2) Realize real-time diagnosis through automatic monitoring and data acquisition system, and improve the diagnosis efficiency;
(3) Verified the feasibility and superiority of this method by comparing with the judgment of experienced oilfield operators.

3 Pattern Recognition Method Based on BP Neural Network

Artificial neural network (ANN) is a kind of distributed information processing algorithmic mathematical model which imitates the behavior of animal neural network. The BP neural network is a multi-layer feedforward neural network trained according to the error back propagation algorithm (BP algorithm), which is the most widely used neural network at present.

BP neural network is a nonlinear dynamic system, which solves the problem of learning the connection weights of hidden layers in multilayer neural networks, and enhances the ability of classification and recognition of networks, especially suitable for the problems where there is a nonlinear relationship between the input of the research object and the output of the system. When there is no obvious linear correlation between the input and output of the control object, the neural network can be used to simulate y = g(u). Although its form is unknown, the connection weight in the neural network can be adjusted by using a suitable excitation function between the actual output and the expected output of the system, so that the neural network can learn itself, and its variance will decrease continuously until it approaches 0 in the learning process. This is the basic idea of realizing direct control by neural network.

3.1 Analysis of Neural Network Structure and Training Process

Neural network is a hierarchical network, which is composed of input layer, hidden layer and output layer. The training process of neural network is as follows: the data is input into the input layer, transformed by sigmod function in the hidden layer, and transmitted to the output layer. Linear transformation function can be used in neurons in the output layer. After a transmission, the results of the forward neural network are transmitted back to the hidden layer and the input layer through feedback adjustment, and the weight coefficient in the nodes of the hidden layer is adjusted by the excitation function, and then reaches the output layer according to the forward transmission process. The back propagation neural network is more accurate and more reliable compared with the forward propagation neural network due to the weight adjustment by back propagation neural network. BP neural network includes the forward propagation of signals and the backward propagation of errors, which makes the errors decrease along the gradient direction. After the repeat training and learning, the training is terminated when the network parameters leading to the minimum error are determined. The structure of neural network and the flow chart of training and learning are shown in Fig. 2.

3.2 Advantages and Main Application of BP Neural Network

BP neural network is mature in both network theory and performance with strong nonlinear mapping ability and flexible network structure. The structure of BP neural network can be adjusted according to the needs of users, by adjusting the number of hidden layers and the number of nodes in a single layer in the model, an ideal balance between the learning speed and the learning accuracy of the neural network can be achieved. Based on the advantages of BP neural network, it is mainly applied to function approximation, pattern recognition, classification and data compression.

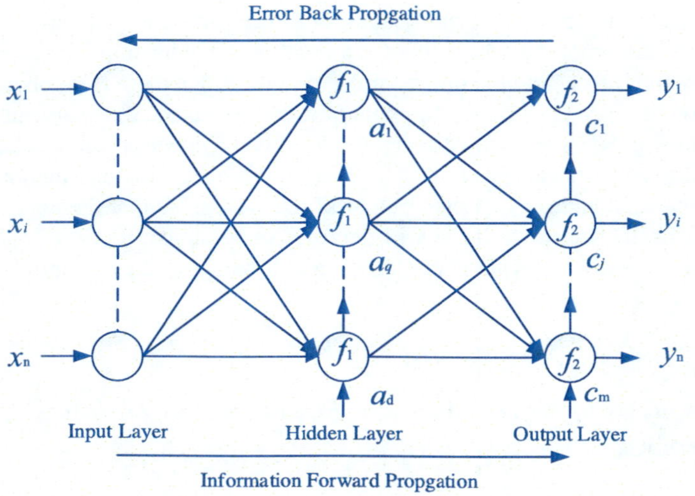

Fig. 2 Learning process of neural network

4 Current Card Pattern Recognition Based on BP Neural Network

The relationship between current and working condition of ESP is a typical nonlinear relationship, and there are strong characteristics among the same type of current cards. The rapid and accurate identification of current cards is the basis of real-time working condition diagnosis of ESP wells. BP neural network is good at solving nonlinear problems, which can be used in pattern recognition and classification. It can achieve the expected learning accuracy and speed by adjusting the model structure of neural network, and can realize real-time discrimination and output after the model is established. Based on this analysis, the problem of rapid recognition of current card can be solved by the advantages of BP neural network, and it is feasible to use BP neural network for pattern recognition of ESP current card. The steps of pattern recognition of current card using BP neural network generally include the following steps: (1). The establishment of sample database; (2). The extraction of Failure feature; (3). The establishment and training of BP neural network; (4). The pattern recognition of new current data with the trained model.

4.1 The Establishment of Sample Library

The training of BP neural network is based on a large number of sample data. As far as the case of pattern recognition of current cards via BP neural network is concerned, it needs a large number of current cards labeled with specific failure types. The common failure types that can be identified by current cards include: gas influence, gas locking, sand containing, sand sticking, insufficient liquid supply, overload pump shut, frequent start and shut down, crude oil emulsification, current fluctuation and normal working conditions, etc. The sources of these current cards can be as follows: (1). Experience accumulated in the industry; (2). Collected in oil field; (3) Generated with simulation software [2].

The experience accumulated in the industry means that experienced engineers qualitatively describe the current card when a certain specific working condition occurs according to their working experience, and quantify it into a specific current card according to the characteristics of the described current card.

Field collection means that after a certain working condition actually occurs in the oil field, after the conclusion of working condition diagnosis is given by pump inspection and maintenance, the current situation in a period of time before the failure occurs is marked as a sample that will happen in this working condition.

Simulation software generation refers to the use of numerical simulation software, taking the characteristic condition of the pump in a certain state as the input condition of the software, reflecting the current fluctuation in this condition as a current card, and taking it as a sample under this working condition to participate in the subsequent neural network training.

4.2 *Extraction of Failure Feature Values*

The current feature values are numerical values used to represent the symbolic attribute under a specific working condition. The selection criterion is that the feature values must reflect the characteristic states of the current card under a certain working condition which is different from other working conditions. The current cards under several typical working conditions are shown in Fig. 3 [7].

The current feature value refers to the characteristic of current value and current fluctuation amplitude in current cards, that is, the shape characteristics of each current card are different from those of other types of current cards. Therefore, current and current fluctuation amplitude are the main sources of current card feature values extracted.

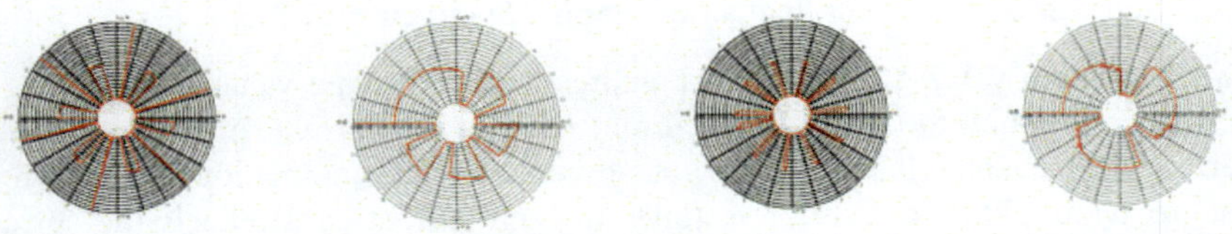

a. Unreasonable start after pump depletion

b. pump depletion

c. Frequent short cycle operation

d. Gas influence

Fig. 3 Current cards under several typical working conditions

As the motors may have different rated currents in the samples obtained, the same failure may occur to motors with different rated currents, so it is necessary to normalize the currents in motors with different rated currents. With the rated current as the standard, the real-time current is normalized [8]. The normalization method of current is:

$$I_{\mathrm{ni}} = \frac{I_{\mathrm{ai}}}{I_{\mathrm{r}}} \tag{1}$$

where, I_{ni} is the normalized current of current of real-time current, dimensionless.

I_{ai} is real-time current, A.

I_{r} is rated current, A.

By analyzing the characteristics of different current cards, we can see that there are many parameters that can reflect the current situation and current fluctuation. By analyzing the standardized current value, the current average value, summation of current fluctuation amplitude, maximum value of the single current fluctuation, solitary peak degree of single current fluctuation maximum value, ratio of the summation of current fluctuation value to current average value in current non-zero time, time when current is zero and time when current is non-zero are obtained. Defined as follows:

(1) current average value of non-zero current in a single current card (eigenvalue 1).

After the normalization method of current values mentioned above, the values recorded in the current card are values between 0 and 1. So for a single current card, the average value of non-zero current is the average value of all normalized current values. The average value of these currents is equivalent to the normalized area of the closed shape surrounded by current cards. It is expressed as follows:

$$S_{\mathrm{n}} = \sum_{I_{\mathrm{ni}} \neq 0} I_{\mathrm{ni}} \tag{2}$$

where, S_{n} is the normalized area, dimensionless;

I_{ni} is the normalized current of real-time current, dimensionless.

(2) Summation of current fluctuation amplitude (eigenvalue 2).

The fluctuation value of current is an expression of absolute value after making a difference between the normalized current value at the current time point and the normalized current value at the previous time point. The summation of the current fluctuation amplitudes at different times in each current card reflects the current fluctuation in the whole current card. Equivalent to the perimeter of the closed shape enclosed by the current card. Expressed as follows:

$$I_{\mathrm{fi}} = \left| I_{\mathrm{ni}} - I_{\mathrm{n(i-1)}} \right| \tag{3}$$

$$C_{\mathrm{n}} = \sum I_{\mathrm{fi}} \tag{4}$$

where, I_{fi} is the current fluctuation at point i, reflecting the difference between the normalized current value at current time point and the normalized current value at the previous time point, dimensionless;

C_{n} is the normalized perimeter of closed shape enclosed by the current card, dimensionless.

(3) Maximum value of single current fluctuation (eigenvalue 3).

The maximum value of single current fluctuation refers to the one with the largest value among the current fluctuation values. In the current card, it is generally shown when the pump just starting or just stopping. Both starting and stopping the pump show the feature of fluctuation from one current range to another, and the difference lies in which area has the lower current value before and after the maximum value of this single current fluctuation. When the pump is started, the current rises from 0 to around the peak, while when the pump is stopped, the current decreases from around the peak to 0. That is, by judging the sequence of the maximum current fluctuation value and the minimum current value, the pump start-stop state can be judged. The pump starts when there is a minimum current value and followed by a maximum current fluctuation value. While the pump shuts down when there is a maximum current fluctuation value and followed by a minimum current value.

$$I_{\mathrm{fmax}} = \max(I_{\mathrm{fi}}) \tag{5}$$

where, I_{fmax} is the maximum value of normalized current fluctuation value, dimensionless.

(4) The current solitary peak degree at the maximum value of single current fluctuation (eigenvalue 4–7)

It indicates the isolation degree of the maximum value of a single current fluctuation, which reflects the duration of the series of current maximum fluctuation values.

$$T_{\mathrm{a}} = \sum T_{\mathrm{fmax}} \tag{6}$$

where, T_{fmax} is the duration of maximum current fluctuation, min;

T_{a} is the summation of the duration of maximum current fluctuation, min;

In addition, the difference between the minimum and maximum value of current fluctuation at the maximum value of current fluctuation and its surroundings can also reflect the isolation degree of current fluctuation. According to the current data density in the current card, generally, if the duration of a single current fluctuation is less than 6 min, only the peak of one point will be displayed in the graph, and if the duration is longer than 6 min, a staircase shape will appear. Therefore, the current peak, three points before and after, a total of 7 points, and data with a duration of

36 min are selected to judge whether the current fluctuation data is in a corresponding state.

$$\Delta I_1 = I_{\text{fmax}} - \min I_{\text{before}} \tag{7}$$

$$\Delta I_2 = I_{\text{fmax}} - \max I_{\text{before}} \tag{8}$$

$$\Delta I_3 = I_{\text{fmax}} - \min I_{\text{after}} \tag{9}$$

$$\Delta I_4 = I_{\text{fmax}} - \max I_{\text{after}} \tag{10}$$

where, I_{fmax} is the maximum value of current fluctuation, dimensionless;

I_{before} is data of 3 points before the peak value of current fluctuation data, dimensionless;

I_{after} is data of 3 points after the peak value of current fluctuation data, dimensionless;

ΔI_1, ΔI_2, ΔI_3, ΔI_4 are the difference between the peak values of current fluctuation data and the adjacent current fluctuation value, dimensionless.

(5) Current fluctuation intensity of unit current intensity during non-zero current time (eigenvalue 8).

It reflects the intensity of standardized current fluctuation. The ratio of the total fluctuation intensity of current to the average value of current. It is the ratio of the perimeter to the area of the closed figure formed by the current card.

$$R = \frac{S_{\text{n}}}{C_{\text{n}}} \tag{11}$$

where, R is current fluctuation intensity of unit current intensity in non-zero time, dimensionless;

S_{n} is normalized area, dimensionless;

C_{n} is normalized perimeter of closed figure enclosed in current card, dimensionless.

(6) Duration of current zero value and non-zero value (eigenvalue 9–10).

The duration of current zero value and non-zero value in a single current card is different, but the ratio of the duration of current zero value and non-zero value in current cards under the same working condition has certain regularity. The ratio of the duration of current zero value and non-zero value corresponding to a certain working condition in a single current card or the ratio of the two is in a certain numerical range, so the respective duration of current zero value and non-zero value can also be used as the judging condition of working condition [1].

$$T_{\text{is0}} = \sum T_{\text{ni}}(I_{\text{ni}} = 0) \tag{12}$$

$$T_{\text{isn0}} = \sum T_{\text{ni}}(I_{\text{ni}} \neq 0) \tag{13}$$

where, T_{is0} is the total duration of normalized current zero value, min;

T_{isn0} is the total duration of normalized current non-zero value, min.

According to the parameter processing method proposed above, the characteristic parameters extracted from each current card are summarized, and some examples of the characteristic parameter matrix and the corresponding working condition parameter matrix are shown in Table 1 and 2.

The current eigenvalue matrix in Table 1 can be used as the input matrix of BP neural network training, and the working condition parameter matrix in Table 2 can be used as the output matrix of BP neural network training. The input matrix and the output matrix are substituted into BP neural network together, and the weight matrix of BP neural network can be obtained after training.

Table 1 Ammeter characteristic parameter matrix extracted from some Ammeter cards

No	EV1	EV2	EV3	EV4	EV5	EV6	EV7	EV8	EV9	EV10
1	0.5501	10.225	0.9878	0	0.0002	0.9878	0.9875	18.58	100	141
2	0.1511	14.045	1	0	0.0193	1	0.9806	92.901	203	39
3	0.6786	12.599	0.9785	0	0.0033	0.9785	0.975	18.564	60	181
4	0.2153	1.4776	1	0.0049	0	0.9950	1	6.8601	182	60
5	0.1647	17	1	0	0.0533	1	0.9466	103.15	196	46
6	0.9833	2.3535	0.0314	0.0045	0.0009	0.0269	0.0305	2.3933	0	242
7	0.2033	1.6738	1	0.0070	0	0.9929	1	8.2329	182	60

Note EV means eigenvalue

Table 2 The working condition parameter matrix corresponding to Table 1

No	WP1	WP2	WP3	WP4
1	1	0	0	0
2	0	1	0	0
3	1	0	0	0
4	0	0	0	1
5	0	1	0	0
6	0	0	1	0
7	0	0	0	1

Note WP means working parameters

4.3 The Establishment and Training of BP Neural Network

According to the accuracy and speed requirements of the model, a three-layer BP neural network is established, with the extracted eigenvalues as the input layer and the corresponding failure indication as the output layer, and a neural network model is established for training. Therefore, it is necessary to process a current card of 241 points in 1 440 min into qualified neural network input data. According to the method mentioned above, 10 different eigenvalues are extracted, so the number of input layer nodes of BP neural network is 10. The neural network model can be used to make judgments in 4 working conditions, so the number of nodes in the output layer is 4; To determine the number of nodes in the middle layer and the number of learning steps, it is necessary to run the model several times in combination with the actual data, and try to select the neural network model structure with shorter time and higher accuracy.

The length of time required for neural network training is related to the number of hidden layer nodes in the neural network structure, as well as the number of iteration steps during training. By adjusting the number of iteration steps and the number of hidden layer nodes and observing the time when the program runs the results, the relationship among learning times, learning difficulty and learning time can be known.

According to the convergence of the learning curve, the training completion of the model is determined, and the accuracy of the model is determined by verifying the results with test set data. It should be noted that because of its own algorithm characteristics, BP neural network may sometimes fall into the local optimum in the training process. In this case, it needs to be retrained, and the training ends when it meets the requirements.

By modifying the number of hidden layer nodes and iteration steps several times to run the program, observing the running time and result accuracy of the program, and considering both factors comprehensively, the appropriate number of hidden layer nodes and iteration steps can be determined.

4.4 The Use and Prediction of BP Neural Network Model

The neural network model established according to the above steps is tested by the test data set, and the confusion matrix of the test data is obtained. When the error of the confusion matrix of the test data reaches a precision high enough, it is considered that the accuracy of the model meets the requirements, that is, the trained model is an available model, and then the model can be used for new prediction of the data. A confusion matrix form of the training model is shown in Table 3.

There are many evaluation methods of confusion matrix, and the evaluation indexes include various mathematical concepts such as precision and recall rate, and their emphasis is also different. However, the overall evaluation index is that the

Table 3 An example of an obfuscation matrix for test data

	Predicted WP1	Predicted WP2	Predicted WP3	Predicted WP4
Actual WP1	1	0	0	0
Actual WP2	0	1	0	0
Actual WP3	0.6667	0.3333	0	0
Actual WP4	0	0	0	1

elements on the main diagonal are closer to 1, and the other elements are closer to 0, which means that the test results of the model are closer to the actual marking results, and the training results of the model are better. This means that the higher the correlation degree between a predicted working condition parameter and the actual working condition parameter, the lower the correlation degree between the predicted working condition parameter and other working condition parameters, and the better the training result.

According to the above analysis, when all the elements on the main diagonal of the confusion matrix of the test data of working conditions are 1, it shows that the neural network has completely correctly classified the parameters of different working conditions. The larger the value of elements on the non-principal diagonal, the larger the error, the worse the classification. The mean square deviation of elements on the non-principal diagonal was calculated as the error rate for error analysis.

$$\varepsilon = \sqrt{\frac{1}{N}\sum_{i \neq j} \lambda_{ij}^2} \tag{14}$$

where, λ_{ij} is the elements in the matrix, dimensionless;
ε is the mean square deviation, dimensionless;
N—is the number of test samples.

5 Case Analysis and Evaluation

According to the above method, the current cards under four working conditions of pump evacuation, frequent short-period operation, overload shutdown and normal operation are extracted from the sample database for data preprocessing, the input parameters and their corresponding working condition output parameters are sorted out, and the input data is sorted out for 44 wells, which is used to build a BP neural network model for learning and training. According to the training results, some data separated in advance after data preprocessing but not participating in training and without fault markers are predicted for failure, and compared with the failure types marked by itself to verify the accuracy of the model.

Here, a unified learning step number of 300 steps is set, and 11–18 are selected as the values of hidden layer nodes to try, so as to obtain a low error rate. The change of error rate with the number of intermediate hidden layer nodes is shown in Fig. 4, and the change of learning time with the number of hidden layer nodes is shown in Fig. 5.

It can be seen from Fig. 4 that the test error will change with the number of nodes in the middle layer. within the range of 11–18, when the number of nodes in the hidden layer is set to 15, the lower test error will be obtained. Furthermore, as can be seen from Fig. 5. The nonlinear degree of test time varying with the number of hidden layer nodes is more obvious when there are more nodes, and the number of hidden layer nodes is determined to be 15 by comprehensive consideration.

After the number of hidden layer nodes is determined, the learning steps are modified respectively, and the learning steps are selected from 200 to 3000 unequal intervals for testing. The relationship between the test error rate and the change with the number of learning steps is shown in Fig. 6, and the change of training time with the number of learning steps is shown in Fig. 7.

It can be seen from Fig. 6 that when the number of iteration steps is small, the test error will greatly decrease with the increase of iteration steps, and when the number of iteration steps is large, the test error will not change much. It can be seen from Fig. 7 that there is almost a linear relationship between the total time of

Fig. 4 The variation of error rate with the number of nodes in the hidden layer

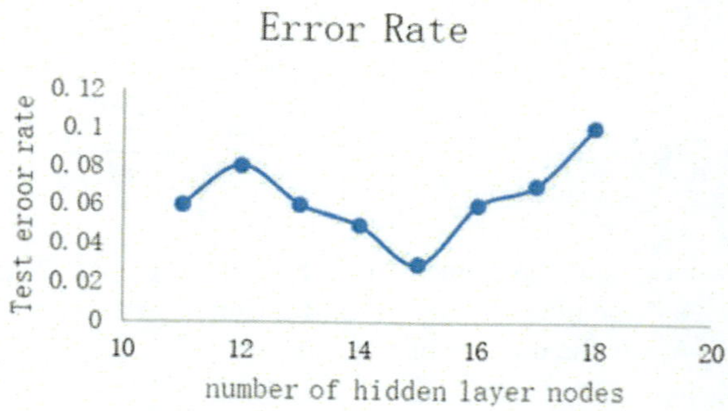

Fig. 5 The relationship of training time with the number of nodes in the hidden layer

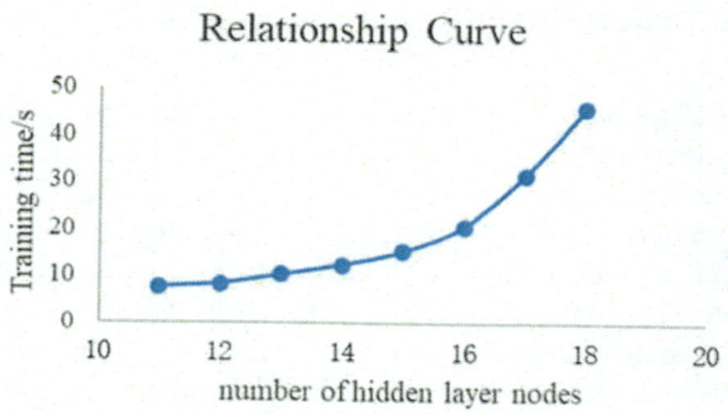

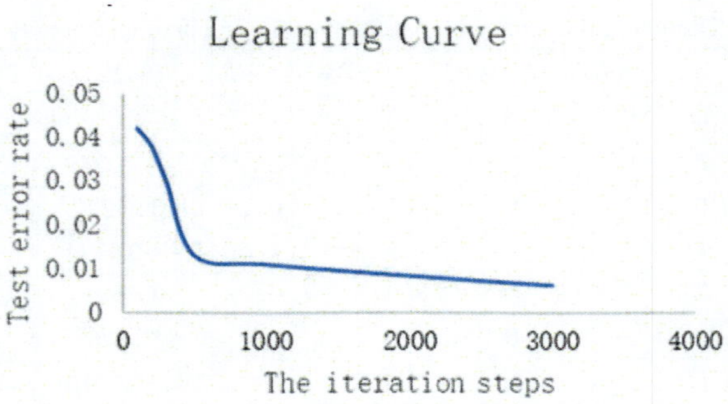

Fig. 6 Learning curve of test data

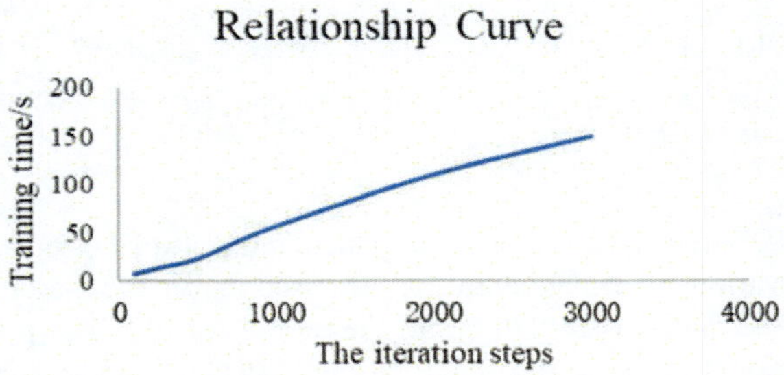

Fig. 7 The relationship of training time with the iteration steps

program calculation and the number of iteration steps, which means that there is a linear relationship between the difficulty of program calculation and the number of iteration steps. After the comprehensive consideration, choosing 300 as time step can satisfy fast and accurate pattern recognition.

According to the previous analysis, when the number of hidden layer nodes is 15, the error rate is low. When the number of time steps is selected as 300, the test accuracy has reached a considerable level, so there is no need to further increase the number of learning steps to avoid unnecessary waste of running resources. Therefore, for the whole network model, it is determined that the selected time step is 300 steps and the hidden layer nodes are 15, which can meet the requirements of accuracy and algorithm complexity. The confusion matrix obtained after training the obtained test data is shown in Table 4.

Table 4 Confusion matrix of test data

	Predicted WP1	Predicted WP2	Predicted WP3	Predicted WP4
Actual WP1	1	0	0	0
Actual WP2	0	1	0	0
Actual WP3	0	0	1	0
Actual WP4	0	0	0	1

Table 5 Comparison of expected results and predicted results

No	AWP1	AWP2	AWP3	AWP4	PWP1	PWP2	PWP3	PWP4
1	0	0	0	1	−0.0557	−0.0122	−0.0316	1.0342
2	0	1	0	0	−0.2249	1.2685	−0.0007	−0.0433
3	1	0	0	0	0.9922	−0.0202	−0.0160	0.0159
4	0	0	0	1	−0.0122	−0.0703	−0.0950	0.9823
5	1	0	0	0	0.8340	0.2191	−0.0354	−0.0090
6	1	0	0	0	1.1246	−0.0845	0.0545	−0.0539
7	0	0	0	1	−0.0010	−0.0775	−0.1017	0.9741
8	1	0	0	0	0.9541	0.0234	−0.0800	0.0129
9	0	0	1	0	−0.1306	0.0163	1.0479	0.0444
10	0	0	1	0	−0.0797	0.0085	1.0471	0.0235

Note AWP means actual working condition parameter; PWP means predicted working condition parameter

After the training is completed, multiple pieces of data separated in advance under various working conditions are used to remove the failure type indication, and then the data are imported into the model for test. The predicted value of the model is compared with the actual failure indication value, and the predicted result matrix is shown in Table 5.

Each line in the table represents a piece of input data to be predicted, the first four columns represent the actual marked working conditions, and the last four columns represent the working condition data calculated by the model. For each row of data, only one of the first four columns is 1, and the rest are all zero, which means that for this data, the marked working condition is the working condition corresponding to the parameter sequence that this parameter is 1 and other parameters are 0. When the data series in the last four columns are closed to the data series in the first four columns, it is considered that the model is successful in prediction. When the predicted results are obviously separated from the actual results, it is considered that there is an error in the model prediction.

Through the comparison and verification of the above model and the actual data with the output data, the model has realized accurate and correct identification of 10 working condition data, and the root mean square error of identification is 0.0747, which means the prediction accuracy reaches 92.53%, meeting the reasonable requirements. It is considered that the BP neural network model has achieved the expected effect in the diagnosis of electric pump working condition.

6 Conclusion

Through the theoretical analysis and practical verification of pattern recognition of current card using BP neural network, the following aspects are obtained:

(1) The correspondence between current card and working condition is a nonlinear correlation, and the traditional manual empirical discrimination method has disadvantages in accuracy and timeliness. BP neural network has strong applicability to pattern recognition and classification of nonlinear problems. The advantages of BP neural network have a strong correspondence with solving the problems in current card recognition, and it is feasible to use BP neural network for pattern recognition of current cards.

(2) After training, it is found that the learning curve converges quickly, the prediction quality is high, the accuracy is good, and it meets the requirements of reasonable error range.

(3) After the establishment of the current card pattern recognition model based on BP neural network, once the data of the current card is imported, the fast pattern recognition can be realized, and the real-time diagnosis of the working condition of the electric submersible pump can be realized under the condition that the incoming speed of the data can be guaranteed, which will become an important part of building a smart oilfield.

References

1. Chen, Z., Feng, D., Zhu, M., et al.: Study on feature extraction method of current card based on pattern recognition. China Pet. Mach. (02), 38–41+61 (2004)
2. Yu, J., Feng, D., Lu, Y., et al.: Diagnosis of submersible oil pump based on neural network. Mech. Eng. (05), 54–56 (2005)
3. Gan, L., Wang, Y., Wang, B.: Application of BP neural network in failure diagnosis of ESP production wells. Oil Drill. Prod. Technol. **33**(02), 124–127 (2011)
4. Han, G., Chen, M., Zhang, H.: Real-time monitoring and diagnosis of electrical submersible pump. Paper Present at the SPE Annual Technical Conference and Exhibition Held in Houston, Texas, USA, 28–30 Sept 2015. SPE-174873-MS (2015). https://doi.org/10.2118/174873-MS
5. Gu, X., Han, G., Zhu, B.: Working condition diagnosis of ESP well based on wellhead pressure-out analysis. Oil Drill. Prod. Technol. **38**(04), 514–518 (2016)
6. Awaid, A., Al-Muqbali, H., Al-Bimani, A., et al.: Alastair Baillie. ESP Well Surveillance using pattern recognition analysis, oil wells, petroleum development Oman. Paper prepared for Presentation at the International Petroleum Technology Conference held in Doha, Qatar, 20–22 Jan 2014. IPTC 17413 (2014)
7. Peng, K.: Fault diagnosis of electric submersible pump based on BP neural network. J. Petrochem. Univ. **29**(01), 76–79 (2016)
8. Zhang, R., Yi, Y., Xu, L., et al.: A new method for intelligent diagnosis of operating conditions of electric submersible pump wells based on feature recognition. J. Shengli College China Univ. Pet. **32**(04), 41–45 (2018)

The Application of Encryption Algorithm in Information Security Reflected

Chuanyue Li

Abstract Computer mass data transmission and information security has always been the focus of scientific research scholars and industry economic innovation. The file system is an important part of the operating system to manage and store file information. File access must be performed through the file system. Therefore, encryption technology is integrated during the system running to ensure the physical security of file data. Based on the understanding of network file transfer encryption key technology on the basis of the proposed hybrid encryption based on AES and ECC algorithm, to solve the previous single password encryption algorithm efficiency is too low or loopholes in management, and design system for the function and performance test and analysis, the final result shows that this article research institute to build file encryption system, can meet the demand of practical work.

Keywords Encryption algorithm · Information security · AES. ECC · The information entropy

1 Introduction

In the context of the information age, the Internet platform with computer technology as the core has brought new changes to the development of modern society and economy [1–3]. While people enjoy the technological achievements, they also face the threat of information being tampered with or forged. Especially for information in a file system, it is easy to compromise information security during management operations. Faced with this phenomenon, data encryption technology has been widely used. Many data encryption technologies can be quickly realized by using some software, but a more perfect encryption algorithm will not affect the system performance, but can improve the key performance of the system. For example, the encryption algorithm can compress the data at a certain rate during operation. The integrated encryption service can be applied to the file system to solve problems such

C. Li (✉)
Jiangsu University of Science and Technology, Zhenjiang, China
e-mail: 1375236797@qq.com

H. Dai (ed.), *Computational and Experimental Simulations in Engineering*,
Mechanisms and Machine Science 119,
https://doi.org/10.1007/978-3-031-02097-1_5

as data destruction and malicious access. In this way, authorized users can access encrypted files randomly without compromising encryption keys, which improves file transfer security. Through reading existing information security and encryption algorithm literature, it can be seen that the research direction of information security technology and cryptography technology at home and abroad has always been the focus of scientific research. This paper mainly explores the encryption technology of advanced Encryption Standard (AES) and elliptic Curve cryptosystem (ECC), which are widely used, and studies the encryption algorithm technology of digital signature, symmetric key and public key. Based on The Windows operating platform, the application software of file encryption system is constructed. In this way, network file information can be safely and stably transmitted [4–6].

2 Methods

2.1 Algorithm Analysis

On the one hand, the basic function of DES algorithm is to ensure the security of information transmission, communication parties need to use the same key during encryption and decryption, although this algorithm is widely used, but because the key length is too short, it is difficult to meet the requirements of open network technology at this stage. Therefore, in the late 1990s, the U.S. government chose the AES cryptography algorithm, a new encryption standard proposed by the National Institute of Standards and Technology. From the practical point of view, as a new data encryption standard, it has the advantages of flexibility, high efficiency, security, convenience and operation. The 128-bit key contained in it is 10 times stronger than the original algorithm's 56-bit key. According to the different encryption methods of plaintext messages, symmetric encryption algorithms can be divided into two types: one is the block cipher and the other is the stream cipher. The AES algorithm studied in this paper refers to the block cipher algorithm. Both the input and output packets and the packets during encryption and decryption are 128 bits, and the key length K reaches 128, 192 and 256 bits. The specific decryption and encryption flow chart is shown in Fig. 1 [7, 8].

On the other hand, ECC is a conclusion proposed by et al. based on the study of the difficulty of solving the discrete logarithm problem on the point group of elliptic curves. In the already defined work cryptosystem, ECC has low overhead, high processing, strong security and other performance in practical application. Up to now, this kind of algorithm is still the best application algorithm to solve logarithmic problem. ECC only needs to use a smaller key length to achieve the same security as RSA.

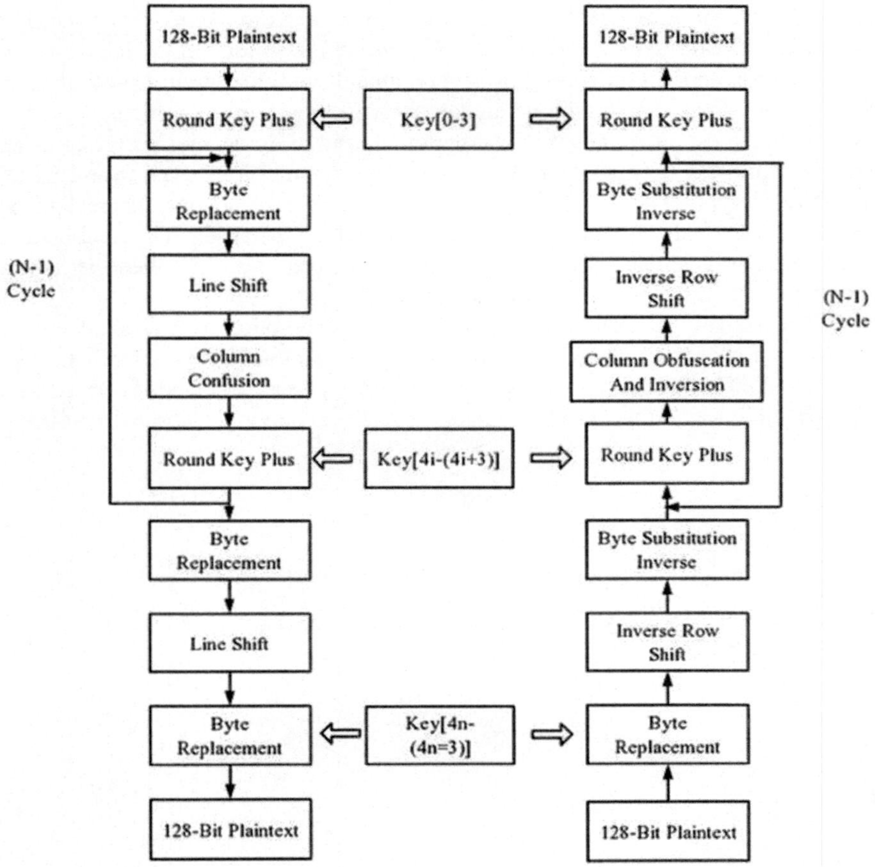

Fig. 1 Flow chart of encryption and decryption based on algorithm

2.2 Design and Analysis of Encryption System

In view of the various requirements of information security in recent years, the encryption system involved in network files should comply with the following principles: first of all, the index data confidentiality, after the completion of the system design, to ensure the security of the transmitted data design function, so as to avoid external attackers at will monitor. Secondly, it is necessary to realize the integrity of data. The data files transmitted by network platform may be lost due to malicious damage or network delay, which requires the use of digital signature to ensure the perfection of the information obtained. Finally, in line with the reliability of the real data, assuming that the file receiver cannot accurately judge the authenticity of the information, it is difficult to obtain effective data information, so it is necessary to encrypt the summary of the information digital, so as to avoid being forged or tampered with during transmission, and thus ensure the validity of the file data.

Symmetric encryption algorithm is very suitable for encryption chunk of data, but because of the distribution of the key management is more difficult, so I can and combined use of asymmetric encryption algorithm, in which the former as an encrypted message data, while the latter generating symmetric encryption and digital signature encryption keys required for digital envelope, such not only can realize the digital signature, can also improve data security during file transfer. The combination of the two forms a new hybrid cryptosystem, which can also be called the hybrid encryption algorithm based on AES and ECC.

Combined with the analysis of encryption and decryption flow chart of sender and receiver shown in Figs. 2 and 3, it can be seen that this paper comprehensively analyzes all kinds of encryption algorithms currently used, and fuses MD5, private key and public key cryptosystems together, effectively solving the problem of using

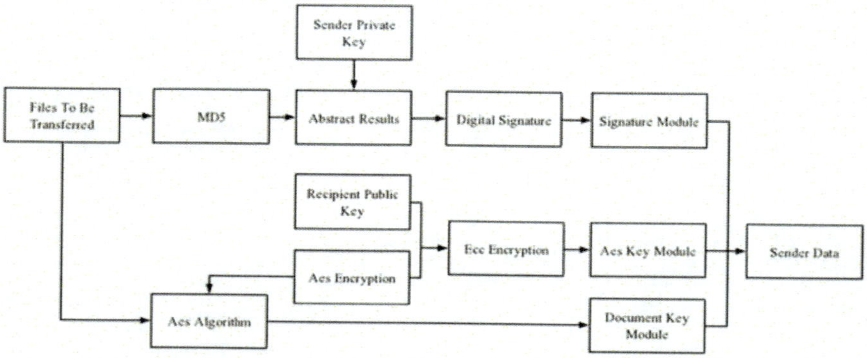

Fig. 2 Encryption and decryption flow chart of sender

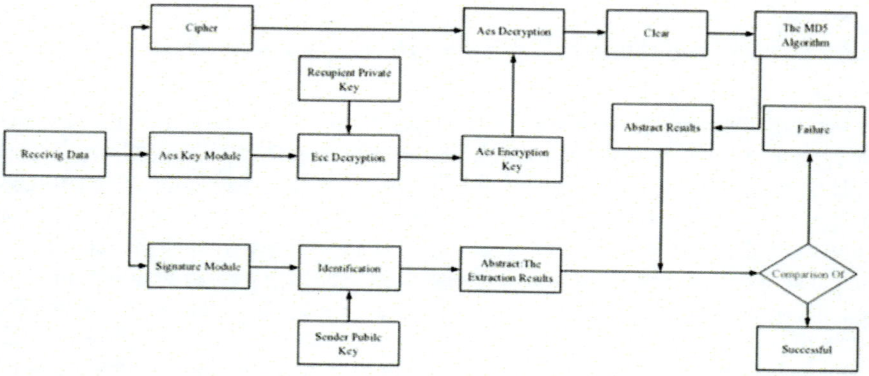

Fig. 3 Encryption and decryption flow chart of the receiver

Fig. 4 Network encryption architecture diagram

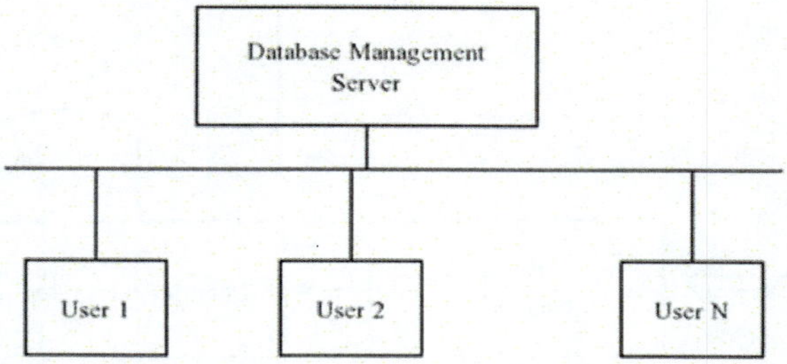

only one encryption algorithm traditionally. And according to the database hierarchical management scheme to deal with the system user rights management work, to ensure that all kinds of users can design rights according to their own needs. This hybrid encryption system not only effectively utilizes the advantages of symmetric key for convenient operation and easy implementation, but also fully demonstrates the advantages of public key cryptosystem, convenient management and strong security. Therefore, it has a very broad development prospect in network information security transmission [9, 10].

From a practical point of view, the management goal of network information security is to transmit data quickly and effectively. Therefore, the corresponding network encryption structure is shown in Fig. 4.

Combined with the above analysis, it can be seen that the network file encryption system includes a server and multiple user computers. The computer of any node can encrypt and decrypt the transmitted files, and the interaction between the user terminal and the server can make the network platform use E-mail or some form of transmission. In this process, users only need to log in to the network system after authentication and encrypt and decrypt files according to their permissions. From the perspective of the client, it is the basic condition for legitimate users to access encrypted files. Users can use the client to send identity authentication requests to the server. Assuming that the user has the assignment permission, the user can directly process the ciphertext. Under normal circumstances, the program used by the client involves multiple modules such as identity authentication, key management and file decryption, and the relationship among them is shown in Fig. 5.

For example, the file decryption module is mainly to decrypt the encrypted file transmitted by the server, so as to restore the original file information, the specific operation involves the following points: first of all, the ID number and D5 digital summary information in the file is accurately extracted; Secondly, the private key is used to provide file serial number, user and password to the server. Again to receive the relevant file information returned by the server and decrypt processing; Finally, the decrypted source file and the digital abstract are compared and analyzed. Assuming that the two are identical, it is proved that the encrypted file has not been arbitrarily changed during transmission. The key management module mainly controls the public key file of the server scientifically, thus provides the public key and private key information needed for the client to decrypt, and uses the public

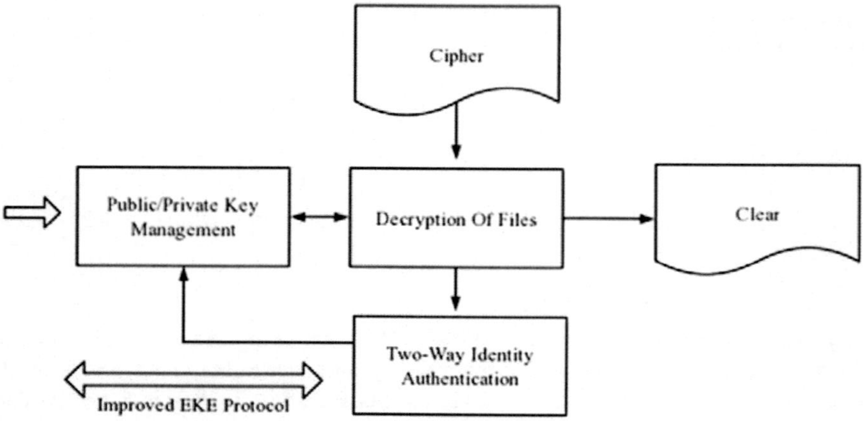

Fig. 5 Relational structure diagram of client module

key of the server to encrypt the sent information, and the private key to decrypt the received information. In the module of identity authentication, not only the service function of user authentication can be realized, but also the login password can be changed scientifically.

Encryption algorithm and key management, as the core content of encryption system design, are proposed by using relevant formulas and rules, which not only accurately master the transformation rules between plaintext and secret, but also further guarantee the security of information transmission with the continuous innovation of computer technology.

3 Result Analysis

This paper chooses the mode of mixing two encryption algorithms, designs and develops the corresponding network system, and uses Visual C++ language to write test programs, selects the broadband network used by a certain enterprise to transmit information files, encrypts and decrypts data in Windows system. In order to verify the network transmission data identity authentication and digital signature and other functions.

Before the test algorithm is executed, a random file text is made and stored in the root directory of drive D of the computer. Select the test file in the interface of adding system files, click the button of encryption file, encrypt the transferred file according to the above research, and store it in the folder set as expected. Open the encrypted folder with Notepad, and the display result is different from the original information. After selecting the test secret file for decryption processing, it is found that the relevant content is consistent with the source file. In the process of this test, a buffer-size M byte area should be found in the internal storage as a Buffer, and random value should be assigned to this data block. T-begin, the record system time, should be regarded as the time value before encryption, and n cycle encryption of data in this area should be carried out by the call of encryption function. The system time after the End of the operation is t-end, which is regarded as the time value after encryption. Then the speed value in Mbps unit can be calculated by using the following formula:

$$(Buffer\_size * n * 8)/\left[(t\_End - t\_Begin) * 10^{-3}\right]$$

The data speed test results of the memory buffer in this paper are shown in Table 1.

Table 1 Test results

Number of data block cycles		The amount of data (M)	Time and speed				Average value (MBPS)
Data block (K)	1	100	Time (MS)	32,327	32,312	32,328	23.07
Cycles	102,400		Speed (MBPS)	23.08	23.07	23.06	
Data block (K)	512	1	Time (MS)	20	27	14	51.14
Cycles	2		Speed (MBPS)	49	36.65	67.76	
Data block (K)	512	100	Time (MS)	25,635	25,646	25,620	30.47
Cycles	200		Speed (MBPS)	30.43	30.53	30.46	
Data block (K)	1	100	Time (MS)	25,677	25,656	25,579	30.34
Cycles	100		Speed (MBPS)	30.32	30.43	30.29	
Data block (K)	4	100	Time (MS)	24,989	112,156	24,896	30.44
Cycles	25		Speed (MBPS)	30.42	30.47	30.48	
Data block (K)	16	400	Time (MS)	112,143	112,156	112,149	30.54
Cycles	25		Speed (MBPS)	30.41	30.47	30.73	
Data block (K)	32	800	Time (MS)	199,789	199,781	199,785	30.44
Cycles	25		Speed (MBPS)	30.41	30.48	30.47	

From this analysis, it can be seen that if smaller data is selected for fast encryption processing, the number of calls to the actual encryption system algorithm will increase accordingly, and the influence of other factors on the algorithm will continue to rise, so the speed value will become smaller and smaller. Given that the allocated buffer area is large and does not exceed the actual available memory allocated by the system, memory problems can easily occur during encryption.

The test results of file system encryption and decryption are shown in Table 2.

By comparing the data analysis of the first group and the second group in the table above, it can be seen that the shortest time required for encryption and decryption of transmitted files in the encryption system is 26.37Mbps and 25.63Mbps. By comparing the data, it can be seen that the actual speed can reach 18.95Mbps and 17.74Mbps when processing oversized file data. This proves that the file encryption system is efficient and fast.

Table 2 Results of file system encryption and decryption

File size (m)	Data module size (K)	Encryption/decryption speed (MBPS)						Average speed (MBPS)
58.2	64	Encryption	18.88	20.08	21.63	20.01	16.38	18.86
		Decryption	18.02	18.43	18.01	18.68	18.82	18.01
	512	Encryption	18.68	21.60	18.68	18.22	16.86	18.26
		Decryption	18.82	18.45	16.08	18.88	18.65	18.82
	50	Encryption	21.10	20.15	18.84	22.65	18.23	20.21
		Decryption	18.16	18.81	16.81	18.81	20.34	18.01
	64	Encryption	16.88	21.62	16.65	16.81	18.38	18.53
		Decryption	18.32	16.62	18.02	15.83	16.62	16.48
	512	Encryption	16.81	18.01	20.15	22.88	18.46	18.88
		Decryption	18.06	18.62	14.55	18.80	16.88	16.40
	50	Encryption	16.84	18.84	20.05	22.11	18.83	18.56
		Decryption	18.04	18.63	14.53	18.08	16.56	16.31
	64	Encryption	16.36	18.23	16.43	16.66	18.83	16.62
		Decryption	15.55	16.60	18.38	15.18	16.56	16.88
	512	Encryption	16.66	16.83	18.18	18.08	16.56	16.60
		Decryption	18.36	16.66	15.88	15.46	16.41	16.88
	50	Encryption	16.80	16.82	16.46	18.35	16.50	16.63
		Decryption	18.44	16.66	16.44	16.03	16.66	16.26

4 Conclusion

To sum up, in the innovation of computer technology, encryption and decryption technology for network information is also facing new development opportunities. Therefore, researchers should focus on improving the security of using keys on the basis of guaranteeing the operation efficiency of encryption system, which is also the focus of research on encryption algorithms at this stage. In view of the advantages and disadvantages of symmetric key and asymmetric encryption system, this paper proposes a hybrid application mode. Although it provides a basic guarantee for the effective transmission of network files, there are still many problems in practice and need to be further improved.

References

1. Zhang, J., Guo, X., Fu, X.: Analysis of AES encryption algorithm and its application in information security. Inf. Netw. Secur. **05,** 31–33 (2011)
2. Liu, Y., Liu, Y.: Application of public-key encryption algorithm in network information security. Inf. Sci. **21**(001), 93–94 (2003)
3. Liang, W.: Application of composite chaotic encryption algorithm in power system information security. Coast. Enterp. Sci. Technol. (009), 21–23 (2012)
4. Zhou, K., Zhu, R.: Application of composite chaotic encryption algorithm in power system information security. East China Sci. Technol. (Academic Edition) (009), 193–193 (2013)
5. Xin, M.: Explore the application of hybrid encryption algorithm in Internet of Things information secure transmission system. Bus. Econ. (07), 80–81 (2015)
6. Ji, X., Zhang, H.: Application of DES data encryption algorithm in computer communication. Inf. Rec. Mater. **V.20**(08), 90–91 (2019)
7. Zheng, J., Zhou, L.: Inf. Commun. (002), 209–210 (2016) (in Chinese)
8. Wang, W., Wang, H.: Application of AES encryption algorithm in air defense system. Inf. Res. **04**, 73–75 (2011)
9. Kim, S.S., Kim, Y.J., Carayannis, E., et al.: The effect of compliance knowledge and compliance support systems on information security compliance behavior. J. Knowl. Manag. 986–1010 (2017)
10. Qin, J.: Application of MD5 encryption algorithm in information security of campus network. Mod. Shopp. Malls **000**(020), 188 (2008)

Multiscale Finite Element Technique for Mathematical Modelling of Multi-physics Processes in Heterogeneous Media

E. P. Shurina, N. B. Itkina, D. A. Arhipov, D. V. Dobrolubova, A. Yu. Kutishcheva, S. I. Markov, N. V. Shtabel, and E. I. Shtanko

Abstract In real-life applications, it is not always possible to decouple multiple physical processes occurring simultaneously in the medium, thus indirect methods based on, for example, electromagnetic measurements may be used to adequately describe the macroscopic behavior of the studied objects. We consider a multiscale problem of modelling the electromagnetic field in the hydrocarbon-bearing rock under the thermal and mechanical effects. The goal of this study is to develop a unified approach to modelling the multi-physical processes in geological media, such as oil reservoirs and carbon deposits. Geological media are characterized by the presence of multiple geometrical scales and highly heterogeneous physical properties. We use upscaling techniques and the effective characteristics of the media that adequately reflect its macroscopic behavior. We carry out mathematical modelling of the processes of heat and mass transfer, elastic deformation and mesoscale electromagnetic interactions with regard to the complex internal structure of the media. Our algorithm for the macroscopic description of the behavior of the geological media is based on the methods of the effective medium theory and numerical homogenization techniques. We present the results of the mathematical modelling of heat and mass transfer, elastic deformation and electromagnetic field in a heterogeneous medium, obtained using the computational schemes based on the multiscale finite element methods.

Keywords Heterogeneous medium · Multi-physics processes · Multiscale non-conforming finite element methods · Numerical homogenization

---

E. P. Shurina (✉) · D. A. Arhipov · D. V. Dobrolubova · A. Yu. Kutishcheva · S. I. Markov · N. V. Shtabel · E. I. Shtanko
Trofimuk Institute of Petroleum Geology and Geophysics, SB RAS, Koptug ave. 3, 630090 Novosibirsk, Russia
e-mail: www.sim91@list.ru

E. P. Shurina · N. B. Itkina · D. V. Dobrolubova · A. Yu. Kutishcheva · S. I. Markov · N. V. Shtabel
Novosibirsk State Technical University, Karl Marx ave. 20, 630073 Novosibirsk, Russia

N. B. Itkina
Institute of Computational Technologies, SB RAS, Academician M.A. Lavrentiev ave. 6, 630090 Novosibirsk, Russia

H. Dai (ed.), *Computational and Experimental Simulations in Engineering*, Mechanisms and Machine Science 119,
https://doi.org/10.1007/978-3-031-02097-1_6

1 Introduction

As the deposits of retrievable hydrocarbon in the world are rapidly decreasing, the development of the unconventional hydrocarbon reserves is of great importance. The technologies commonly used in the development of the high-viscosity oil reservoirs are hydraulic fracturing and thermal gas treatment [1, 2]. Both imply sufficient changes in thermal, mechanical and transport properties of the hydrocarbon-reservoir rock. The coal mining implies rock deformation and destruction. Therefore, heat and mass transport, as well as solid deformation, are the focus of the study as the primal physical processes. Under reservoir conditions, it is impossible to measure the thermal, mechanical and transport properties of the medium directly. To address this problem, non-contact electromagnetic sensing methods are employed [3]. The reservoir rock being highly heterogeneous, with intricately interdependent characteristics, multi-physical mathematical modelling is often required to provide the correct interpretation of the sensing data [4–9].

In natural hydrocarbon reservoirs and coal deposits, multiple physical phenomena, such as phase change due to thermal effects, mechanical deformations, fluid transport occur on a variety of time and length scales. A careful choice of an adequate model is, thus, of utmost importance [6]. A multi-physical problem is formulated as a system of partial differential equations with special interface conditions coupling mathematical models of physical processes. Solving such problems in a unified approach is a challenging task. As a natural medium, hydrocarbon reservoirs are also essentially heterogeneous, involving multiple geometric scales, high contrast and anisotropy of physical properties. The discretization method should take into account the specifics of the problem and preserve the global regularity of the discretized mathematical models.

There are general components in the mathematical models of heat and mass transfer processes, elastic deformation of a solid and electromagnetism: the div-grad and curl-curl operators. The unifying core is a spatial mesh including all internal boundaries. Note, the mathematical models should contain physically relevant conjugation conditions for the coupling processes.

Multiscale non-conforming finite element methods are popular for solving multiphysics problems [10, 11]. Today, there is a limited number of publications in which this method of mathematical modelling is implemented. As a rule, the authors limit themselves to a significant reduction of mathematical models and consideration of one-dimensional and, less often, two-dimensional problems [12, 13].

Our goal is to develop a strategy for the mathematical modelling of the multiphysical processes in 3D geological media specific to oil and gas reservoir and coal deposits.

We propose a unified framework for multi-physical modelling of electromagnetic, thermal and stress–strain fields in heterogeneous media, based on the hierarchy of finite element mesh models reflecting the hierarchy of the multiple studied scales. We consider the oil-saturated media and coal-bearing rock as a geometrical model of a two-phase medium consisting of the matrix (sandstone) and inclusions filled with oil or coal. The multiscale nature of the problems is addressed by constructing a hierarchy of the mesh models of the medium, and the computational schemes developed are based on the multiscale finite element approach.

To adequately describe the physical phenomena in natural hydrocarbon reservoirs and coal-bearing rock, it is important to include the process of mechanical deformation in the scope of the multi-physical simulations [7, 8]. For example, a change in the stress–strain state of the medium may result in a change in the thermal properties, fluid saturation, porosity of the medium, etc. To build a dynamic picture of the evolution of temperature or electromagnetic fields in the geological environment at various depths, it is also important to take into account the change in reservoir conditions.

In this paper, we consider a model of elastic deformation of a solid with complex geometry, which is described by an elliptic equation for the displacements. The computational technique is based on the multiscale finite element approach, which accounts for the multiscale and multi-physical nature of the global problem [14, 15].

To describe the temperature field, we use the multiscale thermal conductivity model. To discretize the thermal conductivity model in the matrix, the computational scheme of the heterogeneous multiscale finite element method is applied. The thermal conductivity model in the inclusions is discretized using the discontinuous Galerkin method. Heat transfer conditions on the interface boundary between matrix and caverns are used to match heat conduction models at different levels of the hierarchy.

The process of fluid transport under mechanical and thermal effects is typical of oil and gas reservoirs. Therefore, the fluid dynamics equations are added to the multi-physical problem statement.

Fluid dynamic in caverns filled with oil under thermal and mechanical effects is governed by the Navier–Stokes system of equations. For this problem, we employ the computational technique based on the discontinuous Galerkin method. A special projection technique is used to solve the system of linear equations arising from the discretization. The elastic deformation and hydrodynamics models are coupled via the continuity condition on the normal component of the stress tensor at the interface boundaries between solid and caverns filled with fluid.

Thermo-mechanical and transport characteristics, although adequately capturing the changes in state and structure of the heterogeneous media, may not be measured directly. Electromagnetic measurements, on the other hand, are non-intrusive and are sufficiently sensitive to thermal effects, porosity changes and fluid saturation. Since the direct or alternating current used in electromagnetic measurements can be especially sensitive to different parameters, mathematical modelling is required to give recommendations on the measurement techniques.

Electromagnetic field propagation is described by Maxwell's system of equations. We consider the harmonic electromagnetic field described by the vector Helmholtz equation. To overcome the challenges of the highly heterogeneous structure of the media, we use the computational technique based on the multiscale vector finite element method. The method relies on a hierarchical approach aimed at capturing micro-scale information in the shape functions defined independently at each element

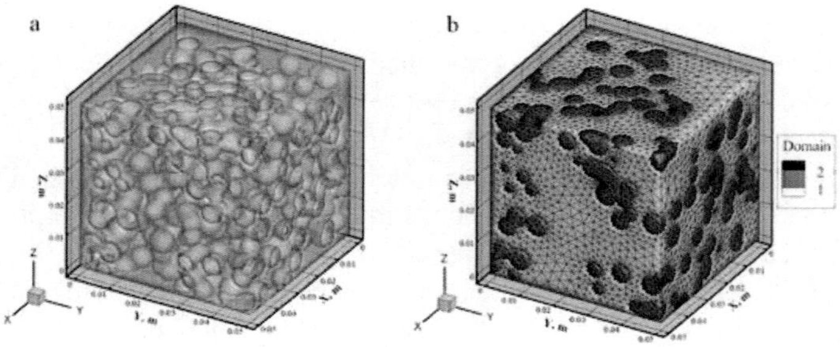

Fig. 1 Ω^{REV}—representative elementary volume (**a**), mesh: 292,783 tetrahedra (**b**)

of the relatively coarse grid. The shape functions are obtained by solving the micro-scale problems on the fine grid inside each coarse element.

Based on the solution of the direct problem, we can compute the effective electromagnetic characteristics, namely, the effective electric tensor for the harmonic electric field. The effective electric tensor is a complex-valued dense second-rank tensor, capturing both conductive and dielectric properties of the media.

The effective characteristics allow one to capture the changes in the heterogeneous media leading to anisotropic effects. Calculated for the relatively small representative elementary volume (REV), the effective electromagnetic tensor may then be used to describe the properties of the reservoir at large scales.

2 Problem Statement

Let Ω^{REV} be a representative volume (see Fig. 1) of a multiscale geological medium Ω. We assume that Ω^{REV} is a heterogeneous object consisting of the matrix (sandstone) and inclusions filled with oil or coal. We consider the problem of mathematical modelling of the heat and mass transfer, elastic deformation and electromagnetic interactions occurring in Ω^{REV} and Ω. To discretize the mathematical models, we apply computational schemes of the multiscale conforming and non-conforming finite element methods. Figure 1b. shows the computational tetrahedral mesh.

The physical properties of the heterogeneous object are presented in Table 1.

3 Thermoelastic Deformation Problem

In this section, we formulate the thermoelastic deformation problem in a fluid-saturated heterogeneous medium. We consider oil as a Newtonian and incompressible fluid. The dependence of the fluid viscosity on temperature is expressed by the Walter

Table 1 Physical properties of representative elementary volume

	Sandstone	Coal	Oil
Viscosity, μ [Pa sec]	–	–	0.03
Density, ρ [kg/m^3]	1300	1200	840
Thermal conductivity, c [W/m K]	1.8	0.45	0.15
Heat capacity, c [J/K]	840	1300	2500
Elastic modulus E, [MPa]	2.50E+03	3.43E+03	–
Poisson's ratio, ν	0.3	0.4	–
Conductivity, σ [S/m]	0.005	0.05	2E−12
Dielectric permittivity, ε_r	10	1.1	2

formula. The pressure is insignificant and does not affect the change in the thermal conductivity and heat capacity of the fluid.

3.1 Mathematical Models

We study the coupled processes of the heat transfer, deformation of a fluid-saturated medium, and fluid-dynamics with thermal and mechanical external influences. Further, we denote the matrix of the medium as Ω_m, and inclusions with oil as Ω_i.

Heat Transfer. The heat transfer process in a fluid-saturated medium is initiated by a temperature gradient along the height of the object $\Omega^{REV} \subset R^3$. The temperature field inside the matrix Ω_m is described by the heat conduction equation:

$$\rho c \frac{\partial T}{\partial t} = \nabla \cdot \lambda \nabla \mathrm{T} \text{ in } \Omega m, \tag{1}$$

where c—heat capacity [J/K], T—temperature [K], λ—thermal conductivity [W/m K], ρ—density [kg/m^3].

At the initial time moment, the temperature field is equal to T_0:

$$T|_{t=0} = T_0. \tag{2}$$

Constant temperatures are set on the upper and lower faces of the sample Ω^{REV}, the side surfaces Γ_{N0} are assumed to be thermally insulated:

$$T|_{\Gamma_{up}} = T_{up},\ T|_{\Gamma_{down}} = T_{down},\ \lambda \nabla T \cdot \mathbf{n}|_{\Gamma_{N0}} = 0, \tag{3}$$

where T_{up} and T_{down}—temperatures on the upper and lower faces, β—heat transfer coefficient [W/m^2 K].

On the interfaces Γ_{in} between the matrix and inclusions, the heat transfer conditions are set:

$$\lambda \nabla T \cdot \mathbf{n}|_{\Gamma_{in}} + \beta\left(T|_{\Gamma_{in}} - T_{incl}\right) = 0, \tag{4}$$

where $\mathbf{n}$—outer unit normal vector, T_{incl}—temperature in inclusions as a result of solving the microscale forced heat transfer problem.

The microscale forced heat transfer problem in the inclusions Ω_i is described by the equation:

$$\rho c\left(\frac{\partial T}{\partial t} + \mathbf{v} \cdot \nabla T\right) = \nabla \cdot \lambda \nabla T + \mu \Psi \text{ in } \Omega i, \tag{5}$$

$$T|_{t=0} = T_0, \tag{6}$$

$$\lambda \nabla T \cdot \mathbf{n}|_{\Gamma_{in}} + \beta\left(T|_{\Gamma_{in}} - T_{matrix}\right) = 0, \tag{7}$$

where $\mathbf{v}$—transfer velocity [m/sec], $\Psi = 2(\nabla^s \mathbf{v} : \nabla^s \mathbf{v})$—dissipation energy due to viscous friction [J], T_{matrix}—temperature in the matrix.

Thermoelastic Deformation. The time-dependent boundary value problem for nonstationary thermoelastic deformation of a solid under the action of external loads and in the absence of internal forces such as gravity is formulated as:

$$\rho \frac{\partial^2 \boldsymbol{u}}{\partial t^2} = \nabla \cdot \boldsymbol{\sigma} \text{ in } \Omega_0, \tag{8}$$

$$u|_{t=0} = 0, \quad \left.\frac{\partial u}{\partial t}\right|_{t=0} = 0, \tag{9}$$

$$\boldsymbol{\sigma} = D(T) : (\nabla_S u - \boldsymbol{\alpha}[T - T_0]) \text{ in } \Omega_0, \tag{10}$$

$$u|_{\Gamma_{down}} = 0,\ \mathrm{u}|_{\Gamma_{\mathrm{up}}} = \mathrm{u_f}, \tag{11}$$

$$\sigma \cdot n|_{\Gamma_{in}} = -p \cdot n, \tag{12}$$

$$\sigma \cdot n|_{\Gamma_{N_0}} = 0, \tag{13}$$

where $\mathrm{u} = \left(\mathrm{u_x}, \mathrm{u_y}, \mathrm{u_z}\right)^T$—displacement [m], $\boldsymbol{\sigma}$—stress tensor [Pa], $\mathbf{D}$—stiffness tensor [Pa], $\nabla_S(\cdot)$—symmetrical part of gradient, $\boldsymbol{\alpha}$—thermal expansion tensor [K^{-1}], p—fluid pressure on the pore boundary Γ_{in} [Pa].

Fluid Flow. The viscous flow of an incompressible fluid inside cracks (pores) Ωi is described by the system of Navier–Stokes equations:

$$\rho\left(\frac{\partial \mathbf{v}}{\partial t} + \mathbf{v}\cdot\nabla\mathbf{v}\right) = \nabla\cdot\boldsymbol{\sigma} + \mathbf{F}\,\text{in}\,\Omega i, \tag{14}$$

$$\nabla\cdot\mathbf{v} = 0\,\text{in}\,\Omega i, \tag{15}$$

$$\mathbf{v}|_{t=0} = \mathbf{v}_0, \tag{16}$$

$$\mu\left(\nabla\mathbf{v} + (\nabla\mathbf{v})^T\right)\cdot\mathbf{n}\big|_{\partial\Omega_i} - p\mathbf{n}|_{\partial\Omega_i} = \boldsymbol{\sigma}\cdot\mathbf{n}|_{\partial\Omega_i}\,\text{in}\,\Omega i, \tag{17}$$

where **v**—flow velocity [m/sec], **F**—volume density of mass forces [N/m^3] (it is assumed that only gravity is significant, therefore, $\mathbf{F} = \rho\mathbf{g}$), $\mathbf{v}_0$—initial flow velocity, **n**—outer unit normal vector at the pore face.

3.2 *Finite Element Discretization*

To discretize the thermoelastic deformation problem, we use the heterogeneous multiscale finite element method. The discontinuous Galerkin method is applied to discretize the mathematical models of fluid dynamics and forced heat transfer.

Thermoelastic Deformation. Solution of the problem (1–4) and (8–13) is a pair (**u**, *T*). We introduce functional spaces for the displacement vector and temperature, respectively:

$$\mathrm{V_u} = \left\{\mathrm{w}|\mathrm{w}\in \mathrm{H}^1(\Omega)\subset\left[\mathrm{L}^2(\Omega)\right]^3\right\}, \tag{18}$$

$$\mathrm{V_T} = \left\{\mathrm{w}|\mathrm{w}\in \mathrm{H}^1(\Omega)\subset \mathrm{L}^2(\Omega)\right\}. \tag{19}$$

Computational schemes based on the heterogeneous multiscale finite element method are used to solve the problem [16, 17]. Let $\Pi^{\mathrm{H}}(\Omega)$ be conforming polyhedral macroelement mesh in domain Ω [18]. Let us denote the conforming tetrahedral mesh as $\widetilde{\mathrm{T}}\,(\mathrm{K^p})$. We define non-polynomial multiscale shape functions $\boldsymbol{\varphi}_i^u = \left\{\varphi_i^x, \varphi_i^y, \varphi_i^y\right\}, \forall i = \overline{1, N_{node}}$ and $\psi_i^T, \forall i = \overline{1, N_{node}}$ associated with polyhedral mesh nodes.

In the heterogeneous multiscale finite element method (FE-HMM), the macroelement is the quadruple $\{\mathrm{K^p}, \Phi, \Sigma, \Lambda\}$ [17]. Note, $K^p \in \Pi^H(\Omega)$ is a geometric element (poly-hedron) with p vertices, $\Phi = \left\{\varphi_i^u, i = \overline{1, p}\right\} \oplus \left\{\psi_i^T, i = \overline{1, p}\right\}$ is a space of local heterogeneous mul-tiscale non-polynomial shape functions defined in K^p, $\dim(\Phi) = 2p$, Σ is a subspace of degrees of freedom dual to Φ, $\dim(\Sigma) = 2p$,

Λ is a numerical integration formula de-fined in K^p:

$$\int_{K^p} f(x)dx = \sum_{t\in\tilde{T}(K^p)} \sum_{l=1}^{n_t} \omega_l f(x_l), \tag{20}$$

where t—tetrahedron in $\tilde{T}(K^p)$, n_t—number of integration points defined in the tetrahedron t, ω_l and x_l—weights and nodes of numerical integration according to the Gauss cubature formula for the tetrahedron.

Taking into account the introduced finite element mesh, we define discrete subspaces:

$$V_H^u\left(\Pi^H(\Omega)\right) = \text{span}\left\{\varphi_i^u = \left\{\varphi_i^x, \varphi_i^y, \varphi_i^y\right\},\ \forall i = \overline{1, N_{node}}\right\} \subset V_u(\Omega), \tag{21}$$

$$V_H^T\left(\Pi^H(\Omega)\right) = \text{span}\left\{\psi_i^T,\ \forall i = \overline{1, N_{node}}\right\} \subset V_T(\Omega). \tag{22}$$

We are using difference schemes for time derivative:

$$\frac{\partial T}{\partial t} = \frac{T_k - T_{k-1}}{\Delta t},\ \frac{\partial^2 u}{\partial t^2} = \frac{u_k - 2u_{k-1} + u_{k-2}}{\Delta t^2},$$

where $k = \overline{1 \dots K}$ is a number of a time iteration, Δt is a time step.

Variational formulation of the heterogeneous multiscale finite element method with polyhedral supports for the problem of nonstationary thermoelastic deformation: find $u^H \in V_H^u\left(\Pi^H(\Omega)\right) + u_0$, $T^H \in V_H^T\left(\Pi^H(\Omega)\right) + T_0$, such that

$$\forall w^H \in \left\{w^H \in V_H^u\left(\Pi^H(\Omega)\right):\ \ w^H\big|_{\partial\Omega} = 0\right\}, \tag{23}$$

$$\forall w^H \in \left\{w^H \in V_H^T\left(\Pi^H(\Omega)\right):\ \ w^H\big|_{\partial\Omega} = 0\right\}, \tag{24}$$

$$\begin{aligned}\int_\Omega \lambda \cdot \nabla w^H . \nabla T_K^H d\Omega + \int_\Omega \frac{c\rho}{\Delta t} w^H T_K^H d\Omega + \int_{\Gamma_{in}} \beta \cdot w^H . T_K^H d\Gamma \\ = \int_\Omega \frac{c\rho}{\Delta t} w^H T_{K-1}^H d\Omega + \int_{\Gamma_{in}} \beta \cdot w^H d\Gamma,\end{aligned} \tag{25}$$

$$\begin{aligned}&\int_\Omega \nabla \mathbf{w}^H : \mathbf{D} : \nabla \mathbf{u}_k^H d\Omega + \int_\Omega \frac{\rho}{\Delta t^2} \mathbf{w}^H \mathbf{u}_k^H d\Omega \\ &= \int_\Omega \frac{\rho}{\Delta t^2} \mathbf{w}^H \mathbf{u}_{k-2}^H d\Omega - \int_\Omega \frac{2\rho}{\Delta t^2} \mathbf{w}^H \mathbf{u}_{k-1}^H d\Omega \quad + \int_{\Gamma_{in}} \mathbf{w}^H \cdot \nabla \cdot \left(\mathbf{D} : \alpha\left[T^H - T_0\right]\right) d\Omega\end{aligned} \tag{26}$$

The discrete analogue of the variational formulation (23–26) can be written as:

$$M^{\mathrm{global}} Q^{\mathrm{H}} = b^{\mathrm{global}}, \tag{27}$$

where Q^H—decomposition weights of u_k^H and T_k^H by non-polynomial multiscale shape functions φ_i^u and ψ_i^T, respectively, M^{global} and b^{global}—the global matrix and vector of the right part of the SLAE obtained by assembling local matrices $M^{K,local}$ and the right-hand sides $b^{K,local}$ for each macro-elements $K^p \in \Pi^H(\Omega)$.

The conjugate gradient method is used to solve system (27).

Fluid Flow Problem and Heat Transfer with Convection. On a non-empty bounded subset $\Omega \subset R^3$, we define a finite cover $M_h(\Omega) = \bigcup_i K_i$ and n-degree polynomial space $\Im_n(K_i)$. In the $M_h(\Omega)$, we introduce discrete functional spaces for temperature, velocity vector and pressure, respectively:

$$Q^h = \left\{ q^h | q^h \in L_0^2(\Omega) : q^h \in \Im_n(K_i) \forall K_i \in M_h(\Omega), q^h|_{\partial K_i} = 0 \right\}, \tag{28}$$

$$\mathbf{V}^h = \left\{ \mathbf{v}^h | \mathbf{v}^h \in \mathbf{H}_0(\mathrm{div}, \Omega) : \mathbf{v}^h \in [\Im_n(K_i)]^3 \forall K_i \in M_h(\Omega), \mathbf{v}^h \cdot \mathbf{n}|_{\partial K_i} = 0 \right\}, \tag{29}$$

$$P^h = \left\{ p^h | p^h \in L_0^2(\Omega) : p^h \in \Im_{n-1}(K_i) \forall K_i \in M_h(\Omega), \int\limits_{K_i} p^h dK_i = 0 \right\}. \tag{30}$$

We will use the non-conforming discontinuous Galerkin method to discretize the problem of fluid dynamics and forced heat transfer. It is necessary to determine the traces of functions on the interelement faces. Let $\Gamma = \bigcup_i \partial K_i$ denote the set of external and internal faces. On the set Γ we introduce the space of traces $\mathrm{Tr}(\Gamma) = \prod_{\Omega_i \in M_h(\Omega)} L^2(\partial K_i)$. The set of internal boundaries is denoted as $\Gamma_0 = \Gamma \backslash \partial\Omega$.

Traces of functions are associated with the mean $\{.\}$ and jump $[.]$ of ambiguous functions on the interfragmentary faces. For the functions $\mathbf{v} \in [\mathrm{Tr}(\Gamma)]^3$, $p \in \mathrm{Tr}(\Gamma)$ and $q \in \mathrm{Tr}(\Gamma)$ the mean $\{.\}$ and jump $[.]$ on the external faces $\partial\Omega$ are determined as [19]:

$$\begin{aligned} &[\underline{\mathbf{v}}]|_{\partial\Omega} = \mathbf{v} \otimes \mathbf{n}, [\mathbf{v}]|_{\partial\Omega} = \mathbf{v} \cdot \mathbf{n}, \{\mathbf{v}\}|_{\partial\Omega} = \mathbf{v}, \\ &[p]|_{\partial\Omega} = p\mathbf{n}, \{p\}|_{\partial\Omega} = p, [q]|_{\partial\Omega} = q\mathbf{n}, \{q\}|_{\partial\Omega} = q, \end{aligned} \tag{31}$$

on the internal faces $\Gamma_0 = \partial K_i \cap \partial K_j$ [19]:

$$\begin{aligned} &[\underline{\mathbf{v}}]|_{\Gamma_0} = \mathbf{v}_i \otimes \mathbf{n}_i + \mathbf{v}_j \otimes \mathbf{n}_j, [\mathbf{v}]|_{\Gamma_0} = \mathbf{v}_i \cdot \mathbf{n}_i + \mathbf{v}_j \cdot \mathbf{n}_j, \\ &\{\mathbf{v}\}|_{\Gamma_0} = (\mathbf{v}_i + \mathbf{v}_j)/2, [p]|_{\Gamma_0} = p_i\mathbf{n}_i + p_j\mathbf{n}_j, \{p\}|_{\Gamma_0} = (p_i + p_j)/2, \\ &[q]|_{\Gamma_0} = q_i\mathbf{n}_i + q_j\mathbf{n}_j, \{q\}|_{\Gamma_0} = (q_i + q_j)/2. \end{aligned} \tag{32}$$

The time derivatives of the velocity and temperature can be approximated as:

$$\mathbf{v}_t^h = \frac{\mathbf{v}_k^h - \mathbf{v}_{k-1}^h}{\Delta t}, T_t^h = \frac{T_k^h - T_{k-1}^h}{\Delta t}. \tag{33}$$

The variational formulation of the discontinuous Galerkin method for the Navier–Stokes problem: find $\mathbf{v}_k^h \in \mathbf{V}^h \times [0, T]$, $p^h \in P^h \times [0, T]$, $\forall \mathbf{w}^h \in \mathbf{V}^h$ and $q^h \in P^h$ [20]:

$$\begin{aligned}
&\frac{1}{\Delta t}\int_\Omega \rho \mathbf{v}_k^h \cdot \mathbf{w}^h d\Omega + \int_\Omega \mu(\nabla \mathbf{v}_k^h + \nabla^T \mathbf{v}_k^h) : \nabla \mathbf{w}^h d\Omega - \int_\Gamma \{\mu(\nabla \mathbf{v}_k^h + \nabla^T \mathbf{v}_k^h)\} : [\underline{\mathbf{w}^h}] \\
&\quad + \{\mu(\nabla \mathbf{w}^h + \nabla^T \mathbf{w}^h)\} : \left[\underline{\mathbf{v}_k^h}\right] d\Gamma + \tau^{DG} \int_\Gamma [\underline{\mathbf{w}^h}] : \left[\underline{\mathbf{v}_k^h}\right] d\Gamma - \int_\Omega \nabla \cdot \mathbf{w}^h p_k^h d\Omega \\
&\quad + \int_\Gamma [\mathbf{w}^h]\{p_k^h\} d\Gamma + \int_\Omega \rho(\mathbf{v}_{k-1}^h \cdot \nabla \mathbf{v}_k^h) \cdot \mathbf{w}^h d\Omega + \frac{1}{2}\int_\Omega \rho(\nabla \cdot \mathbf{v}_{k-1}^h) \cdot \mathbf{v}_k^h \\
&\quad \mathbf{w}^h d\Omega + \int_\Omega \nabla \cdot \mathbf{v}_k^h q^h d\Omega + \tau_0^{DG} \int_\Gamma [p^h] \cdot [q^h] d\Gamma = \int_\Omega \rho \left(\mathbf{g} + \frac{\mathbf{v}_{k-1}^h}{\Delta t}\right) \\
&\quad \mathbf{w}^h d\Omega + \int_{\partial\Omega} \sigma \cdot \mathbf{n} \cdot \mathbf{w}^h dS.
\end{aligned} \tag{34}$$

The uniqueness of the solution to the Navier–Stokes problem is determined by the inf–sup condition [19]:

$$inf_{q^h \in P^h} sup_{\mathbf{v}_k^h \in \mathbf{V}^h} \frac{(\nabla \cdot \mathbf{v}_k^h, p^h)}{\|p^h\|_{L^2(\Omega)} \|\mathbf{v}_k^h\|_{[H^1(\Omega)]^3}} \geq \alpha > 0. \tag{35}$$

The second-order basis of the $\mathbf{H}(\text{div}, \Omega)$ is applied to approximate the velocity. To approximate the pressure, we use the first-order hierarchical basis of the $H^1(\Omega)$. To solve the discrete analogue, we apply a special projection technique for the Navier–Stokes problem, see [21].

The variational formulation of the discontinuous Galerkin method for the heat transfer problem has the form: find $T_k^h \in P^h \times [0, T]$, $\forall q^h \in Q^h$ [20]:

$$\begin{aligned}
&\int_\Omega \rho c \left(\frac{1}{\Delta t} T_k^h + \mathbf{v} \cdot \nabla T_k^h\right) q^h d\Omega + \int_\Omega \lambda \nabla T_k^h \cdot \nabla q^h d\Omega \\
&\quad + \int_{\Gamma_0} \lambda([T_k^h] \cdot \{\nabla q^h\} + [q^h] \cdot \{\nabla T_k^h\} + \tau [T_k^h] \cdot [q^h]) dS + \int_{\partial\Omega} \lambda \beta T_k^h (\mathbf{n} \cdot \nabla q^h) dS \\
&\quad + \int_{\partial\Omega} \lambda((\mathbf{n} \cdot [T_k^h]) q^h + \tau T_k^h q^h) dS = \int_\Omega \left(\frac{\rho c}{\Delta t} T_{k-1}^h + \Psi^h\right) q^h d\Omega \\
&\quad + \int_{\partial\Omega} \lambda \beta T_g^h (\mathbf{n} \cdot \nabla q^h + \tau q^h) dS,
\end{aligned} \tag{36}$$

where $\tau = \frac{\rho c \|\mathbf{v}\|}{\lambda}$.

To solve the discrete analogue, we use BiCGStab solver with the algebraic multilevel preconditioning technique [22].

4 Electromagnetic Problem

In this section, we describe the model problem of the electric field in the near-wellbore region. We consider the model domain with vertical non-cased well and the horizontal layer of highly heterogeneous structure (see Fig. 2).

Since it is impossible to capture all the microscopic features when solving the problem in the entire region, we use the two-step upscaling technique. At the initial stage, we choose a certain representative volume (see Fig. 1a) and perform computational homogenization. At the next step, the electromagnetic problem is solved in the entire near-wellbore region using the obtained homogenized characteristic to describe the electric properties of the layer (see Fig. 2). Throughout the section, we will refer to the scale of the near-wellbore region as a macroscopic scale; the scale of the representative volume as a mesoscopic scale; the pore scale as a microscopic scale.

4.1 Mathematical Models

The electric field $\mathbf{E}$ in the computational domain Ω is determined by solving the Helmholtz equation numerically:

$$\nabla \times \mu^{-1} \nabla \times \mathbf{E} + k^2 \mathbf{E} = -i\omega \mathbf{J} \text{ in } \Omega, \tag{37}$$

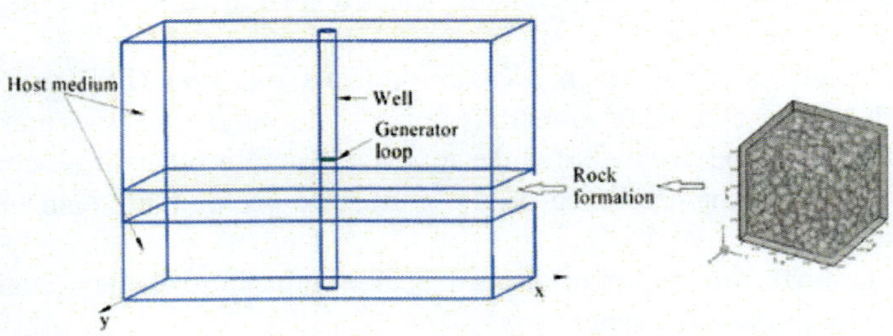

Fig. 2 Computational domain: near-wellbore region

where **J**—current density vector [A/m^2] $k^2 = i\omega\sigma - \omega^2\varepsilon$, $\omega = 2\pi f$—angular frequency [Hz], f—frequency [Hz], $\varepsilon = \varepsilon_r\varepsilon_0$—dielectric permittivity [F/m], ε_r—relative dielectric permittivity, $\varepsilon_0 = 8.85 \times 10^{-12}$ F/m; $\mu = \mu_r\mu_0$ – magnetic permeability [H/m], μ_r—relative magnetic permeability, $\mu_0 = 4\pi \times 10^{-7}$ H/m; σ—electrical conductivity [S/m], Ω - computational domain with boundary $\partial\Omega = \Gamma_m \cup \Gamma_e$.

An experimental measuring procedure is usually held on a small sample with source electrodes applied to the top and bottom surfaces, so we will introduce the following field excitation scheme. The electromagnetic field in the representative volume Ω^{REV} (see Fig. 1) is excited by the following boundary condition:

$$\mathbf{n} \times \mathbf{E}|_{\Gamma_e} = \mathbf{E}_0, \tag{38}$$

where Γ_e is the top surface of the sample. The Perfect Electrical Conductor (PEC) boundary condition is set on the bottom surface. Homogeneous magnetic boundary conditions are set on the lateral boundary:

$$\mu^{-1}\nabla \times \mathbf{E} \times \mathbf{n}\big|_{\Gamma_m} = 0,$$

where $\mathbf{n}$ is a unit outward normal to the lateral boundary Γ_m.

For the electromagnetic problem in the representative volume Ω^{REV}, the right-hand side in Eq. (37) is $\mathbf{J} = \mathbf{0}$.

In the near-wellbore region, the electromagnetic field is excited by a generator loop. The homogeneous electric boundary conditions of the form (38) are set at all boundaries of the computational domain. Here $\mathbf{E}_0 = \mathbf{0}$ in (38).

4.2 *Finite Element Discretization*

Here, we consider a variational formulation of the vector finite element method for discretizing the mathematical model of the electric field distribution in the REV and in the near-wellbore region.

Electromagnetic problem in a heterogeneous medium (REV). Since REV is highly heterogeneous with numerous small-scale pores, it is essential that the computational scheme accounts for the multiscale nature of the problem. We use the computational schemes based on the multiscale vector finite element method [23, 24].

The variational formulation of the multiscale vector finite element method is set in the following functional spaces:

$$\mathbf{H}(\mathbf{curl}, \Omega) = \left\{\mathbf{u} \in L^2(\Omega) : \nabla \times \mathbf{u} \in L^2(\Omega)\right\}, \tag{39}$$

$$\mathbf{H}_0(\mathbf{curl}, \Omega) = \{\mathbf{u} \in \mathbf{H}(\mathbf{curl}, \Omega) : \mathbf{u} \times \mathbf{n}|_{\partial\Omega} = 0\}. \tag{40}$$

Let $\Pi^H(\Omega)$ be a conforming discretization of the domain Ω into polyhedral elements K^P, diam(K^P) = H [25, 26]. We will refer to $\Pi^H(\Omega)$ as a mesoscopic or coarse mesh. Let us introduce the discrete space $\mathrm{V}^{\mathrm{H}}(\Pi^{\mathrm{H}}(\Omega)) \subset \mathbf{H}(\mathbf{curl}, \Omega)$ spanned by the nonpolynomial multiscale shape functions $\{\mathbf{v}_i^H, i = \overline{1, N_e}\}$ associated with the edges of the elements $K^P \in \Pi^H(\Omega)$, where N_e is the number of edges. Let $T^h(K^T)$ be a conforming discretization of the element $K^T \in T^H(\Omega)$ into tetrahedral elements K^t, diam(K^t) = h << H. We will refer to $T^h(K^T)$ as a microscopic or fine mesh. Nonpolynomial multiscale shape functions $\{\mathbf{v}_i^H\}$ are constructed independently for each element of the coarse mesh as a linear combination of the Whitney functions [27, 28] defined on the fine mesh, so that their tangential trace is continuous across the inter-element boundaries of the coarse mesh [25].

Thus, the problem in the REV is solved at the mesoscopic level using coarse polyhedral mesh, while the microscale features are captured via the multiscale basis functions.

The discrete variational formulation of the multiscale finite element method at the mesoscopic level is set:

find $\mathbf{E}^H \in V^H(\Pi^H(\Omega)) + \mathbf{E}_0^H$, such that for all $\mathbf{v}^H \in \{\mathbf{v} \in V^H(\Pi^H(\Omega)) : \mathbf{v} \times \mathbf{n}|_{\partial\Omega} = 0\}$:

$$\sum_{K^P \in \Pi^H(\Omega)} \int_{K^P} \mu^{-1} \nabla \times \mathbf{E}^H \cdot \nabla \times \mathbf{v}^H dV + \sum_{K^P \in \Pi^H(\Omega)} \int_{K^P} k^2 \mathbf{E}^H \cdot \mathbf{v}^H dV = 0. \tag{41}$$

To solve the discrete analogue, both for macro- and micro-level problems, we use Krylov subspace conjugate gradient method for complex-valued matrixes [29].

Effective tensor characteristic. Due to the highly heterogeneous nature of the medium, its characteristics at the macroscopic level are, in general, anisotropic. We consider nonmagnetic media. The effective electric characteristics of the representative volume are described as a complex-valued symmetric second-rank tensor [3]:

$$\mathrm{Z}^{\mathrm{eff}} = \begin{bmatrix} z_{11} & z_{12} & z_{13} \\ z_{21} & z_{22} & z_{23} \\ z_{31} & z_{32} & z_{33} \end{bmatrix} = \underbrace{Re(\mathrm{Z}^{\mathrm{eff}})}_{\sigma^{\mathrm{eff}}} + i\omega\varepsilon_0 \underbrace{Im(\mathrm{Z}^{\mathrm{eff}})}_{\varepsilon^{\mathrm{eff}}}. \tag{42}$$

The computational technique for the effective tensor (42) is based on solving the direct problem in the representative sample Ω^{REV}. This effective characteristic captures the anisotropy effect in a medium with complex internal structure. The effective tensor $\mathrm{Z}^{\mathrm{eff}}$ may then be used to describe the entire anisotropic layer in the near-wellbore region.

Electromagnetic problem in a near-wellbore region. The macroscale problem in the near-wellbore region is solved by the classical vector finite element method on an adaptive tetrahedral partition $T^h(\Omega)$, which accounts for the generator loop geometry. The discrete subspace $\mathrm{V}^h(\mathrm{T}^h(\Omega)) \subset \mathbf{H}_0(\mathbf{curl}, \Omega)$ is spanned by Whitney functions. The discrete variational formulation of the vector finite element method is set: for $\mathbf{J} \in \mathbf{L}^2(\Omega)$, find $\mathbf{E}^h \in \mathrm{V}^h\left(\mathrm{T}^h(\Omega)\right)$, such that for all $\mathbf{v}^h \in \mathrm{V}^h\left(\mathrm{T}^h(\Omega)\right)$:

$$\sum_{K \in T^H(\Omega)} \int_K \mu^{-1} \nabla \times \mathbf{E}^h \cdot \nabla \times \mathbf{v}^h dV + \sum_{K \in T^H(\Omega)} \int_K Z^{\mathrm{eff}} \mathbf{E}^h \cdot \mathbf{v}^h dV = -i\omega \sum_{K \in T^h(\Omega)} \int_K J \cdot \mathbf{v}^h dV, \quad (43)$$

where Z^{eff} is the effective electrical tensor (42).

The source is represented as a known distribution of the current density in the loop:

$$J = \delta(x - x_i, y - y_i, z - z_i) \cdot \mathbf{N}_{i1} \cdot |\mathbf{J}|$$

where $\mathbf{J}$ is a current density, N_{i1} is first order incomplete Nedelec basis associated with the edges of the mesh partition, $\delta(x - x_i, y - y_i, z - z_i)$ is the Dirac function, (x_i, y_i, z_i) are points on the generator loop partition edges.

Nedelec basis functions of the first order, first type ensure the zero-divergence condition is met automatically, as $\nabla \cdot \mathrm{N}_{i1}, \forall i = 1, \ldots, 6$. The discrete analogue is solved by a multilevel solving technique specifically tailored to that class of problems [30].

5 Numerical Results

In this section, we present the results of computational experiments performed for a heterogeneous sample (REV) and for the near-wellbore region.

5.1 *Heterogeneous Medium (REV)*

The results of computational experiments are given for a cross section (see Fig. 3) of a sample with pores forming cluster structures (see Fig. 1), the physical properties of the media are indicated in Table 1.

Thermoelastic Deformation Problem. The sandstone sample is a cube (edge size is 0.05 m). The pores filled with oil. The inclusions are randomly placed and can form clusters. At the initial time, the sample temperature is +5 °C. The bottom surface of the cube is fixed and a temperature is +5 °C. Along the Z axis (see Fig. 1), the sample is compressed. On the top surface, the temperature is +50 °C. The side surface is assumed to be thermally insulated. Figures 4 and 5 show the distributions

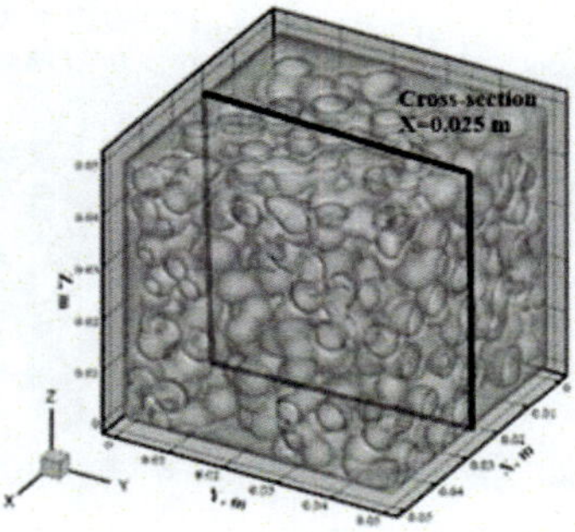

Fig. 3 Location of the cross-section X = 0.025 m of the computational domain

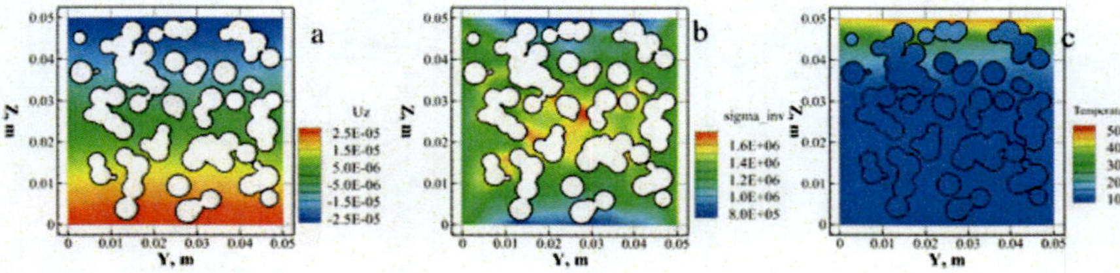

Fig. 4 Solution in cross-section X = 0.025 m: Z-component of deformation of the sample matrix (**a**), invariant of the stress tensor of the sample matrix (**b**), temperature distribution in the matrix and pores (**c**)

of deformation, temperature, stress and pressure fields in the cross-section passing through the center of the sample (see Fig. 3).

The above results of the computational experiment demonstrate the correctness of the proposed algorithm for coupling the problem of thermoelastic deformation of the matrix and heat and mass transfer of fluid in the cavities. Namely, we see the transfer of heat from the matrix to the pores (Fig. 4c), according to the law of heat exchange (10). The stress–strain state of the matrix (Fig. 4a, 4b) generates a change in fluid pressure and fluid movement in the cavities (Fig. 5).

Elastic Deformation Problem. The sandstone sample is a cube (edge size is 0.05 m), and the pores are filled with coal. The sample temperature is assumed to be constant (+20 °C). The sample is compressed along the Z axis (see Fig. 1). Figure 6 shows the distributions of the deformation components and stress tensor components, and pressure in the cross-section passing through the center of the sample (see Fig. 3).

The results of the computational experiment demonstrate that the coupling conditions at the matrix-pore interfaces, namely, the equality of displacements (Fig. 6a) and jumps of the field of tension corresponding to the contrast of the elastic properties of the matrix material (sandstone) and pores (coal) (Fig. 6b) are fulfilled. Thus, these results are physically relevant and can be used in the future, for example, to perform modeling of the electromagnetic field in a deformed sample.

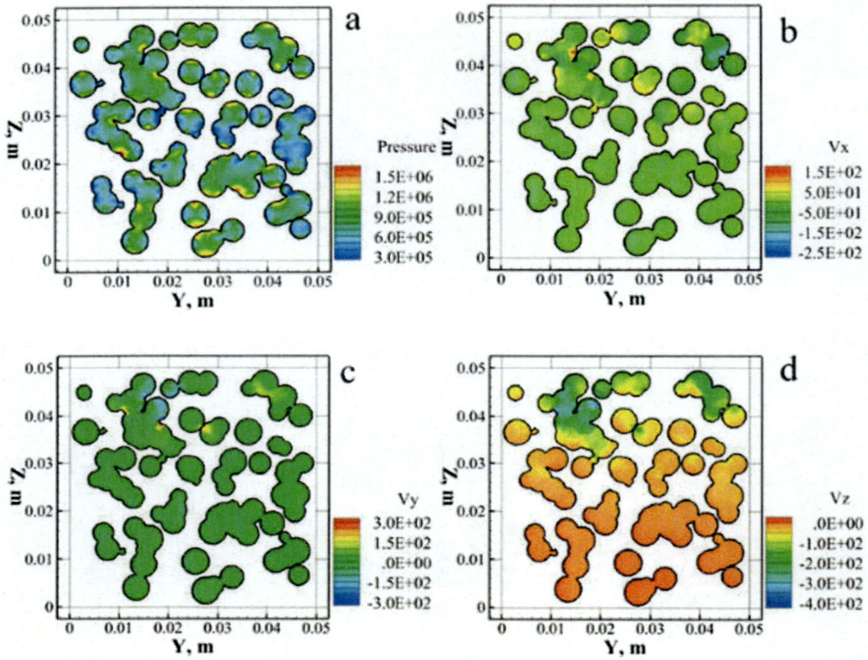

Fig. 5 Solution in section X = 0.025 m: pressure in pores (**a**), fluid velocity in pores (**b–d**)

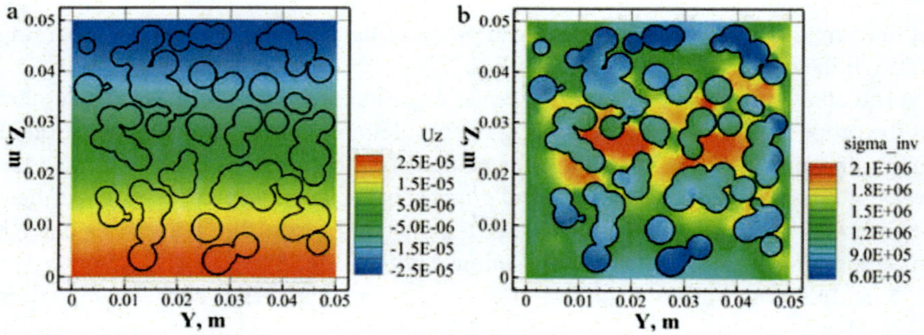

Fig. 6 Solution in the cross-section X = 0.025 m: Z-component of the deformation of the sample (**a**), invariant of the stress tensor of the sample (**b**)

Electromagnetic Problem. The sample is a cube (edge size is 0.05 m). It contains 44% of pores filled with coal. The material of the matrix is sandstone. Figure 7 shows the distribution of the real and imaginary parts of the Ey component in the cross-section YZ.

The effective electric tensor characteristic of the sample (see Fig. 1) has the form:

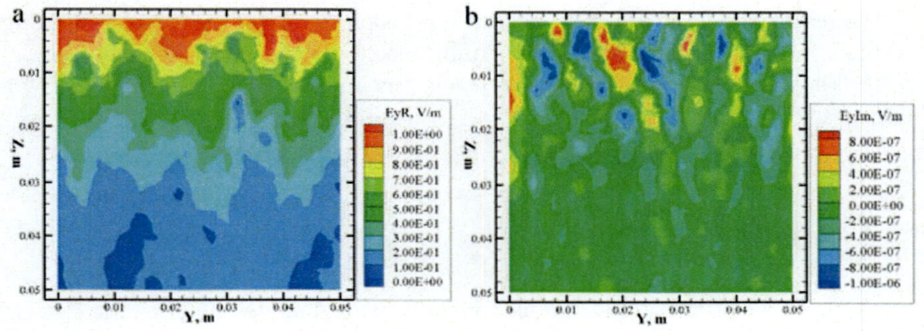

Fig. 7 The distribution of real (**a**) and imaginary (**b**) parts of the Ey component in the cross-section YZ ($X = 0.025$ m) of REV

$$Z^{\text{eff}} = \begin{bmatrix} 3.25E-02 & -3.63E-04 & 3.03E-07 \\ -3.63E-04 & 3.97E-02 & 8.63E-03 \\ 3.03E-07 & 8.63E-03 & 3.11E-02 \end{bmatrix} + i\omega\varepsilon_0 \begin{bmatrix} 4.57E+00 & -1.42E-01 & 1.72E-02 \\ -1.42E-01 & 2.41E+00 & 3.08E-01 \\ 1.72E-02 & 3.08E-01 & 4.25E+00 \end{bmatrix}.$$

The electromagnetic wave interacts with the internal microstructure of the sample, as a result, the distribution of the real and imaginary components of the electric field is significantly different from the distribution in a homogeneous medium with the characteristics of a matrix or inclusions. Since the contrast is set both for σ and ε_r, this difference is essential in both the real and imaginary parts of the field E.

The calculated effective tensor reflects the onset of anisotropy in the sample. The diagonal elements differ from each other. Diagonal dominance is noted, but off-diagonal elements are not small enough to be considered equal to zero.

5.2 The Near-Wellbore Region

The near-wellbore region model is the following: well radius is 0.108 m, generator loop radius is 0.05 m. The current in the loop is 1 A and the source frequency is 10 kHz. The heterogeneous layer thickness is 0.5 m. Electrophysical characteristics of the drilling fluid: $\varepsilon_r = 1$, $\mu_r = 1$, $\sigma = 0.5$ S/m; of the host medium: $\varepsilon_r = 1$, $\mu_r = 1$, $\sigma = 10^{-1}$ S/m. The electrophysical properties of the layer are described by the effective characteristic $\varepsilon_r = \text{Im}(Z^{\text{eff}})$, $\sigma = \text{Re}(Z^{\text{eff}})$, the layer is nonmagnetic $\mu_r = 1$.

Figures 8 and 9 show the non-zero components of the electric field in the near-wellbore region with a horizontal layer with the electrophysical characteristics described by the tensor given above.

The results shown in Fig. 8 reflect the anisotropic nature of the rock constituting the horizontal layer. The Ez component of the electromagnetic field, which is normal to the " layer-host medium" interface, reacts on an anisotropic layer, undergoing a jump at the interface (see Fig. 9).

6 Conclusion

In this paper, we proposed a technique for mathematical modelling of multi-physical processes in multiscale media based on the computational schemes of conforming and non-conforming multiscale finite element methods. We demonstrate the stages of this technique in the context of solving the problem of modelling electromagnetic field in a rock specific to hydrocarbon reservoirs and coal deposits under thermal and mechanical effects.

We considered two scale levels of spatial description of the objects under study. At the mesoscale, we capture all structural features of the heterogeneous media. At the macroscopic level, we deal with a homogenized medium characterized by the physical properties that, although invariant to spatial coordinates, preserve its macroscopic behaviour.

To describe the physical properties of the medium at the mesoscale level, we choose a representative elementary volume (REV) sample with an idealized structure. REV consists of a matrix and microinclusions. Using the REV sample, we performed

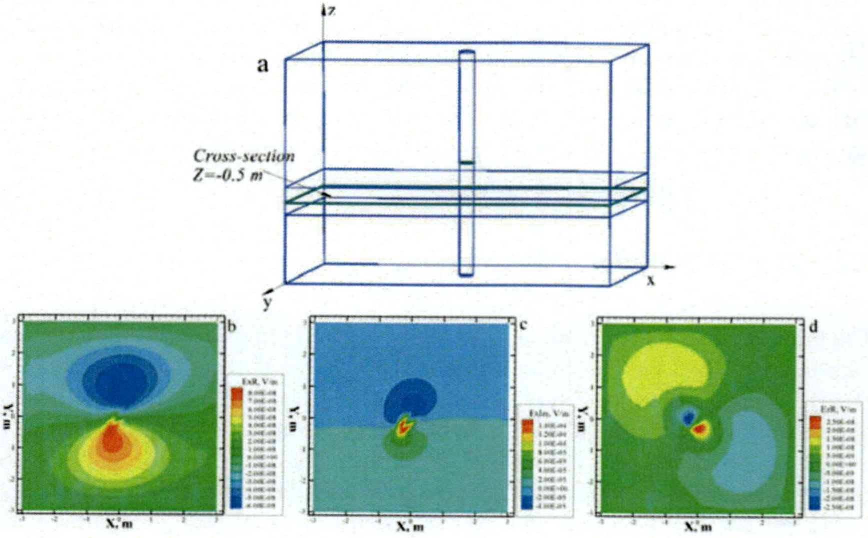

Fig. 8 Cross-section XY $Z = -0.5$ m (**a**); Ex component in the cross-section XY (**b**, **c**); real component Ez in the cross-section XY (**d**)

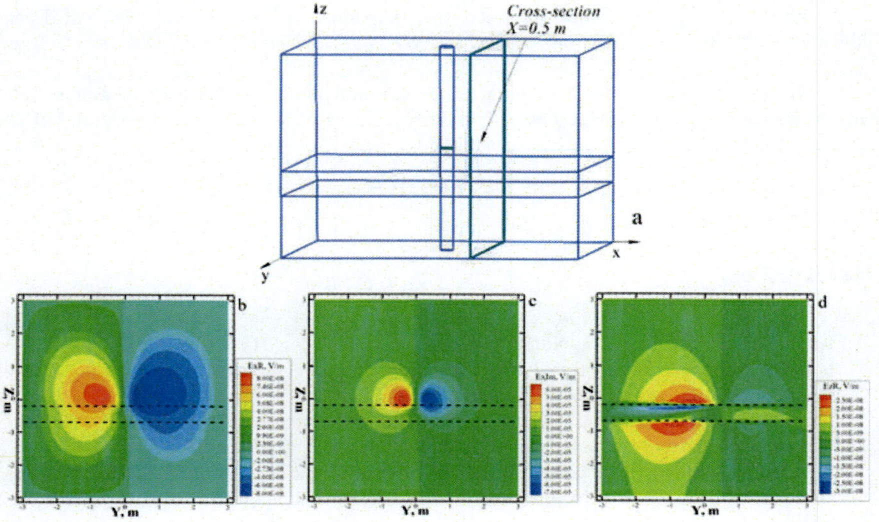

Fig. 9 Cross-section YZ $X = 0.5$ m (**a**); Ex component in the cross-section YZ (**b**, **c**); real component Ez in the cross-section YZ (**d**)

the mathematical modelling of the thermoelastic deformation process, taking into account the fluid saturation of the rock or the heterogeneity in the distribution of Young's modulus and Poisson's ratio. To reduce computational costs, we employed a multilevel procedure. The problems of elastic deformation and heat transfer in the medium matrix, and the problems of heat transfer and fluid dynamics in microinclusions, are solved independently. The coupling of physical fields is achieved through conjugation conditions at the pore-matrix interfaces.

To describe the behaviour of the electromagnetic fields at the macroscopic level, we employed an algorithm for numerical homogenization based on the methods of the effective medium theory. We then performed mathematical modelling of the electric field in the near-wellbore region with the electrophysical properties of the contrasting rock layer described by the calculated effective electric tensor.

The proposed technique for the mathematical modelling of multi-physical processes in multiscale media does not require the reduction of mathematical models and retains the property of their global regularity, which ensures the physical relevance of the obtained modelling results. We should note, however, that the choice of a representative elementary volume, as well as an analysis of the stability and accuracy of the numerical homogenization algorithm, requires additional research. These problems are beyond the scope of this paper and determine the path for our future research.

Acknowledgements 1. This work has been supported by the grants the Russian Science Foundation, RSF 20-71-00,134 (modelling of the thermal and fluid dynamics fields).

2. The research was carried out within the state assignment of Ministry of Science and Higher Education of the Russian Federation, Project No. FWZZ-2022-0030 (modelling of the stress–strain fields).

3. The research was carried out within the state assignment of Ministry of Science and Higher Education of the Russian Federation, Project No. FWZZ-2022-0025 (modelling of the electromagnetic fields).

References

1. Kiani, S., Jafari, S., Jafari, S., Norouzi-Apourvari, S., Mehrjoo, H.: Simulation study of wormhole formation and propagation during matrix acidizing of carbonate reservoirs using a novel in-situ generated hydrochloric acid. Adv. Geo-Energy Res. **5**(1), 64–74 (2021)
2. Xue, H., Huang, Z., Zhao, L.Q., Wang, H., Liu, P.: Wormholing influenced by injection temperature in carbonate rocks. Open J. Yangtze Gas Oil **4**, 12–30 (2019)
3. Shurina, E.P., Epov, M.I., Shtabel, N.V., Mikhaylova, E.I.: The calculation of the effective tensor coefficient of the medium for the objects with microinclusions. Engineering **6**(3), 101–112 (2014)
4. Rin, R., Tomin, P., Garipov, T., Voskov, D.: general implicit coupling framework for multi-physics problems. In: Conference: SPE Reservoir Simulation Conference, pp. 1–16. Society of Petroleum Engineers (2017)
5. Rin, R.: Implicit coupling framework for multi-physics reservoir simulation. Thesis (Ph.D.), Stanford University (2017)
6. Garipov, T., White, J., Lapene, A., Tchelepi, H.: Thermo-hydro-mechanical model for source rock thermal maturation. In: 50th US Rock Mechanics (Geomechanics Symposium), Houston, USA. Society of Petroleum Engineers (2016)
7. Mauthe, S.: Variational Multiphysics Modeling of Diffusion in Elastic Solids and Hydraulic Fracturing in Porous Media. Report No. II-33, Institut f¨ur Mechanik (Bauwesen) Lehrstuhl f¨ur Kontinuumsmechanik Universit¨at Stuttgart, Germany (2017)
8. Epov, M.I., Shurina, E.P., Itkina, N.B., Kutishcheva, A.Y., Markov, S.I.: Finite element modeling of a multi-physics poro-elastic problem in multiscale media. J. Comput. Appl. Math. **352**, 1–22 (2019)
9. Keyes, D.E., et al.: Multiphysics Simulations: Challenges and Opportunities. Tech. Rep. ANL/MCS-TM-321, Argonne National Laboratory, Report of workshop sponsored by the Institute for Computing in Science (ICiS), July 30–Aug. 6, 2011, Park City, Utah (2011)
10. Recent, G.: Advances in splitting methods for multiphysics and multiscale: theory and applications. J. Alg. Comp. Tech. **9**(1), 65–93 (2013)
11. Ghasemi, F., Nordström, J.: Coupling requirements for multiphysics problems posed on two domains. SIAM J. Num. Anal. **55**(6), 2885–2904 (2017)
12. Pantelyat, M.: Magneto-thermo-mechanical analysis of electromagnetic devices using the finite element method. Eng. Tech. Internat. J. Elec. Com. Eng. **10**(5), 652–658 (2016)
13. Bogdanova, M., Belousov, S., et al.: Simulation platform for multiscale and multiphysics modeling of OLEDs. Comp. Manag. Sci. **29**, 740–753 (2014)
14. Steinhauser, M.: Computational Multiscale Modeling of Fluids and Solids: Theory and Applications, 2nd edn. Springer-Verlag, Berlin, Heidelberg (2017)
15. Babaei, M.: Multiscale wavelet and upscaling-downscaling for reservoir simulation. Imperial College London (2013)
16. Eidel, B., Fischer, A.: The heterogeneous multiscale finite element method for the homogenization of linear elastic solids and a comparison with the FE^2 method. Comput. Methods Appl. Mech. Eng. **329**, 332–368 (2018)

17. Epov, M.I., Shurina, E.P., AYu., Kutischeva: Computation of effective resistivity in materials with microinclusions by a heterogeneous multiscale finite element method. Phys. Mesomech. **20**(4), 407–416 (2017)
18. Jeong, K.-L., Seo, D.-W.: Automatic polyhedral mesh generation for ship resistance based on the locally refined cartesian cut-cell method. J. Mar. Sci. Technol. **28**(4), 3 (2020)
19. Riviere, B.: Discontinuous Galerkin Methods for Solving Elliptic and Parabolic Equations: Theory and Implementation. Society for Industrial and Applied Mathematics (2008)
20. Li, B.Q.: Discontinuous Finite Elements in Fluid Dynamics and Heat Transfer. Springer-Verlag, London Limited (2006)
21. Donea, J., Huerta, A.: Finite Element Methods for Flow Problems. John Wiley & Sons Ltd., Chichester (2003)
22. Xu, J., Chen, L., Nochetto, R.: Optimal Multilevel Methods for H(grad), H(curl), and H(div) Systems on Graded and Unstructured Grids. Multiscale, Nonlinear and Adaptive Approximation. Springer (2009)
23. Shurina, E.P., Dobrolyubova, D.V., Shtanko, E.I.: Modified multiscale vector finite element method on polyhedral meshes for the time-harmonic electric field. In: 14th International Scientific—Technical Conference on Actual Problems of Electronic Instrument Engineering: Proceedings, vol. 1, no. 4, pp. 283–286 (2018)
24. Shurina, E.P., Mikhaylova, E.I.: Modified multiscale discontinuous Galerkin method in the function space H(curl). In: 13th International Scientific and Technical Conference on Actual Problems of Electronic Instrument Engineering: Proceedings, vol. 1, no. 2, pp. 398–402 (2016)
25. Veiga, L.B., Brezzi, F., Marini, L., Russo, A.: H (div) and H (curl)-conforming virtual element methods. Numer. Math. **133**, 303–332 (2016)
26. Veiga, L.B., Brezzi, F., Dassi, F., Marini, L.D., Russo, A.: A family of three-dimensional virtual elements with applications to magnetostatics. SIAM J. Numer. Anal. **56**(5), 2940–2962 (2018)
27. Nedelec, J.C.: Mixed finite elements in R^3. Numer. Math. **35**(3), 315–341 (1980)
28. Nedelec, J.C.: A new family of mixed finite elements in R^3. Numer. Math. **50**(1), 57–81 (1986)
29. Saad Y.: Iterative methods for sparse linear systems. In: Society for Industrial and Applied Mathematics (2003)
30. Shurina, E.P., Arkhipov, D.A.: Multilevel algebraic methods of modeling the 3D electromagnetic field. In: 12th International Conference on Actual Problems of Electronic Instrument Engineering: Proceedings, vol. 7040757, pp. 603–610 (2014)

Multilevel Modeling of Woven Composite Shell Structure

Eva Kormanikova and Lenka Kabosova

Abstract The focus of this paper is the multilevel modeling of the woven carbon-reinforced-fiber composite shell structure. Through the steps of (1) Micro-level modeling, (2) Meso-level modeling, (3) Macro-level modeling, and (4) Structural-level modeling, an experimental complex-shaped pavilion is designed. The mechanical behavior of fiber bundles is determined on the micro-level; subsequently, the mechanical properties of the 2-D woven fabric composite are set on the meso level. The response of the double-curved laminated shell considering the 1st order shear deformation theory is examined on the macro level. The resulting material is applied to a shell structure, which is vertically loaded. Employing the linear FEM approach and considering the shear deformation theory of a simply supported shell structure, the mechanical response of the 3D shell structure is accurately predicted.

Keywords Woven composite · Shell structure · Micro-meso-macro modeling level

1 Introduction

Textile structural composites are, recently, gaining increasing technological importance [1–3]. Forms used as reinforcements for composites can be designed to accommodate diverse requirements, including dimensional stability, subtle conformability, and deep-draw shape ability. Woven composites are essentially two-dimensional structures that are firm in the mutually orthogonal warp and fill direction. They have more balanced properties in the fabric plane when compared to unidirectional laminae. The bidirectional reinforcement in the single layer of fabric implies excellent impact resistance. Such composites are attractive for structural applications due

E. Kormanikova (✉)
Technical University of Kosice, Vysokoskolska 4, 04200 Kosice, Slovakia
e-mail: eva.kormanikova@tuke.sk

L. Kabosova
Center for Research and Innovation in Construction, Technical University of Kosice, Park Komenskeho 10, 04200 Kosice, Slovakia
e-mail: lenka.kabosova@tuke.sk

H. Dai (ed.), *Computational and Experimental Simulations in Engineering*, Mechanisms and Machine Science 119,
https://doi.org/10.1007/978-3-031-02097-1_7

to their serviceability and low fabrication costs [4–7]. Triaxially woven fabrics, made from three sets of yearns, interlacing at 60-degree angles offer improved isotropy and higher in-plane shear rigidity. The three-dimensional integrated structural geometry enables the creation of complex shapes which exhibit a high transverse shear strength and impact resistance [8–12].

Moreover, it is possible to design composites with considerable flexibility in performance, ranging from complete directional stability to engineered directional elongation. Incorporating CFRP polymers as a reinforcement, thin, lightweight, durable, and visually attractive structures can be designed.

The fabric considered in this paper is composed of two sets of mutually orthogonal bundles of the same material.

It is well-known that the overall performance and response of such structures are highly dependent on the micromechanical behavior of the composite. This kind of research direction involves analyses on different scales, ergo multi-scale modeling, as is suggested in Fig. 1, showing a sample of a graphite fiber fabric–polymer matrix composite system. The significance of multi-scale modeling is demonstrated on a large-scale structural element, i.e., composite ply (Fig. 1a), which has specific effective, macroscopic properties derived for the geometry specified on the mesoscale (Fig. 1b). A classical problem of determining the effective elastic properties becomes even more complex when taking the level of constituents, microscale (Fig. 1c), into consideration. Such a step creates demands for specific techniques, enabling the determination of the effective properties of disordered media [13–17].

The Mori–Tanaka homogenization method determines the effective elastic properties of the composites, which consist of several phases. The HELP (Heat and Elasticity Properties) software is employed to determine effective elastic properties utilizing

Fig. 1 A fiber fabric polymer matrix composite [16]

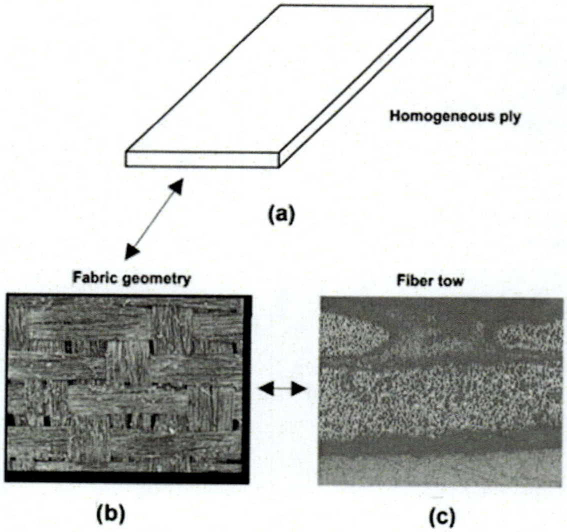

the Mori–Tanaka method [4, 5]. Through this software, ellipsoidal inhomogeneities embedded in a generally anisotropic medium can be solved. Homogenization with orientational averaging is also implemented.

Success of composite materials is due to their excellent mechanical properties. Maybe even more important is the design freedom they offer, allowing tailoring of mechanical properties [18, 19] and creating totally new shapes [22] by using plate and shell elements [23, 24].

2 Micro-level Modeling

For a composite with a random microstructure, it is suitable to use the periodic microstructure model. The model for long cylindrical fibers, regularly arranged in a square microstructure, is illustrated in Fig. 2.

The representative volume element (RVE), approximating the real material statistics as close as possible, is generally used for periodic microstructures. Therefore, the RVE is assumed to be surrounded by periodic replicas of itself. This step allows substituting the complicated microstructure by the periodic unit cell, consisting of a small number of reinforcements, while the resulting material possesses similar statistical properties as the original.

Developing a multi-scale numerical model requires starting from the microscale with the fiber and matrix mixture forming a fiber tow. Up to date, simple averaging techniques were mostly used to generate effective properties of this basic building element.

The Mori–Tanaka method can determine the elastic properties of the composites, made of several phases. The principle lies in the approximation of the effect of the phase interaction to local stresses, provided that the stress at each phase is equal to the individual inclusion inserted into an unbounded matrix subjected to unknown average strain.

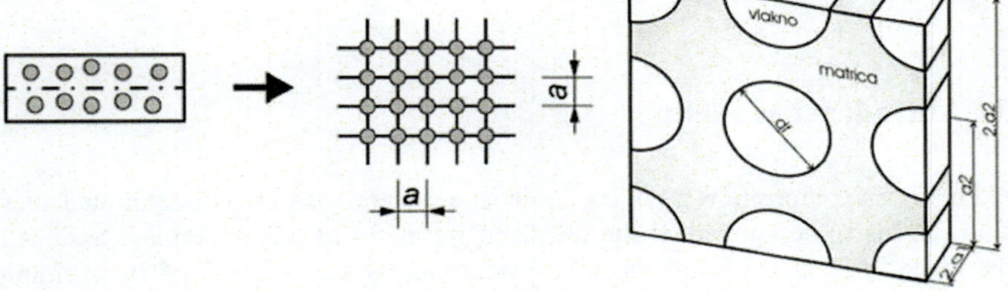

Fig. 2 A periodic microstructure model for effective elastic properties of a bundle

Stress–strain relation valid for transversely isotropic materials with axis x_1 in the direction of the fibers can be defined using the assumption of Hill moduli:

$$\begin{Bmatrix} \sigma_1 \\ \sigma_2 \\ \sigma_3 \\ \sigma_4 \\ \sigma_5 \\ \sigma_6 \end{Bmatrix} = \begin{bmatrix} m & l & l & 0 & 0 & 0 \\ l & (k+m) & (k-m) & 0 & 0 & 0 \\ l & (k-m) & (k+m) & 0 & 0 & 0 \\ 0 & 0 & 0 & m & 0 & 0 \\ 0 & 0 & 0 & 0 & p & 0 \\ 0 & 0 & 0 & 0 & 0 & p \end{bmatrix} \begin{Bmatrix} \varepsilon_1 \\ \varepsilon_2 \\ \varepsilon_3 \\ \varepsilon_4 \\ \varepsilon_5 \\ \varepsilon_6 \end{Bmatrix} \tag{1}$$

where

$$\begin{aligned} k &= \frac{k_2 k_1 + m_1(c_2 k_2 + c_1 k_1)}{c_2 k_1 + c_1 k_2 + m_1} \\ l &= \frac{c_2 l_2(k_1 + m_1) + c_1 l_1(k_2 + m_1)}{c_2(k_1 + m_1) + c_1(k_2 + m_1)} \\ n &= c_2 n_2 + c_1 n_1 + (1 - c_2 l_2 - c_1 l_1)\frac{l_2 - l_1}{k_2 - k_1} \\ m &= \frac{m_2 m_1(k_1 + 2m_1) + k_1 m_1(c_2 m_2 + c_1 m_1)}{k_1 m_1 + (k_1 + 2m_1)(c_2 m_1 + c_1 m_2)} \\ p &= \frac{2c_2 p_2 p_1 + c_1\left(p_2 p_1 + p_1^2\right)}{2c_2 p_1 + c_1(p_2 + p_1)} \end{aligned} \tag{2}$$

The effective elastic properties of the carbon-fiber composite bundle were investigated. Carbon fiber reinforced polymer (CFRP) was considered with the following characteristics: $E_f = 294$ GPa; $E_\mathrm{m} = 2$ GPa; $\nu_f = 0.4$; $\nu_m = 0.24$; $G_f = 11.8$ GPa; $G_m = 0.85$ GPa; fiber volume fraction $\xi_f = 0.65$, pores in bundles volume fraction $\xi_{pb} = 0.01$. The material characteristics were calculated through HELP software, which allows the calculation of the effective elastic properties of multi-phase composites with pores. Figure 3 depicts the electron microscope shot for the identification of the fiber volume fraction of the woven composite bundle. Voids and cracks can be understood as special cases of inhomogeneities.

The calculated material properties of one composite bundle are given in Table 1.

3 Meso-level Modeling

Plain woven composite is a typical composite system where the orientational averaging needs to be applied. If the idealized geometry of this material is assumed, the centerlines of the warp and fill systems of tows are described by a simple trigonometric form [4].

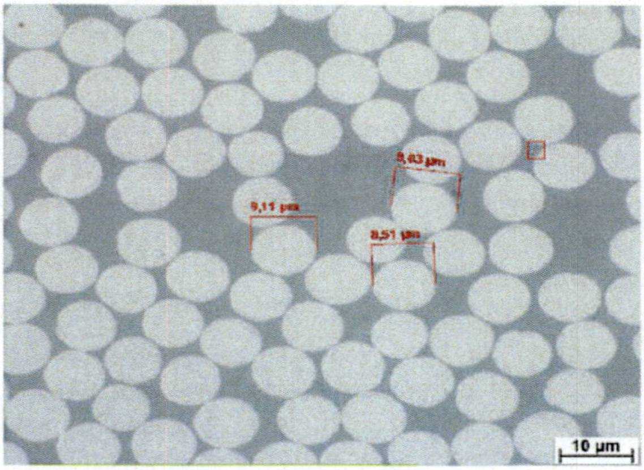

Fig. 3 Electron microscope shot for the identification of the fiber volume fraction

Table 1 Effective material characteristics of the composite bundle

Material characteristics	Fibers-matrix bundle	Bundle with pores
E_1 (GPa)	203.5	201.9
E_2 (GPa)	6.1	4.7
G_{12} (GPa)	2.3	2.1
G_{23} (GPa)	2.7	2.3
ν_{12}	0.26	0.24
ν_{23}	0.43	0.36

The dimensions of the periodic unit cell (PUC) are required for the calculation (Fig. 4). The matrix material properties are assigned in the same way as above. It is necessary to specify the effective properties of perpendicular bundles. The standard volume fraction for the fill and warp system of tows is assumed. The shape of the

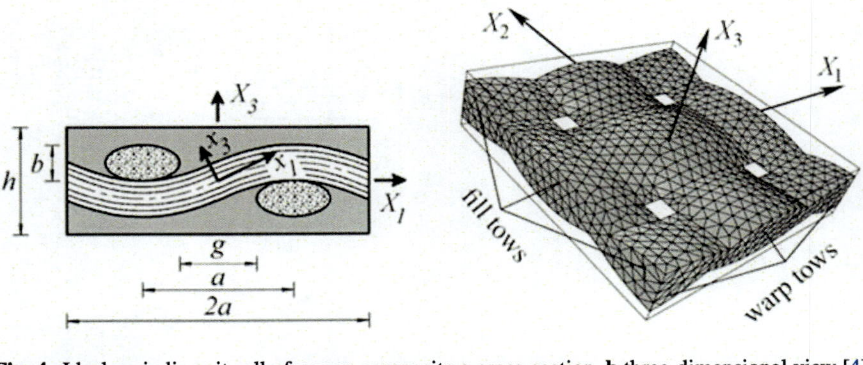

Fig. 4 Ideal periodic unit cell of woven composite **a** cross-section, **b** three-dimensional view [4]

Table 2 Effective material characteristics of the CFRP woven composite

Material characteristics	Bundles-matrix composite	Woven composite with pores
E_1 (GPa)	49.8	37.3
E_2 (GPa)	49.8	37.3
G_{12} (GPa)	1.3	1.24
G_{23} (GPa)	1.5	1.33
ν_{12}	0.28	0.11
ν_{23}	0.46	0.16

bundle is defined through its semi-axes and orientation, concerning the global coordinate system, using the Euler angles. The required material parameters are introduced next, again depending on the selected type of material symmetry in the bundle local coordinate system [15].

The effective elastic properties of the woven carbon-fiber-reinforced composite were investigated. Geometrical parameters of the periodic unit cell of the woven composite are the following: $a = 2035.58$ μm, $b = 146{,}828$ μm, $g = 464.855$ μm, $h = 313.508$ μm. Bundles volume fraction $\xi_b = 0.47$, pores in composite volume fraction $\xi_{pb} = 0.119$. The material characteristics were calculated employing the HELP software.

The calculated material properties of the woven composite are presented in Table 2.

A similar procedure used to obtain the RVE at the microscale level can be used to analyze laminate on the mesoscale level. In this case, the RVE represents a laminate (Fig. 5). The through-thickness direction should remain free to expand with the thickness.

The tensile effective elastic properties of a woven carbon-fiber-reinforced laminated composite consisting of [0/90], [45/−45], [0/45/−45] were investigated.

Fig. 5 Laminated RVE

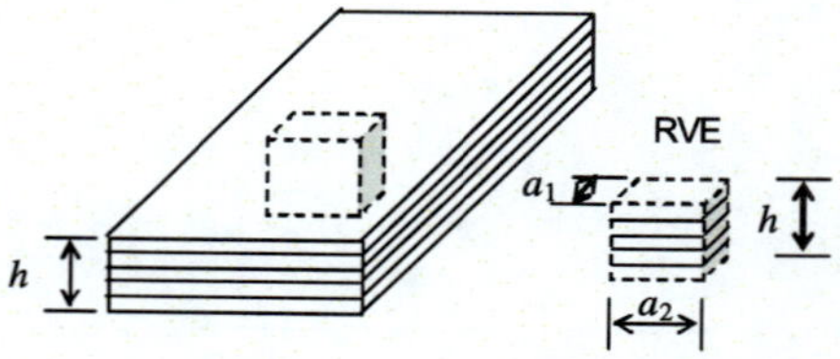

4 Macro-level Modeling

4.1 Shell Laminate Theory

Laminate shells can be also modeled as two-dimensional structural elements, with single or double-curved reference surfaces (Fig. 6). Figure 7 shows a laminated double-curved panel of the rectangular platform, of a total thickness of h. The coordinates x_1 and x_2 represent the directions of the lines of curvature of the middle surface, while the x_3-axis is a straight line perpendicular to the middle surface. R_i (i = 1, 2) denotes the principal radii of curvature of the middle surface.

The displacement field, based on first-order shear deformation theory, is given by

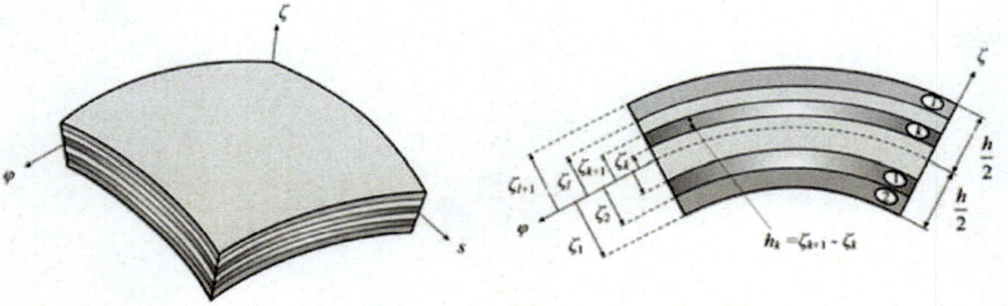

Fig. 6 Double curved laminated shell and layout of layers [20]

Fig. 7 Double-curved laminated shell [21]

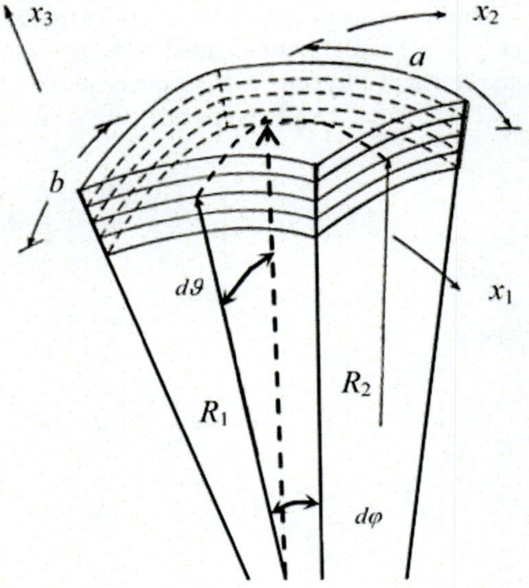

$$u_1 = (1 + x_3/R_1)\overline{u}_1 + x_3\frac{\partial u_3}{\partial x_1} \quad u_2 = (1 + x_3/R_2)\overline{u}_2 + x_3\frac{\partial u_3}{\partial x_2} \\ u_3 = \overline{u}_3 \tag{3}$$

in which u_i ($i = 1, 2, 3$) represents the components of displacement at a point x_i ($i = 1, 2, 3$), while $\overline{u}_i$ denotes the same for the corresponding point at the mid-surface.

Assumptions of shallowness, vanishing geodesic curvatures, transverse inextensibility, and the strain displacement relations for a double-curved shell, based on first-order deformation theory, are given by

$$\varepsilon_1 = \overline{\varepsilon}_1 + x3\kappa_1 \quad \varepsilon_2 = \overline{\varepsilon}_2 + x_3\kappa_2 \quad \varepsilon_4 = \overline{\varepsilon}_4 \\ \varepsilon_5 = \overline{\varepsilon}_5 \quad \varepsilon_6 = \overline{\varepsilon}_6 + x_3\kappa_6 \tag{4}$$

where

$$\overline{\varepsilon}_1 = \frac{\partial u_1}{\partial x_1} + \frac{u_3}{R_1} \quad \overline{\varepsilon}_2 = \frac{\partial u_2}{\partial x_2} + \frac{u_3}{R_2} \quad \overline{\varepsilon}_4 = \frac{\partial u_3}{\partial x_1} - \frac{u_1}{R_1} \\ \overline{\varepsilon}_5 = \frac{\partial u_3}{\partial x_2} - \frac{u_2}{R_2} \quad \overline{\varepsilon}_6 = \frac{\partial u_2}{\partial x_1} + \frac{\partial u_1}{\partial x_2} \tag{5}$$

$$\kappa_1 = \frac{\partial^2 u_3}{\partial x_1^2} \quad \kappa_2 = \frac{\partial^2 u_3}{\partial x_2^2} \\ \kappa_6 = 2\frac{\partial^2 u_3}{\partial x_1 \partial x_2} - \frac{1}{2}\left(\frac{1}{R_1} - \frac{1}{R_2}\right)\left(\frac{\partial u_2}{\partial x_1} - \frac{\partial u_1}{\partial x_2}\right) \tag{6}$$

The curved shell geometry, illustrated in Fig. 7 [21], is described by the coordinates (x_1, x_2, x_3), and it is subdivided into angular segments with the apex angles $d\phi$, $d\vartheta$ and constant curvature radii of the centerline R_1 and R_2.

The internal forces can be written in the following form

$$\boldsymbol{N} = \begin{pmatrix} N_1 \\ N_2 \\ N_6 \end{pmatrix} \quad \boldsymbol{M} = \begin{pmatrix} M_1 \\ M_2 \\ M_6 \end{pmatrix} \quad \boldsymbol{V} = \begin{pmatrix} V_1 \\ V_2 \end{pmatrix} \tag{7}$$

where

$$N_1 = \int_{-\frac{h}{2}}^{+\frac{h}{2}} \sigma_1 dz \quad M_1 = \int_{-\frac{h}{2}}^{+\frac{h}{2}} \sigma_1 z dz \\ N_2 = \int_{-\frac{h}{2}}^{+\frac{h}{2}} \sigma_2 dz \quad M_2 = \int_{-\frac{h}{2}}^{+\frac{h}{2}} \sigma_2 z dz \tag{8}$$

$$N_6 = \int_{-\frac{h}{2}}^{+\frac{h}{2}} \tau_6 dz \quad M_6 = \int_{-\frac{h}{2}}^{+\frac{h}{2}} \tau_6 z dz$$
$$V_1 = \int_{-\frac{h}{2}}^{+\frac{h}{2}} \tau_{xz} dz \quad V_2 = \int_{-\frac{h}{2}}^{+\frac{h}{2}} \tau_{yz} dz \tag{9}$$

$$\begin{aligned} \boldsymbol{N} &= \int_{-h/2}^{+h/2} \boldsymbol{E}(z) dz \overline{\boldsymbol{\varepsilon}} + \int_{-h/2}^{+h/2} \boldsymbol{E}(z) z dz \boldsymbol{\kappa} \\ \boldsymbol{M} &= \int_{-h/2}^{+h/2} \boldsymbol{E}(z) z dz \overline{\boldsymbol{\varepsilon}} + \int_{-h/2}^{+h/2} \boldsymbol{E}(z) z^2 dz \boldsymbol{\kappa} \\ \boldsymbol{V} &= (k^*) \int_{-h/2}^{+h/2} \boldsymbol{E}^t(z) dz \boldsymbol{\gamma} \end{aligned} \tag{10}$$

where $\boldsymbol{N}$ is the membrane force resultant vector, $\boldsymbol{M}$ is the moment resultant vector and $\boldsymbol{V}$ is the transverse shear force resultant vector.

The internal forces can be written in hypermatrix form

$$\begin{pmatrix} \boldsymbol{N} \\ \boldsymbol{M} \end{pmatrix} = \begin{pmatrix} \boldsymbol{A} & \boldsymbol{B} \\ \boldsymbol{B} & \boldsymbol{D} \end{pmatrix} \begin{pmatrix} \overline{\boldsymbol{\varepsilon}} \\ \boldsymbol{\kappa} \end{pmatrix}$$
$$\boldsymbol{V} = \boldsymbol{k}\overline{\boldsymbol{A}}\boldsymbol{\gamma} \tag{11}$$

where $\boldsymbol{N}$ is the membrane force resultant vector, $\boldsymbol{M}$ is the moment resultant vector and $\boldsymbol{V}$ is the transverse shear force resultant vector. In addition, $\boldsymbol{A}$, $\boldsymbol{D}$, $\boldsymbol{B}$ denote the classical extensional stiffness matrix, bending stiffness matrix, and bending-extensional coupling stiffness matrix, respectively, whereas $\overline{\boldsymbol{A}}$ is the shear stiffness matrix [22].

The components of $\boldsymbol{A}$, $\boldsymbol{B}$, $\boldsymbol{D}$, $\overline{\boldsymbol{A}}$ matrix, are written as

$$\begin{aligned} \boldsymbol{A} &= \int_{-h/2}^{+h/2} \boldsymbol{E}(z) dz = \sum_{n=1}^{N} \int_{n-1_z}^{n_z} {}^n\boldsymbol{E} dz = \sum_{n=1}^{N} {}^n\boldsymbol{E}^n h \\ \boldsymbol{B} &= \int_{-h/2}^{+h/2} \boldsymbol{E}(z) z dz = \sum_{n=1}^{N} \int_{n-1_z}^{n_z} {}^n\boldsymbol{E} z dz = \sum_{n=1}^{N} {}^n\boldsymbol{E} \frac{{}^nZ^2 - {}^{n-1}Z^2}{2} \\ \boldsymbol{D} &= \int_{-h/2}^{+h/2} \boldsymbol{E}(z) z^2 dz = \sum_{n=1}^{N} \int_{n-1_z}^{n_z} {}^n\boldsymbol{E} z^2 dz = \sum_{n=1}^{N} {}^n\boldsymbol{E} \frac{{}^nZ^3 - {}^{n-1}Z^3}{3} \\ \overline{\boldsymbol{A}} &= \int_{-h/2}^{+h/2} \boldsymbol{E}^t(z) dz = \sum_{n=1}^{N} {}^n\boldsymbol{E}^{t^n} h \end{aligned} \tag{12}$$

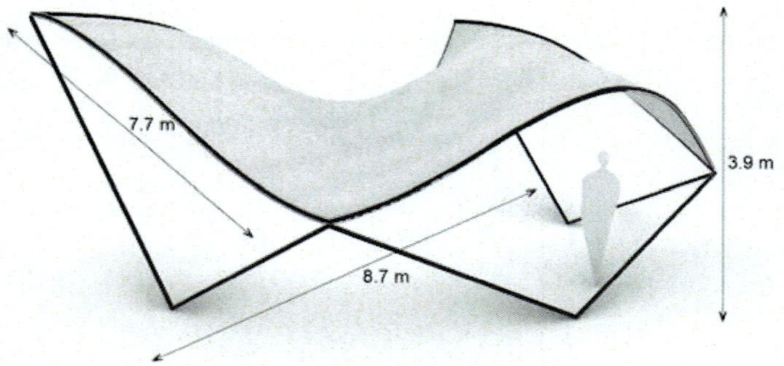

Fig. 8 The carbon-fiber shell pavilion

5 Structural-Level Modeling

A pavilion, made of shell structure elements, was designed to demonstrate the structural modeling level (Fig. 8). The Institute of Structural Engineering and the Center for Research and Innovation in Construction of the Technical University of Kosice tested carbon-fiber-reinforced composite in the collaborative concept of a double-curved pavilion. The pavilion is simply supported ($u = v = w = 0$) by a stiffening frame and thin, V-shaped columns and is 3.9 m high, 8.7 m long, and 7.7 m wide. Because the material is lightweight, it enables a design setup for a long-span fiber composite construction, which very thin. The material properties were taken from Tables 1 and 2. The input load was set to 5 kPa, uniformly loading the designed structure. The response of the structure was investigated for three different types of fiber orientation of woven and non-woven [0/90], [45/−45] and [0/45/−45] laminates. The designed shell structure pavilion was subsequently analyzed in ANSYS Mechanical software to obtain an idea of the deformation of the pavilion under vertical load.

The meshing of the structure was realized by SHELL99 finite elements. Then mesh modifying, with refine it in 7 iteration steps, was analyzed (Fig. 9). Advanced option with fifth level of refinement and cleanup plus smooth postprocessing was adopted. This option was only valid with an all-quadrilateral mesh.

The deflection analysis shows that the optimal arrangement of fibers in the designed pavilion is [0/45/−45], using the non-woven material (Fig. 10). The maximum deflections of CFRP laminate plate made of woven and non-woven material in various lay-ups are shown in Table 3.

6 Conclusion

Using the multilevel modeling approach, a woven, shell structure pavilion made of the carbon-fiber-reinforced laminated composite (CFRP), was designed. Woven and

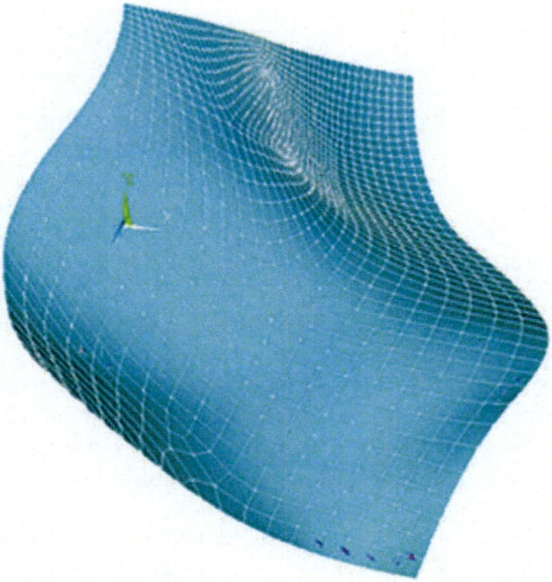

Fig. 9 The mesh modifying of the mid-plane and boundary conditions

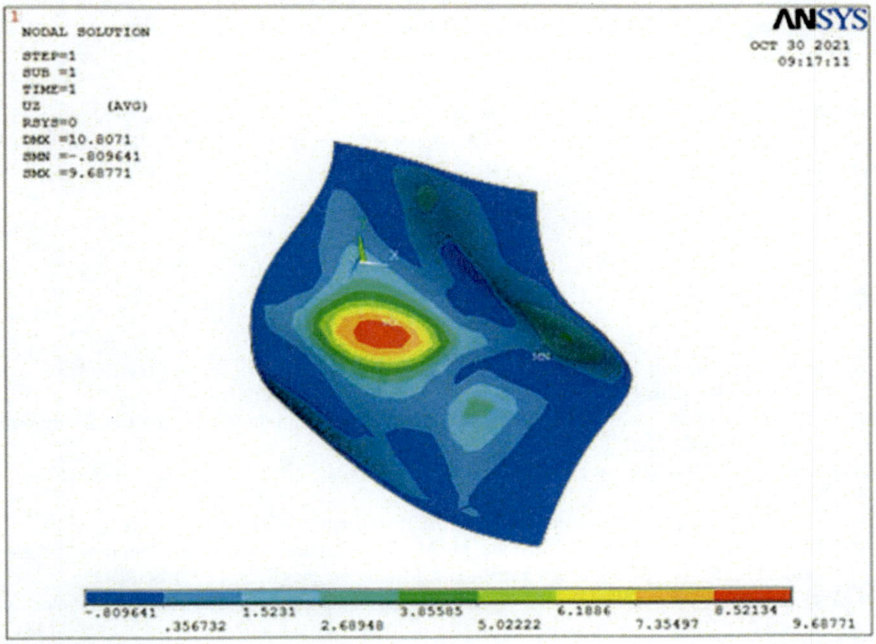

Fig. 10 Contour plot of deflection of pavilion with legend for non-woven [0/45/−45] laminate

Table 3 Maximum deflection of the woven and non-woven CFRP laminates

Material characteristics	Woven [0/90]	Non-woven [0/90]	Woven [45/−45]	Non-woven [45/−45]	Woven [0/45/−45]	Non-woven [0/45/−45]
w (mm)	12.3204	10.8999	27.8116	26.8200	10.2555	9.6877

non/woven alternatives of the CFRP pavilion, with the fiber orientation of [0/90], [45/−45], and [0/45/−45] were vertically loaded and investigated in the Ansys Mechanical software. The Mori Tanaka method, combining periodic unit cell models with randomly distributed ellipsoidal-shaped pores, was utilized in the approach. The double-curved laminated shell element was applied to the pavilion structure, employing the shell laminate theory.

The analysis in ANSYS shows that the optimal arrangement of fibers in the designed pavilion is [0/45/−45], using the non-woven material (Fig. 10). However, the difference between the maximal deflection of the woven and non-woven pavilion shell structure is low, 6.14%. The fiber orientation of [45/−45] caused the highest values of deflection of designed composite pavilion.

This paper focused on micro-level variations in fiber orientation, which influenced the behavior of the proposed pavilion on the structural level. In future research, we aim to focus on experimenting at the meso level and vary the dimensions of the periodic unit cell while observing the change in the overall performance of the CFRP pavilion.

Combining the untethered freedom of the shape with the advanced composite, offered the potential for various material fiber arrangements, and unique structural performance.

Acknowledgements This work was supported by the Scientific Grant Agency of the Ministry of Education of Slovak Republic and the Slovak Academy of Sciences under Projects VEGA 1/0374/19 and VEGA1/0363/21.

References

1. Ma, P., Jiang, G., Gao, Z.: The three dimensional textile structures for composites. In: Advanced Composite Materials: Properties and Applications, pp. 497–526 (2017)
2. Bright, M., Kurlin, V.: Encoding and topological computation on textile structures. Comput. Graph. **90**, 51–61 (2020)
3. Lombardi, S., Canobbio, R.: Textile structures for climate control. Proc. Eng. **155**, 163–172 (2016)
4. Vorel, J., Sejnoha, M.: Documentation for HELP program, CVUT Praha (2008)
5. Stransky, J., Vorel, J., Zeman, J., Sejnoha, M.: Mori-Tanaka based estimates of effective thermal conductivity of various engineering materials. Micromachines **2**(2), 129–149 (2011)
6. Wang, P., Wang, B.L., Wang, K.F., Xi, L.: Effective behaviors of anisotropic thermoelectric composites containing ellipsoidal inclusions. Compos. Struct. **267**, Article Number 113817 (2021)
7. Kulovana, T., et al.: Mechanical, durability and hygrothermal properties of concrete produced using Portland cement-ceramic powder blends. Struct. Concr. **17**(1), 105–115 (2016)

8. Melcer, J., Merčiaková, E., Kúdelčíková, M., Valašková, V.: Response to kinematic excitation, numerical simulation versus experiment. Mathematics **9**(6), 678 1–23 (2021)
9. Major, M., Major, I., Kucharova, D., Kulinski, K.: Reduction of dynamic impacts in block made of concrete-rubber composites. Civil Environ. Eng. **14**(1), 61–67 (2018)
10. Krejsa, M., et al.: Parallelization in DOProC method and its using in probabilistic modelling of fatigue problems. In: AIP Conference Proceedings, 2116, Article Number 120007 (2019)
11. Kralik, J., Kralik, J. Jr.: Nonlinear analysis of NPP safety against the aircraft attack. In: AIP Conference Proceedings, 1738, Article Number 480079 (2016)
12. Kormanikova, E., Kotrasova, K.: Resonant frequencies and mode shapes of rectangular sandwich plate. Chem. Listy **105**, 535–538 (2011)
13. Kormanikova, E., Mamuzic, I.: Buckling analysis of a laminate plate. Metalurgija **47**(2), 129–132 (2008)
14. Kormanikova, E., Kotrasova, K., Harabinova, S., Panulinova, E.: Elastic mechanical properties of random oriented short fiber composites. In: AIP Conference Proceedings, 2293, Article Number 130008 (2020)
15. Vorel, J., Urbanová, S., Grippon, E., Jandejsek, I., Maršálková, M., Šejnoha, M.: Multi-scale modeling of textile reinforced ceramic composites. Ceram. Eng. Sci. Proc. **34**(10), 233–245 (2014)
16. Sejnoha, M., Zeman, J.: Micromechanical Analysis of Random Composites. Czech Technical University (2000)
17. Sladek, J., Novak, P., Bishay, P.L., Sladek, V.: Effective properties of cement-based porous piezoelectric ceramic composites. Constr. Build. Mater. **190**, 1208–1214 (2018)
18. Kotrasova, K., Kormanikova, E.: Two-step scheme for solution of the seismic response of liquid-filled composite cylindrical container. Math. Methods Appl. Sci. **43**(13), 7664–7676 (2020)
19. Kormanikova, E., Kotrasova, K.: Multiscale modeling of liquid storage laminated composite cylindrical tank under seismic load. Compos. B **146**, 189–197 (2018)
20. Tornabene, F., Ceruti, A.: Mixed static and dynamic optimization of four-parameter functionally graded completely doubly curved and degenerate shells and panels using GDQ method. Math. Probl. Eng. **1**, 1–33 (2013)
21. Topal, U.: Frequency optimization of laminated composite spherical shells. Sci. Eng. Compos. Mater. **19**, 381–386 (2012)
22. https://www.icd.uni-stuttgart.de/projects/icditke-research-pavilion-2016-17
23. Garcia-Macias, E., Rodriguez-Tembleque, L., Saez, A.: Bending and free vibration analysis of functionally graded graphene vs. carbon nanotube reinforced composite plates. Compos. Struct. **186**, 123–138 (2018)
24. Liu, D., Kitipornchai, S., Chen, W., Yang, J.J.: Three-dimension buckling and free vibration analyses of initially stressed functionally graded graphene reinforced composite cylindrical shell. Compos. Struct. **189**, 560–569 (2018)

Study on Rock-Breaking Mechanism of Highly Plastic Formations

Fangyuan Shao, Wei Liu, and Deli Gao

Abstract Polycrystalline Diamond Compact (PDC) bits have occupied a large market of unconventional oil and gas drilling. Efficient rock breaking of PDC cutters greatly promotes the wide variety of applications of PDC bits. However, conventional cylinder-shaped PDC cutters have encountered great difficulties in shearing highly plastic mudstone. It is almost impossible to mimic the highly plastic and over-pressured conditions where the mudstone is located, hindering the laboratory investigations on the rock-breaking mechanism of highly plastic mudstone. To solve this issue, this work proposed an approximate cutting test using plastic rubbers to mimic highly plastic mudstone. The rubbers were fixed to a vertical turret lathe (VTL) and cut by various shaped PDC cutters with the purpose of investigating their cutting mechanism for plastic formations. The approximate cutting test was capable of mimic two types of formations: elastic–plastic mudstone and soft mudstone. Elastic–plastic mudstone exhibits highly elastic–plastic deformation during the cutter-rock interaction, resulting in pretty low cutting efficiency of conventional cylinder-shaped PDC cutters. Soft mudstone doesn't exhibit elastic–plastic deformation and can be easily removed by PDC cutters. In this work, three kinds of PDC cutters: cylinder-shaped cutter, axe-shaped cutter and optimized axe-shaped cutter were studied using the approximate cutting test. The optimized axe-shaped cutter was designed by curving the intersected planes of axe-shaped cutter. The tests showed that the cutting efficiency of optimized axe-shaped cutter was much better than the conventional cylinder-shaped cutter on cutting the elastic–plastic mudstone. On the other hand, if the mudstone didn't exhibit elastic–plastic deformation, the conventional cylinder-shape cutter showed highest cutting efficiency than axe-shaped cutter and optimized axe-shaped cutter. The findings in this work will provide valuable and meaningful

F. Shao · W. Liu (✉) · D. Gao (✉)
MOE Key Laboratory of Petroleum Engineering, China University of Petroleum, Beijing 102249, China
e-mail: wei.liu@cup.edu.cn

D. Gao
e-mail: gaodeli_team@126.com

State Key Laboratory of Petroleum Resources and Engineering, Beijing 102249, China

H. Dai (ed.), *Computational and Experimental Simulations in Engineering*, Mechanisms and Machine Science 119,
https://doi.org/10.1007/978-3-031-02097-1_8

guidelines for the drilling of mudstone. According to the lithology, the right cutter shape shall be used to optimize the cutting efficiency.

Keywords Rock-breaking mechanism · Highly plastic mudstone · Optimized axe-shaped · PDC cutter

1 Introduction

PDC bits are designed primarily for soft to medium hard formations [1]. There is a kind of elastic–plastic mudstone formation whose hardness is not high but hard to drill and figuratively called 'rubber layer' in field drilling [2]. The low rate of penetration (ROP) in highly plastic mudstone has been a difficult problem in geological, coal and petroleum engineering for a long time. Many scholars have analyzed the problem of this highly plastic mudstone. The highly plastic mudstone refers to a kind of rock that exhibits obvious elastic–plastic deformation before failure. The cutters slip on the surface of the mudstone due to the highly plastic mudstone has a resistance to be cut [3]. PDC bits with cylinder-shaped cutters in these formations show very low ROP. The mudstone near the bottom hole shows highly plasticity, which mainly caused by the hydrostatic column pressure and deep confining pressure [4].

In response to increasingly complex formations, many shaped cutters have been invited, but many of them have not been solved by the mechanism of rock breaking. The cutters on coring bit were optimized based on reducing the cross-sectional area and contact arc length, which is helpful to make the rock directly to be destroyed over the elastoplastic deformation stage, greatly improving the drilling efficiency in the field [5]. The shaped PDC cutters for effectively cutting hard rock can also be introduced into the cutting problem of plastic mudstone. The v-shaped cutter was designed for interbedded application, which accounting for 16% reduction in contact area [6]. Luo et al. verified v-shaped cutters have stronger aggression and higher crushing efficiency for highly plastic mudstone by numerical simulation [4].

There are two main measures to control efficiency of PDC bits: rock destruction and rock removal [7]. The axe-shaped cutter was originally designed to break hard rock, which combines the shearing action of a conventional PDC bit with the crushing action of a tungsten carbide insert [8, 9]. Since the ridge structure of axe-shaped cutter increases its sharpness, it is possible to invade effectively into highly plastic mudstone by optimizing the geometry. Comparing with the cylinder-shaped cutters and conventional axe-shaped cutter, a new optimized axe-shaped cutter was developed to cutting rubber that mimic cutting highly plastic mudstone. A series of experiments have been carried out on rubber layer to explore the cutting efficiency of different shaped PDC cutters.

2 Experimental Material and Process

2.1 PDC Cutters

As shown in Fig. 1, three types of PDC cutters were selected in this experiment: cylinder-shape cutter, axe-shaped cutter and optimized axe-shaped cutter. The optimized axe-shaped cutter was improved from the following two aspects: (1) The radius of transition arc in the middle was reduced to make the ridge edge sharper. (2) The intersected planes of axe-shaped cutter were designed as concave structure. Table 1 shows geometric dimensions of all cutters. All cutters have the same materials and manufacturing process expect the geometry. The cross-sectional area and contact area of three kinds of PDC cutters were calculated by CREO software under different DOCs. The results in Figs. 2 and 3 show that both the cross-sectional area and contact area of sample #3 and #2 are lower than sample #1 under the same DOC. The cross-sectional area of sample #3 is slightly reduced compared with sample #2. Furthermore, the concave structure of sample #3 makes its contact area increase compared with Sample #2 to a certain extent.

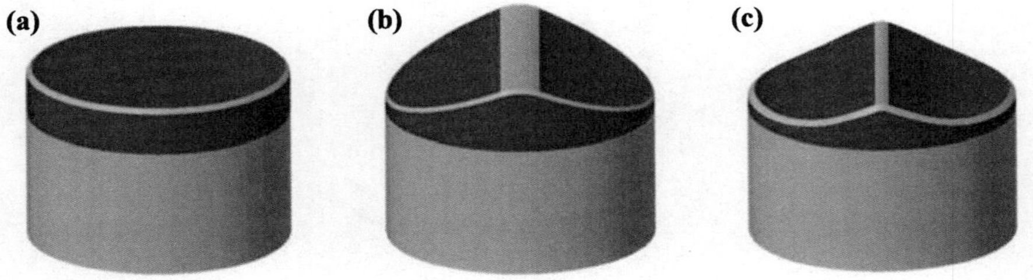

Fig. 1 Schematic diagram of three PDC cutters, **a** Cylinder-shape cutter, **b** Axe-shaped cutter, **c** Optimized axe-shaped cutter

Table 1 Geometric dimensions of PDC cutters

Sample ID	Geometry	Diameter (mm)	Height (mm)	Chamfer size (mm)	Axe shape angle (°)
#1	Cylinder	16.00	13.00	0.40	–
#2	Axe-shape	16.00	13.00	0.40	135
#3	Optimized axe-shape	16.00	13.00	0.40	135

Fig. 2 Cross-sectional area of all samples

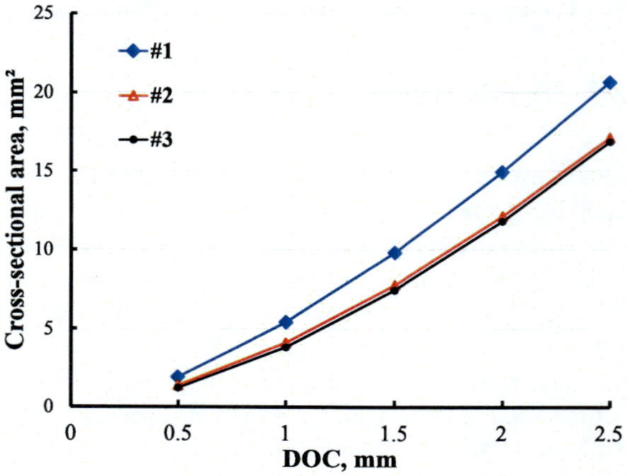

Fig. 3 Contact area of all samples

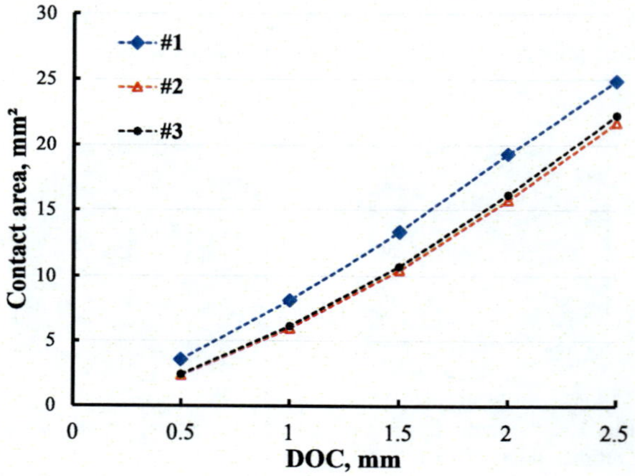

2.2 Experimental Materials

It is difficult to create the cutting conditions for PDC cutter to cut highly plastic mudstone under confining pressure. Luo et al. [4] studied the process of cutting plastic mudstone with V-shaped cutter by numerical simulation software. However, the shape of cuttings cannot be obtained by numerical simulation. Rubber in Fig. 4a with high toughness and elastic-plasticity is used to mimic the highly plastic mudstone, which is expected to obtain the experimental reference of cutting in plastic mudstone. The experiment setup (vertical turret lath, VTL) was used to measure the thermal stability and wear resistance of PDC cutter [10]. And the sample holder of VTL has been upgraded for this experiment. Figure 4b shows a new steel structure fixture, which is used to fix the rubber. The rubber was rotated by the sample holder of VTL during the experiments.

Fig. 4 **a** Rubber layer, **b** Steel structure fixture, **c** Cutting trajectories

Before the formal experiment began, several trial cutting experiments were carried out with different DOCs. After the trial cutting experiments, it was found that there were three states of cutting paths when cutting rubber. (1) Figure 5a shows scratches on the surface trajectory, which is considered an invalid cut. There was no crumbled rubber produced because the PDC cutter did not invade into the rubber layer. This

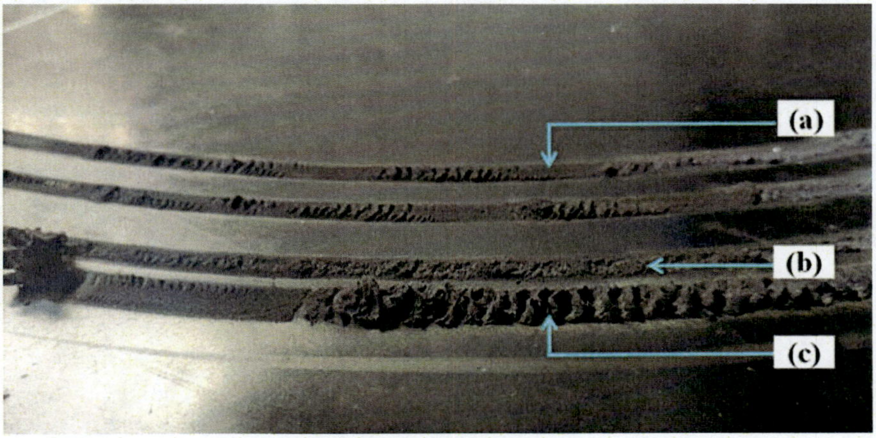

Fig. 5 **a** Scratches on rubber, **b** Cutting trajectory, **c** Serious scratches

Table 2 SCR test's parameter

Test number	Cutter number	Cutting speed (mm/s)	Back rake angle (°)	DOC (mm)
1	#1	250	20.0	2.0
2	#2	250	20.0	2.0
3	#3	250	20.0	2.0
4	#1	250	20.0	3.5
5	#2	250	20.0	3.5
6	#3	250	20.0	3.5

situation may lead to small footage and low ROP in drilling field. (2) Figure 5b demonstrates that the rubber has been cut effectively, which is considered to effective cutting mode. The crumbled rubber was produced at corresponding trajectory. There is no doubt that the effective cutting distance is proportional to the mass of the crumbled rubber finally harvested. (3) Although the PDC cutter invaded into the formation, no slippage is shown in Fig. 5c. There was few crumbled rubbers generated, which was considered an ineffective cutting mode. Since these three phenomena are consistent with the abnormal ROP from the plastic mudstone formation in field drilling, rubber can be used as a similar material for approximate cutting experiments. To make the comparison more significant, the DOC should be set as large as possible.

2.3 Cutting Experiment Process

Samples #1, #2 and #3 were selected for single circle rotating (SCR) tests on VTL. The rubber was rotated only one circle creating a ring of groove. In the cutting experiments, six positions were selected in the flat position on the rubber surface. And the cutting linear velocity and back rake angle were kept unchanged. DOCs of 2.0 and 3.5 mm were selected for these groups of cutting experiments. Detailed parameters are shown in Table 2. In other hand, the second experiment of each sample was carried at the same position with the same DOC.

3 Experimental Results and Analysis

3.1 Analysis of the State After Cutting

Cutter-rubber interaction was captured by a video to analyze the cutting process in detail. The process starts with plunging the cutter in an already rotating rubber layer at the desired cutting speed. The first cutting of each sample on rubber mainly to mimic elastic–plastic mudstone. The ribbon gets thicker and longer with the test continuing and then curls forward possibly. Figure 6 shows that the first cutting process on the rubber layer for all samples. As can be seen from Fig. 6a_1–c_1, sample #3 has the shortest time (0.32 s) to effectively invade into rubber layer. When the cutting time is 1.00 s, the crumbled rubber produced by sample #1 and sample #2 are similar, but both are much smaller than that of sample #3. Figure 6c_3 displays that the ribbon rubber has been fractured, yet the rubber in front of the samples #1 and #2 just begins to curl in Fig. 6a_3, b_3.

Figure 7 demonstrates that the second cutting process of each sample on the rubber layer, which mimics PDC cutter cutting on the soft mudstone. The biggest difference between the second cutting and the first cutting is that there has been scratches on the rubber layer, making it easier for the cutters to invade. It's clear that all the samples cut more ribbon rubber than that of the first cutting at the same moment in from Fig. 7a_1–c_1. At the time of 1.00 s, Fig. 7c_2 shows that the ribbon rubber has broken. While the ribbon rubber in front of sample #2 and sample #3 are still in the process of increasing gradually in Fig. 7a_2, b_2. The ribbon rubber in front of both sample #1 and sample #3 begin to break in Fig. 7a_3, c_3. Figure 7b_3 indicates that the ribbon rubber still grows at 1.20 s. As can be seen from the above, sample #1 produces the most ribbon rubber, while sample #3 has the fastest fracture of rubber cuttings.

Figure 8 shows that the crumbled rubbers in front of the cutters form in the shape of ribbons. Figure 8a_1 shows the state of sample #1 at the end of cutting, in which

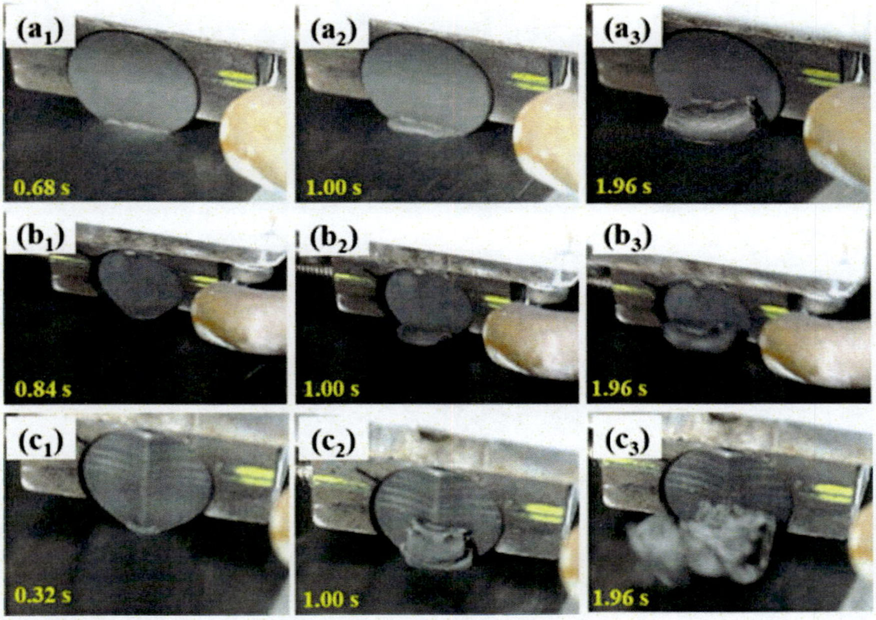

Fig. 6 The first cutting on rubber layer

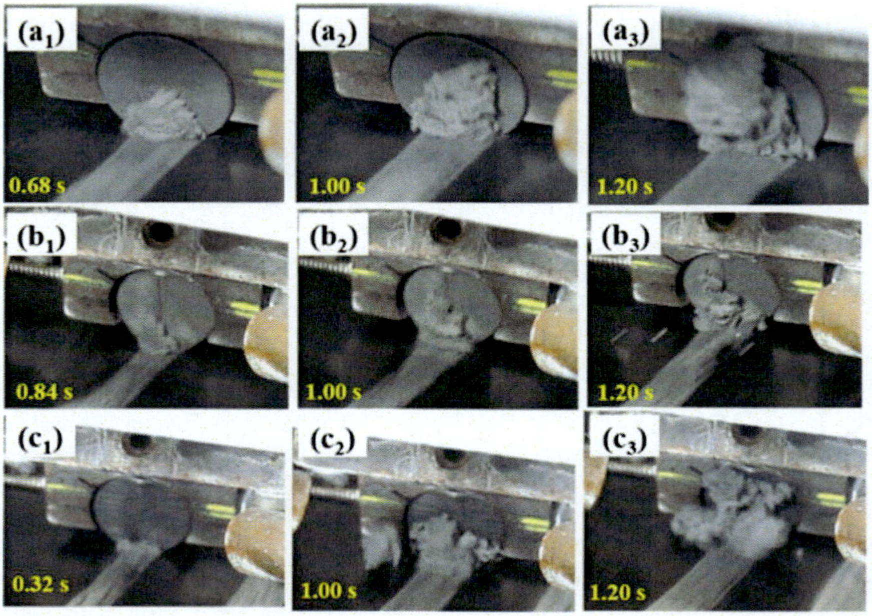

Fig. 7 The second cutting on rubber layer

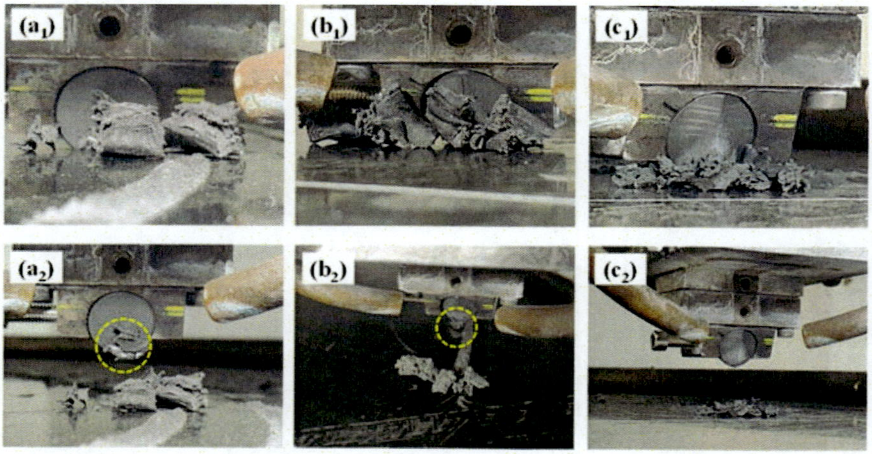

Fig. 8 State after cutting for all samples in first cutting

the ribbon rubber is uniform and flat. Dotted yellow line area in Fig. 8a_2 displays some ribbon crumbled rubber still stuck to the surface when the sample #1 was lifted. Figure 8b_1 shows that the ribbon rubber of sample #2 is concave in the middle and convex on both sides due to the axe-shaped structure. As shown in Fig. 8b_2, there was also some ribbon rubbers stuck to the surface of sample #2. Compared with the previous two samples, sample #3 (Fig. 8c_1) was easier to be invaded into the rubber layer during the process of cutting. The ribbon length of the rubber is shorter than other samples due to the optimized axe-shaped structure and concave cutting face design. Figure 8c_2 indicates that there is no ribbon rubber adhered to the cutter surface. From the above analysis, it can be concluded that the non-planar structure, especially the optimized axe-shaped structure, is conducive to invade effectively into rubber. The concave structure of sample #3 is helpful to the fracture of long ribbon rubber debris.

3.2 *The Trajectory and Mass of Cutting*

Figure 9 shows the collected crumbled rubber of all samples at two DOCs in first and second cutting. The mass of every sample was normalized, which is easy to compare the ability of the sample to cut rubber. Figure 10a illustrates that the ability of sample #1 is higher than that of sample #2, but lower than that of sample #3. It can be found from the cutting trajectory that the ineffective cutting distance was longer when DOC = 2.0 mm than that of DOC = 3.5 mm during the whole cutting process. Figure 10b points out that the mass per meter of sample #1 is the highest due to the largest cross-sectional area in three samples in the second cutting. It is concluded that the cross-sectional area has a great impact on the cutting efficiency after effectively invading into the rubber layer.

With the aim of making ensure that the three tracks at the same position are all in effective cutting state as far as possible, the first cutting trajectories with DOC =

(a) First cutting (b) Second cutting

Fig. 9 Mass of crumbled rubber

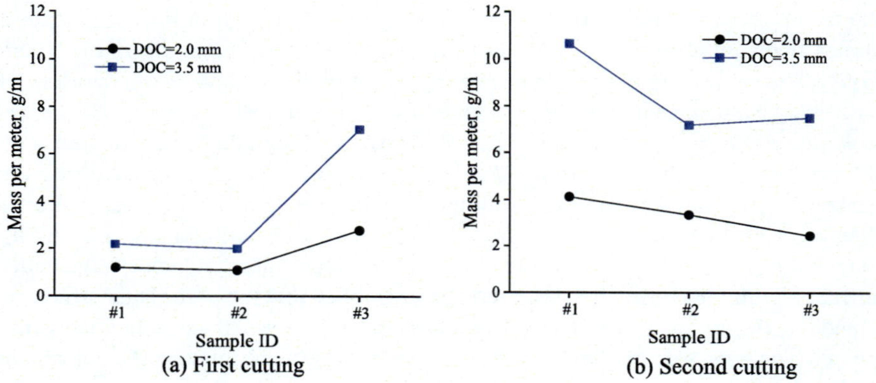

Fig. 10 Mass per meter of samples in different DOCs

3.5 mm were selected for detailed analysis. Figure 11 indicates that all the measured average values of width are lower than theoretical value. And this situation shows that the highly plastic rubber is hard to cut. On the other hand, it is more difficult to achieve volume broken. Based on Figs. 10a and 11, the measured width of sample #2 is larger than that of sample #1, but the mass of crumbled rubber per meter is less than that of sample #1. There is no doubt that the ridge structures of sample #2 helps to invade the rubber layer than sample #1. However, the effective cutting is

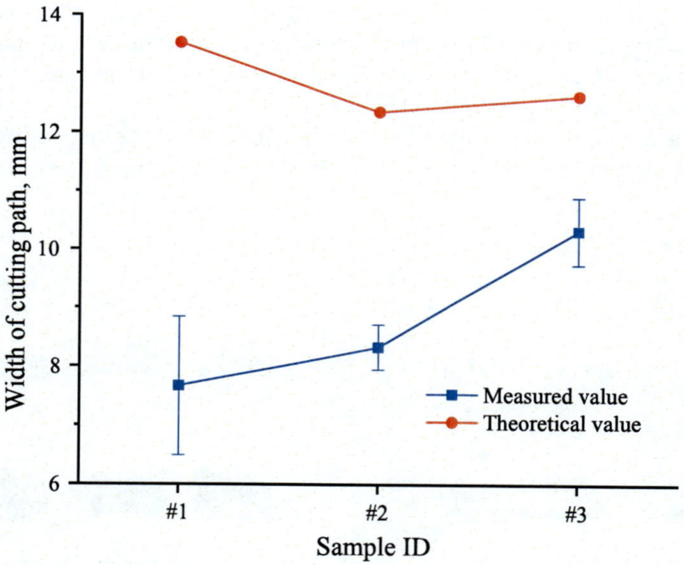

Fig. 11 Average width of the first cutting trajectory, DOC = 3.5 mm

not enhanced due to the large radius of transition arc on the ridge structure. On the contrary, both the measured width and crumbled rubber per meter of sample #3 are higher than that of sample #1. This proves that the optimized axe-shaped structure, especially the reduced transition radius, is conducive to improve the ability to cutting highly plastic formation in the first cutting.

3.3 *Cutting Forces and Special Mass*

In the whole cutting process, the cutting forces of all the samples have been collected. The forces mainly come from the average value of the whole cutting stage. The radial force is too small to be ignored, and the tangential and normal forces are mainly analyzed in this paper. Figure 12 indicates that the tangential forces of all samples increase with the increase of DOC in first and second cutting. And the tangential force of sample #3 is the lowest of all samples at two DOCs mm in Fig. 12a. Figure 12b illustrates that the tangential forces are nearly half lower than that of the first cutting at corresponding DOC. During the second cutting process, the gap between the cutting forces of the three samples was narrowed. Surprisingly, the tangential force of sample #2 is the greatest cutting force, not sample #1. Figure 13a displays that the normal forces of all samples are smaller than that of tangential forces. The increase of normal forces with DOC are lower than that of tangential forces. And the normal force of sample #3 is still the lowest of all samples in two DOCs. Figure 13b denotes that the normal forces are also lower than that of the first cutting at corresponding DOC. The normal force of sample #2 is the greatest cutting force in DOC of 2.0 mm. While the normal forces of three samples are very close at DOC of 3.5 mm. As the DOC of the rubber is large enough, the influence of its geometry diminishes gradually.

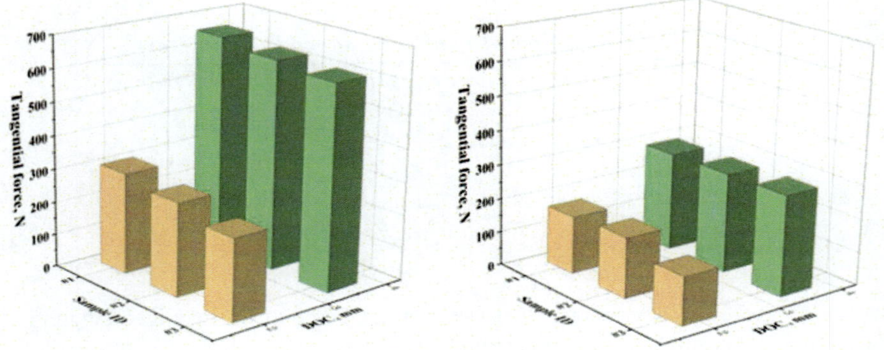

(a) First cutting (b) Second cutting

Fig. 12 Tangential forces of all samples in different DOCs

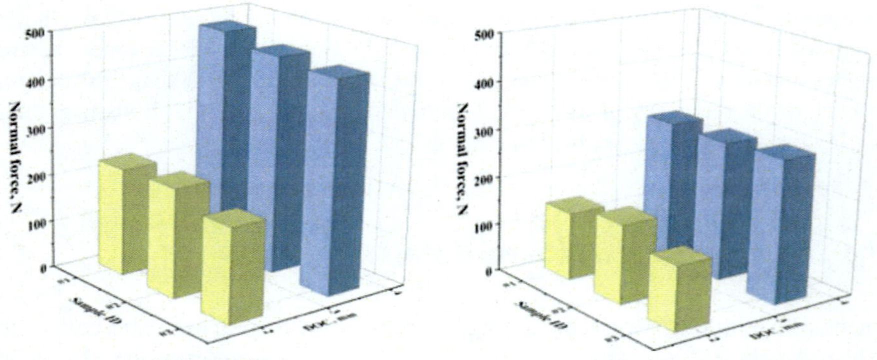

(a) First cutting (b) Second cutting

Fig. 13 Normal forces of all samples in different DOCs

Combined Figs. 12a and 13a, the tangential force and normal forces of the optimized axe-shaped cutter are lower than those of the cylinder-shaped cutter. It is further shown that the optimized structures, including ridge structure and concave structure, are more suitable for cutting highly plastic formation which is hard to invade. Consider Figs. 12b and 13b, the tangential force and normal forces of all cutters are close, which means that the shape of the sample has little effect on the cutting force especially the DOC is large enough.

Hareland et al. [11] introduced specific volume to measure the efficiency of cutting. In this paper a new parameter as shown in Eq. 1, namely specific mass, was used to express to intuitively compare the cutting efficiency of different cutters. The advantage of this new parameter is that it contains three factors: cutting forces, trajectory length and cuttings mass. It also can be defined as mass of cuttings under unit cutting force with per meter, which is shown in Eq. 2.

$$Specific\,Mass = \left(\frac{Mass\,of\,rubber\,removed\,in\,one\,meter}{Average\,force\,required\,to\,remove\,of\,rubber}\right) \tag{1}$$

$$M_0 = \frac{M'}{F} \tag{2}$$

$$F = \sqrt{F_c^2 + F_n^2} \tag{3}$$

where M' is the mass per meter, F is the resultant force of the averaged tangential force (F_c) and normal force (F_n) acting on the cutter.

According to Eq. 2, greater special mass represents higher cutting efficiency. As shown in Fig. 14a, the special mass is higher when the DOC is 3.5 mm than that DOC is 2.0 mm for samples #1 and sample #3. However, the special mass of sample #2 at DOC of 3.5 mm is lower than that of 2.0 mm. This special situation may be

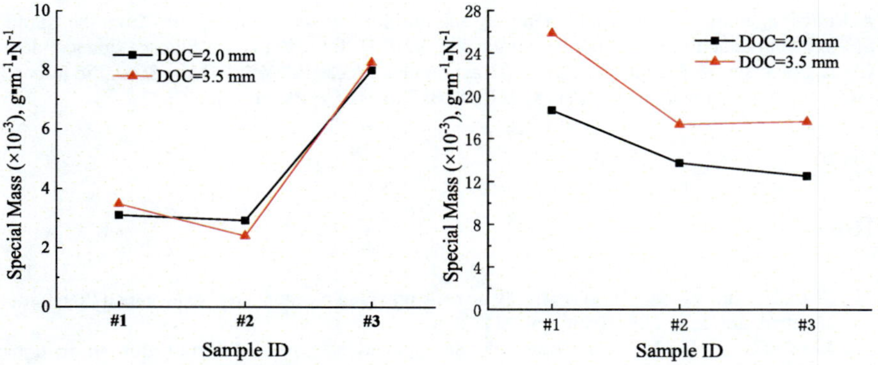

(a) First cutting (b) Second cutting

Fig. 14 Special mass of different samples

the reason that the corresponding crumbled rubber is not produced after the cutting forces increasing. The invalid cutting track of sample #2 at 3.5 mm may be similar with the situation in Fig. 5c. However, the special mass of sample #3 is the highest in both DOCs. Figure 14b demonstrates that the special mass of the second cutting is higher than that of the first cutting at correspond DOC. In particularly, the special mass of sample #1 is the highest at both DOCs in the second cutting. Especially when DOC is 3.5 mm, the advantage of sample #1 is even more obvious.

4 Conclusion

An approximate cutting test was designed with a rubber fixed on a VTL. The rubber was used to simulate the mudstone under downhole. The test results show that the tangential force and normal force of the optimized axe-shaped cutter are smaller than that of cylinder-shaped cutter and axe-shaped cutter. The axe-shaped cutter, especially the optimized one, has a sharp point due to a small radius of the transition arc, which is easier to invade into the elastic–plastic mudstone in the first cutting on rubber layer. In addition, the curved intersected planes of optimized axe-shaped cutter are also helpful for invading in. The axe shape improves the evacuation of cuttings, such as the fracture of long ribbon rubber debris due to the crumbled rubber produced by optimized axe-shaped cutter is relatively small. By introducing cutting force, cutting quality and trajectory length into the model, it can also be concluded that the optimized axe-shaped cutter is more suitable for cutting highly plastic mudstone formations that is hard to invade. The results of the second cutting demonstrate that soft mudstone is much easy to remove than elastic–plastic mudstone under the same DOC. Conventional cylinder-shape cutter has highest cutting efficiency when cutting soft mudstone that didn’t exhibit elastic–plastic deformation.

Acknowledgements The authors gratefully acknowledge the financial supports from the Natural Science Foundation of China (Grant numbers: 51821092, U1762214), the Science Foundation of China University of Petroleum, Beijing (Grant number: ZX20190065) and the Scientific research projects (Grant numbers: HX20191151, ZLZX2020-01-07-01, PRP/indep-1-2002).

References

1. Shao, F., Liu, W., Gao, D., et al.: Study on rock-breaking mechanism of axe-shaped PDC cutter. J. Petrol. Sci. Eng. **205**, 108922 (2021)
2. Li, Y., Cai, J., Jia. M., et al.: Research and application on new PDC bit drilling in hard and compact mudstone. Drill. Eng. 60–61+54 (2006)
3. Yao, E., Zhang, F., Yang, A., et al.: Research and application of bits for drilling in elastic plastic dense mudstone of in-situ leaching mudstone uranium deposit. Drill. Eng. **36**(06), 72–75 (2009)
4. Luo, M., Zhu, H., Liu, Q., et al.: A V-cutter PDC bit suitable for ultra-HTHP plastic mudstones. Nat. Gas. Ind. **41**, 97–106 (2021)
5. Ruan, H., Shen, L., Li, C., et al.: Research and the application of new sharp-tooth PDC bit used in elastic-plastic compact mudstone. Drill. Eng. **41**, 80–83 (2014)
6. Rahmani, R., Pastusek, P., Yun, G., et al.: Investigation of geometry and loading effects on PDC cutter structural integrity in hard rocks. In: IADC/SPE International Drilling Conference and Exhibition. Society of Petroleum Engineers, Galveston, Texas, USA (2020), pp. 1–22
7. Rahmani, R.: Rock customized shaped cutters improve rock cutting efficiency. In: SPE/IADC International Drilling Conference and Exhibition. Society of Petroleum Engineers, The Hague, The Netherlands (2019), pp. 1–15
8. Negm, S., Aguib, K., Karuppiah, V., et al.: The disruptive concept of 3D cutters and hybrid bits in polycrystalline diamond compact drill-bit design. In: Abu Dhabi International Petroleum Exhibition & Conference. Society of Petroleum Engineers, Abu Dhabi, UAE (2016), pp. 1–10
9. Lomov, A., James, B., Konysbekuly, G., et al.: The combination of ridge and conical elements is a new approach for drilling out hard carbonates and mudstones without drop in ROP. In: SPE Russian Petroleum Technology Conference. Moscow, Russia, 15–17 October 2018, SPE-191522–18RPTC-MS
10. Shao, F., Liu, W., Gao, D.: Effects of the chamfer and materials on performance of PDC cutters. J. Petrol. Sci. Eng. **205**(108887), 1–10 (2021)
11. Hareland, G., Yan, W., Nygaard, R., et al.: Cutting efficiency of a single PDC cutter on hard rock. PETSOC-09-06-60, 2007 48, 60–65

The Application of Adaptive Algorithm in the Maximum Power Tracking of Power Photovoltaic System

Haiquan Feng and Yunpeng Wang

Abstract Solar energy, as a kind of clean energy, has played a positive role in the development of today's society. However, the low efficiency of power generation is the main reason for the slow development of this energy. How to accurately track the maximum power of solar energy is the main subject of current research scholars. Therefore, on the basis of understanding the operation of the power photovoltaic system, this paper systematically understands the content of the adaptive algorithm, and conducts experimental exploration and analysis to prove the application performance of the adaptive algorithm in the maximum power tracking of the power photovoltaic system.

Keywords Adaptive algorithm · Power · Photovoltaic system · Maximum power · Tracking

1 Introduction

Solar power generation has the advantages of no noise pollution and low cost in practical application. It is considered as the most promising clean energy in today's development. It has been widely used in many fields, such as battery charging system, water pump, satellite, household appliances, etc., and has achieved excellent results in practical exploration. However, there are certain constraints to this energy source, the most significant of which is climate conditions. Therefore, in order to ensure the efficiency of energy application, energy storage equipment is usually installed, but the related costs and actual efficiency are not as expected. At present, domestic and foreign researchers have proposed a variety of solutions for the study of the efficiency of solar power generation, among which the most common is the disturbance and

General Project of National Natural Science Foundation of Chin (No. 62171060).

H. Feng (✉) · Y. Wang
Beijing Electric Power Corporation, Beijing, China
e-mail: ranyunpower@hotmail.com

H. Dai (ed.), *Computational and Experimental Simulations in Engineering*, Mechanisms and Machine Science 119,
https://doi.org/10.1007/978-3-031-02097-1_9

observation method. In practical application, it has the advantages of simple algorithm and few parameters, but there are also many problems, such as the very large radial artery coefficient at the maximum power point 10 [1, 2]. According to the results proposed by researchers, the current and voltage are regarded as constants, and the reference values of the two as a comparison, the solar array sampling current and voltage can be compared and analyzed, and then the duty cycle adjustment can be implemented. Although this method is simple to operate, it does not consider the influence of temperature and radiation composition, so it is difficult to guarantee the accuracy of actual tracking. This paper starts with the strategic characteristics of solar energy, uses the adaptive principle to study the maximum power tracking of solar array, and analyzes the experimental research results. The final results show that the tracking speed and accuracy of the adaptive algorithm are higher [3–5].

2 Methods

2.1 Adaptive Algorithm

In essence, the adaptive algorithm will automatically adjust the actual processing order, parameters, methods, boundary conditions, constraints and other contents according to the characteristics of data processing, so as to ensure the adaptability of the actual statistical distribution and data structure, and finally get the optimal processing results. According to the current research results, this kind of algorithm mainly uses two ways to operate, one is program control, the other is processing circuit. The former will compile the mathematical model of the algorithm into a program, and then use computer technology to operate; The latter will design related circuits according to the mathematical model of the algorithm. From the point of view of time, there are many types of algorithms that can be used, so the algorithm with high performance and feasibility should be chosen first in the research and design. For example, the common algorithm types include transform domain adaptive algorithm, affine projection algorithm, conjugate gradient algorithm, LMS algorithm and so on. Taking LMS algorithm as an example, the core idea of its application is to update only part of the coefficients in each iteration, so as to reduce the number of calculations. In addition, m-Max NLMS algorithm and Max NLMS algorithm are the same, and the corresponding adaptive coefficient is shown as follows [6–8]:

$$Periodic\ LMS\ Algorithm\colon W_i(n+1)\begin{cases} W_i(n)+\mu e_l x_l - i + 1(n+i)\mathrm{mod} = O\ and\ l = N[n|N] \\ W_i(n) otherwise \end{cases}$$

$$M - MaxNLMS\ Algorithm\colon W_i(n+1)\begin{cases} W_i(n)+\frac{\mu e(n)x(n-i+1)}{X^T(n)X(n)} i\ corresponds\ to\ the\ first. \\ M \cdot \max ls \cdot x(n-i+1)1, i = 1, \ldots, L \\ W_i(n)\ otherwise \end{cases}$$

MaxNLMS Algorithm: $\begin{cases} W_i(n) + \frac{\mu e(n)}{x(n-i+1)} if\ .lx(n-i+1)1 = \max lx(n-j+1)1, j = 1, \ldots, L \\ W_i(n)\ otherwise \end{cases}$

In the application of LMS algorithm, the minimized mean square error is E[e2 (n)], and RLS algorithm can minimize the weighted square sum of errors J (n) = $\sum_{i=1}^{n} \lambda^{n-i} |e(i)|^2$. However, some researchers have proposed the LMF algorithm of minimizing E[E4 (n)] and the RLF algorithm of minimizing $\sum_{i=1}^{n} \lambda^{n-i} |e(i)|^4$ in the study. The convergence of these two algorithms in non-Gaussian environment is stronger than the previous two algorithms. In addition, some researchers have proposed a hybrid LMF algorithm, which has strong stability to noise changes. At present, there are more and more research on the application of adaptive algorithm at home and abroad, and it has gradually become the most critical research topic in signal processing. Therefore, in order to fully show the application advantages of adaptive algorithm in the maximum power tracking of power photovoltaic system, it is necessary to conduct in-depth research from practical research [9, 10].

2.2 *Solar Energy Performance*

The formula of solar battery array is:

$$I = I_g - I_{set}\left\{\exp\left[\frac{q}{AKT}(V + IR_g) - 1\right]\right\} - \frac{V + IR_g}{R_{sh}}$$

In the above formula, I and V on behalf of the solar cell output current and voltage, Ig on behalf of the optical current and Isat on behalf of the diode saturation current, q represents the electronic charge quantity, A representative of the diode characteristic factor, on behalf of the boltzmann constant K, T, on behalf of solar battery temperature, Rs and Rsh represent the resistance of solar cells in series and in parallel.

Figures 1 and 2 represent the I-V curve and P–V curve in the solar cell array respectively.

The above image analysis, including solid lines represent the resistance of the load line, L represents the curve of load, such as virtual lines represent power curve, ISC on behalf of the short circuit current of solar cell, solar battery open circuit voltage Voc representative, Pm on behalf of the maximum power point of solar battery, Vm and Im representing the maximum power point of solar battery current voltage. In the current and voltage path, the solar array with consistent light intensity and temperature is visually presented. Under the condition of studying only solar cells, the corresponding maximum power point will be within the REGION of Pm (Vm, Im). Assuming that a solar cell can be connected to a load using a converter, the cell shop will be defined according to the load. In the case that the load cannot be adjusted, the solar cell should always be at point A according to the load characteristics and solar characteristics. Under the condition that the load can be adjusted, according to

Fig. 1 I–V curve

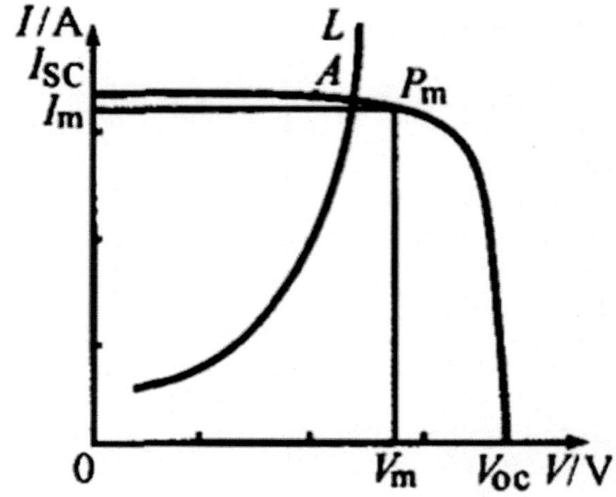

Fig. 2 P–V curve

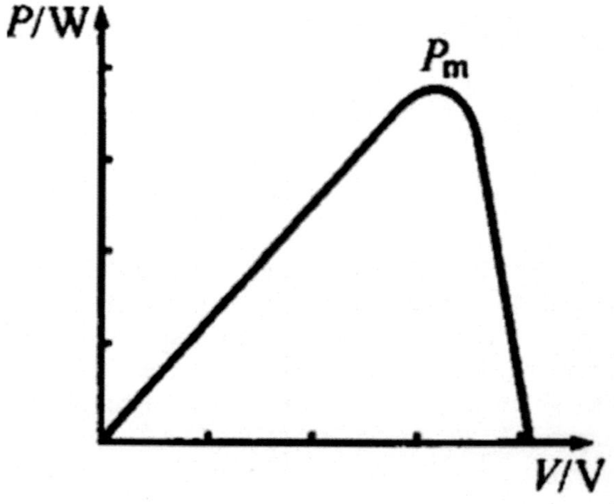

the analysis in the figure above, the output power of the battery at point A is lower than that at point Pm. In the process of adjusting the output voltage, ensure that the load voltage reaches point V, so that the load power can be transferred from point A to point B. At this point, point B and the Pm of the battery will be on the same isopower line, thus the output result of the maximum power of the battery can be obtained. Figure 2 shows the p–V curve of the battery array at the same light intensity and temperature. The characteristics of the battery array will change continuously with temperature and light intensity. Generally speaking, temperature will affect the output voltage and light intensity will affect the output current.

2.3 Experimental Principle

Based on the analysis in Fig. 3, it can be seen that as the schematic diagram of adaptive algorithm, the following expression should be used to apply it to the maximum power tracking analysis of power photovoltaic system:

P = VI

From this analysis, it can be seen that under the condition of dP/dV > 0, the system will run on the left side of the maximum power. In the case of dP/dV < 0, the system will run to the right of the maximum power. With dP/dV = 0, the system will be running exactly at maximum power. Combined with the above condition analysis, after the sampling voltage and current of the battery array are obtained at the (i-1) and I time of the approaching time interval, the corresponding slope is:

$$\frac{dP}{dV}$$

The specific calculation formula is:

$$\frac{\Delta P}{\Delta V}(n) = \frac{P(n) - P(n-1)}{V(n) - V(n-1)}$$

In the above formula, the condition P (n) = V (n) I (n) is met. Meanwhile, the duty cycle D of the switch tube is continuously adjusted until this requirement is met.

The specific operation process is shown in Fig. 4.

In the above flowchart, K1 represents the constant quantity of D regulation, and K2 represents the quantity of adaptive factor regulation.

The adaptive algorithm is used to study the precision E first, so as to judge whether it meets the requirements of voltage and current at the moment I + 1 of Pmo sampling.

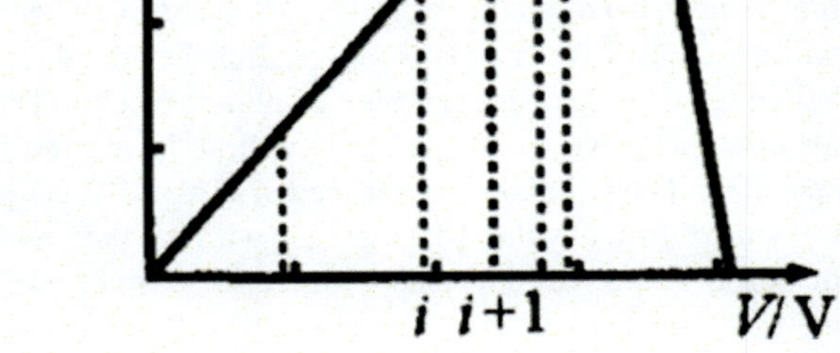

Fig. 3 Schematic diagram

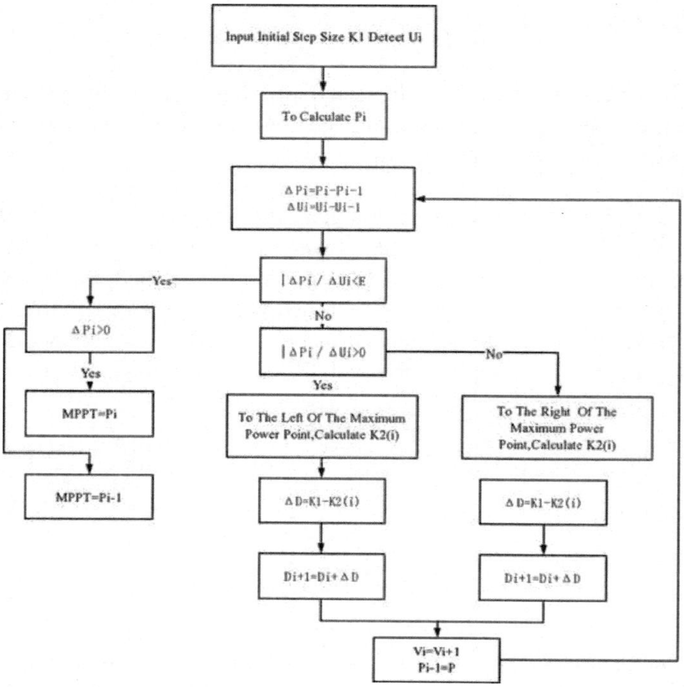

Fig. 4 Tracking flow chart of power photovoltaic system based on adaptive algorithm

The calculation is carried out according to the formula P = VI, and the slope of p–V curve between the moment I and the moment I + 1 is studied. In the case of dP/dV > 0, the value of D needs to be increased to increase the output power. In other words, Di + 1 = Di + (K1 − k2) at the time (I + 1), then on the contrary, under the condition that dP/dV is < 0, D value should be controlled, thus Di + 1 = Di − (k1 − k2).

In this way, after using the adjustable K2, the switching tube is allowed to adjust the D value according to the collected current and voltage. Since (K1 − K2) is proportional to the change in the step size of the sample, adjusting D is essentially to adjust the sampling current and voltage at the next moment, so as to optimize the output power of the battery.

The adaptive factor K2 (I) is used to adjust the D value at the next moment. When the distance between the sampling point and Pm is longer, the value of K2 is smaller, while ΔD = K1 − k2 (I) is larger, and the actual tracking speed is also very fast. When the distance between the sampling point and Pm is short, the value of K2 will also increase, ΔD = K1 − k2 (I) will be lower and lower, and the actual tracking speed can be optimized. Combined with the d-P diagram it can be seen that on the left side of Pm, D value will increase with the increase of Pm, while on the right side of Pm, D value will decrease with the increase of Pm.

3 Result Analysis

Based on the analysis of the system block diagram shown in the following Fig. 5, it can be seen that the main circuit mainly uses Buck converter to connect the power supply and the load, while MOSFET IRF840 is used for the switch tube, the diode VD is used for the fast recovery diode, and the output point sense L is used for winding in the magnetic core with a space gap of iron powder, so as to avoid inductance saturation. Load R0 is a non-adjustable resistive load. Combined with the practical analysis, the control loop is mainly divided into three parts: first, the 80C196KC produced by Intel is selected, the actual power consumption is low, with E2PROM, the experiment chooses E2PROM 2864, latch 74LS737; Second, the interface circuit is mainly designed with hall voltage sensor and current sensor with higher accuracy, which can ensure the operation of the whole system and has stronger overload level and response force, so as to avoid excessive energy consumption of the detected circuit. Thirdly, the selected 80C196KC has three characteristics. Firstly, samples can be collected flexibly according to the specific situation, and the actual collection time is less; Secondly, the frequency of oscillating signal can reach 16 MHz, and the operation speed of instruction is very fast. Finally, the HSO interrupt used in the experiment does not produce continuous BWM, so the driving loop is used to control the switch tube.

As shown in the Fig. 6, to study the performance of words under different temperature and light intensity conditions, the experiment was designed to use an adjustable DC power supply instead of a solar cell. By adjusting the current and voltage of the DC power supply, the influence of external factors on the composition of the solar cell can be changed. Based on the analysis of the linear circuit diagram shown in the following figure, it can be seen that when R0 obtains the maximum power, the voltage on both sides of R0 conforms to the formula V0 = Vi/2. Therefore, the research experiment in this paper uses dc power supply to detect the voltage on both sides of capacitor C1 instead of solar cell.

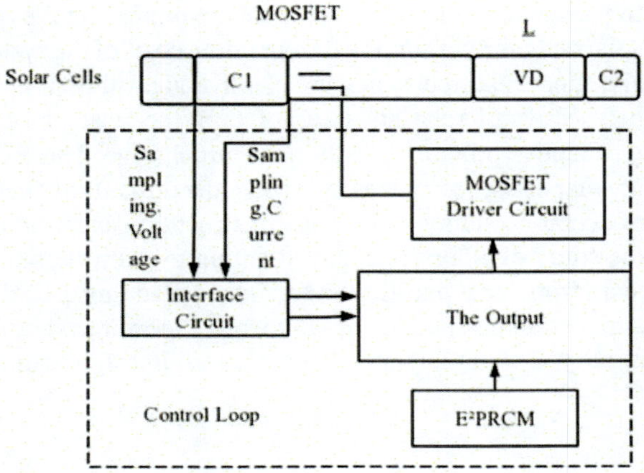

Fig. 5 System block diagram

Fig. 6 Linear circuit diagram

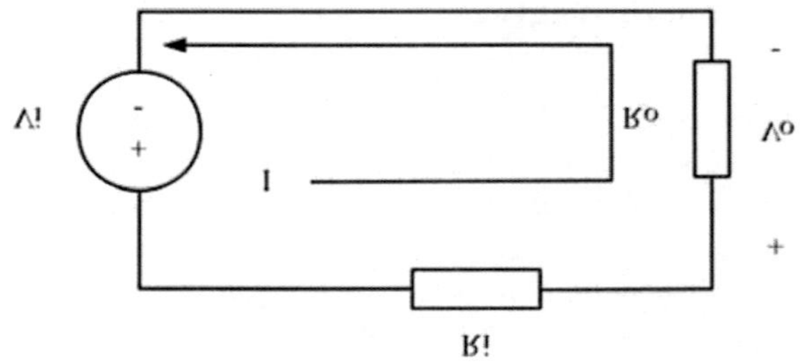

When the power supply voltage reaches Vin = 12.3 V and Vin = 15.46 V, it means that the solar output is affected by external factors, mainly involving radiation, temperature and other contents. The adaptive algorithm is used to study the waveforms on both sides of C1 in Buck circuit. The final results show that the capacitor voltage UC1 = Vin/2 meets this condition, and the Pm output point can be quickly found at about 20 ms. At the same time, because K2 is quoted in the experiment, it can be found that when the capacitance around Pm changes, the actual change amount is small, and the corresponding oscillation is also low. From the overall point of view, the use of adaptive algorithm to track and investigate the maximum power of power photovoltaic system can not only guarantee the actual tracking rate, but also improve the accuracy of the tracking survey, which can solve the nonlinear characteristics of solar cells in a certain sense, and optimize the overall system operation efficiency. In other words, this algorithm is an effective method for the efficient utilization of solar energy resources [11, 12].

4 Conclusion

To sum up, as a technical method for signal processing proposed in recent decades, adaptive algorithm has been reasonably applied in many fields in practical research. In particular, it plays a positive role in improving the power generation efficiency of solar cells and controlling the operating costs of the system. At present, there are more and more research topics on adaptive algorithm at home and abroad, but from the perspective of solar array exploration, the research on tracking investigation based on adaptive principle is still in the initial stage. Therefore, in order to reasonable use of solar energy resources, and solve the difficulties faced by the current development of social economy and urban construction, scientific research scholars to intensify the study of algorithms, pay attention to cultivate more high-quality talent, how to learn from the domestic and foreign outstanding case study, pay attention to from the application perspective of photovoltaic power systems to conduct a comprehensive inquiry, Only in this way can basic resource application be guaranteed.

References

1. Reisi, A.R., Moradi, M.H., Jamasb, S.: Classification and comparison of maximum power point tracking techniques for photovoltaic system: a review. Renew. Sustain. Energy Rev. **19**(1), 433–443 (2013)
2. Zhang, F., Thanapalan, K., Procter, A., et al.: Adaptive hybrid maximum power point tracking method for a photovoltaic system. IEEE Trans. Energy Convers. **28**(2), 353–360 (2013)
3. Shi, J.Y., Xue, F., Qin, Z.J., et al.: Improved global maximum power point tracking for photovoltaic system via cuckoo search under partial shaded conditions. J. Power Electron. **16**(1), 287–296 (2016)
4. Bhatnagar, P., Nema, R.K.: Maximum power point tracking control techniques: state-of-the-art in photovoltaic applications. Renew. Sustain. Energy Rev. **23**(Complete), 224–241 (2013)
5. Kasa, N., Iida, T., Majumdar, G.: Maximum power point tracking with estimation of the capacitance of the capacitor connected to photovoltaic array. Electr. Eng. Jpn. **144**(4), 75–85 (2010)
6. Hussein, A.A., Chen, X., Alharbi, M., et al.: Design of a grid-tie photovoltaic system with a controlled total harmonic distortion and tri maximum power point tracking. IEEE Trans. Power Electron. **35**(5), 4780–4790 (2020)
7. Khan, M.J., Pushparaj: A novel hybrid maximum power point tracking controller based on artificial intelligence for solar photovoltaic system under variable environmental conditions. J. Electr. Eng. Technol. 1–11 (2021)
8. Cid-Pastor, A., Martinez-Salamero, L., Leyva, R., et al.: Design of photovoltaic-based current sources for maximum power transfer by means of power gyrators. Iet Power Electron. **4**(6), 674–682 (2011)
9. Hekss, Z., Abouloifa, A., Lachkar, I., et al.: Nonlinear adaptive control design with average performance analysis for photovoltaic system based on half bridge shunt active power filter. Int. J. Electr. Power Energy Syst. **125**, 106478 (2021)
10. Jyotirmaya, S., Susovon, S., Shamik, B.: Adaptive PID controller with P&O MPPT algorithm for photovoltaic system. IETE J. Res. 1–12 (2018)
11. Verma, P., Garg, R., Mahajan, P.: Smooth LMS-based adaptive control of SPV system tied to grid for enhanced power quality. IET Power Electron. **13**(15) (2020)
12. Cheng, L., Zhou, et al.: Research on improving adaptive variable step length MPPT algorithm. Int. J. Sensor Net. **17**(3), 139–145 (2015)

Forest Environment Association Analysis for the Pandemic Health with Rectified Linear Unit Correlations

Hong Wei Shi, Li Shen Wang, Jiamin Moran Huang, and Jun Steed Huang

Abstract The covariance is a measure of the joint variability of two random variables forming a Cartesian coordinate. Variance is a special case of covariance, when two variables are the same. The root of variance is the standard deviation. The normalization of covariance to standard deviation is called Pearson correlation coefficient. The covariance for the region of first and third quadrant is called upper semi-covariance. The covariance for the region of second and fourth quadrant is called down semi-covariance. Here we present semi-covariance, an accurate ReLU (Rectified Linear Unit) way of measuring the non-linear correlation between variables. Our framework is applied to successfully analyze the association between alternative environment social life factors and the pandemic toll recovery health. The results of our analyses of the 2021 pandemic toll suggest that pesticide residual, annual precipitation, forest coverage, economy development, lifestyle, etc. have different impacts on toll of each country. The pesticide may kill immune system, pandemic lifestyle is impacted. We picked total 8 related factors from 22 countries in order to set up future model for digital twin that predicting pandemic trend.

Keywords Semi-covariance · Forest coverage · Non-linear model · COVID-19 · Digital twin

---

H. W. Shi · J. M. Huang · J. S. Huang
Nanjing Kangbo Health Academy, Nanjing, China
e-mail: 17107@squ.edu.cn

J. M. Huang
e-mail: moran@genieview.com

J. S. Huang
e-mail: steed.huang@visionx.org

L. S. Wang (✉)
Institute for Industrial Technology Research, Suqian University, Suqian, Jiangsu, China
e-mail: sweetsins@vip.qq.com

H. Dai (ed.), *Computational and Experimental Simulations in Engineering*, Mechanisms and Machine Science 119,
https://doi.org/10.1007/978-3-031-02097-1_10

1 Introduction

The COVID-19 pandemic has led to a dramatic loss of human life worldwide and presents an unprecedented challenge to public health, food systems and the world of work. The economic and social disruption caused by the pandemic is devastating, tens of millions of people are at risk of falling into extreme poverty, while the number of undernourished people, currently estimated at nearly 690 million, could increase by up to 132 million by the end of the year. It is time to examine the underline factors that affected our ability to cope with it by establishing digital twin model that is able to predict the next outbreak.

We conduct this research to obtain information that can help predict or confirm a pandemic toll's outcome and provide key insights into why people die for a certain mutation and the demographic environment social information associated with them. Owing to the short window period and data confidentiality, most of the previous researchers focused on study of the randomly sampled community. Therefore, it is urgent to develop a systematic approach to (1) collect meaningful environmental demographic data for analysis; (2) develop an analysis framework to extract insights from the long-term data; (3) establishing ongoing digital twin model to monitor and to predict the future pandemic.

We are concerned about the pesticides that impact human immune system [1]. Epidemiological evidence from countries indicates that the prevalence of diseases associated with alterations in the immune response, such as asthma, certain autoimmune diseases, and cancer, are increasing to such an extent that it cannot be attributed to improved diagnostics. There is a concern that this trend could be partially attributable to the increased patterns of exposures to chemicals, including pesticides, animals. Inspired by this idea, we do further investigation of some of them. They are GDP for each country, COVID-19 death toll for each country, pet situation for each country, life span status and education level implicit environment awareness for each country. The statistical period is chosen to reflect the recent yet within 30 years relevant period. To avoid survivor bias, we picked the data that are the most complete data set for high impacted country by pandemic.

UCSD Professor Harry M. Markowitz, the winner of the 1990 Nobel Prize in Economics, introduced the concept of semi-variance in Portfolio Selection [2] and pointed out that in the case of asymmetric distribution of investment returns, semi-variance can reflect the riskiness of investment returns more accurately than the variance or the covariance between two risk factors. UCSD Professor Robert F. Engle, the winner of the 2003 Nobel Prize in Economics, introduced the concept of realized semi-covariance matrix [3], when he pointed out an unbiased estimator for the ex-post true volatility is the realized volatility. Inspired by these two laurel ideas, we hypothesize that the usage of a single semi-covariance can analyze the causal relationship between factors more precisely yet simpler than entire covariance matrix. Because it not only shows the non-linear interaction between the variables and the response variable, but also reveals the patterns of the response variable. Compared with the Pearson coefficient, which only provides two directions, the semi-covariance approach provides a four-direction measurement between the target and the factors. That is, the target has an upward trend, the factor has an upward trend for first quadrant; the target has an upward trend, the factor has a downward trend for second quadrant; the target has a downward trend, the factor has a downward trend for third quadrant; the target has a downward trend, the factor has a upward trend for fourth quadrant. Overall, the semi-covariance approach provides us with more informative insights and interpretability to measure the non-linearity relationship, where Pearson coefficient can't be due to the contradictive trends' cancellation with each other, while the matrix can but too many calculations.

Considering the complex relationship between different factors and the final pandemic outcome, we hypothesize that the advanced semi-covariance analysis should be more suitable for analyzing the current pandemic situations. Thus, we propose a framework based on it and perform the corresponding study in this research. To evaluate our framework, we also compare the results from our framework with other methods, such as the Pearson correlation coefficient method and analysis of variance (ANOVA). The experiments suggest that our framework can indeed mine more insightful information from the pesticide precipitation forest extra data. In summary, we make the following contributions in this research.

1. We propose a newly derived method that can be used to analyze the non-linear pattern based on a semi-variance idea.
2. We identify, collect, and integrate data for the latest pandemic pesticide association analysis, bearing in mind a digital twins prediction framework is to be set up.
3. We provide a great resource for studying the precipitation map, establish an analysis framework to understand the economic and demographic factors that may affect the pandemic, and identify the potential factors impacting the formation of the pandemic.

2 Results

Our goal is to investigate whether the semi-covariance framework can accurately help us conduct association analysis between the environmental factors and the pandemic outbreak [4] and set up digital twins for city to be prepared for next pandemic. We select factors that may have impacts on the coalition from different fields, such as pesticide, pet, GDP, education, life, forest and precipitation. Finally, we also calculate Pearson correlation coefficients and analysis of variance (ANOVA) as references.

The data we use comes from government agencies and mainstream media. We fill the missing data using regression imputation. The GDP data comes from IMF report 2020. The pandemic data comes from the Center for Diseases Control and Prevention (CDC) data tracker 2021. As of July 7, 2021, the death toll reached 5554 in China. Dog and Cat total Ranking per family is used to assess the possible effects of the national result on culture wellness. Physical and Astronomy scientist (2018) is used as the representative of the College group. We extract the data from the U.S.

Bureau of Labor Statistics. The underlying assumption of choosing this factor is the higher number of physical scientists, the freer academic atmosphere in which people can pursue truth, exercise reasoning, and respect pandemic science. Life span is one of the oldest standard of evaluation for overall quality of life [5]. The data is collected from Wikipedia (2015). We use Forest coverage and precipitation from Wikipedia and Indexmundi for 2014 as the data source of the same year, as the Forest grows with rain. Finally, the pesticide data is collected from United Nation database 1990–2018 summation.

Below, we calculate Semi-Covarince correlation and Pearson correlation and deep dive into each factor for further analysis.

Note that the up Semi plus down Semi equals Pearson, that cross verifies the calculations are all correct. For the value above 0 we note it as "upper side" means positive correlation, data is in 1st or 3rd quadrants, for value below 0 we note it as "down side" means negative correlation, data is in 2nd or 4th quadrants. Excelsheet: https://github.com/steedhuang/Forest-pesticide-covid-19-toll-GDP.

2.1 *Forest Coverage*

Forest cover in general refers to the relative in percent land area that is covered by forests. According to the Food and Agriculture Organization, a forest is defined as land spanning more than 0.5 ha with trees higher than 5 m and a canopy cover of more than 10% [6]. It does not include land that is predominantly under agricultural or urban land use.

Global forest cover, however crucial for soil health, the water cycle, climate and air quality are severely threatened by deforestation, as a direct consequence of agriculture, grazing and mining. Forest cover can be increased by reforestation and afforestation efforts, but it is impossible to restore the full range of ecological services once natural forests are converted to other land uses. We selected 2014 data source listed at the end of the paper as our baseline analysis.

In semi-covariance analysis (Table 1) we can tell that forest coverage is weakly positive related to the pesticide residual, and (Table 2) mildly negative related to pesticide residual [7]. In semi-covariance analysis (Table 3) we can tell that forest

Table 1 Down-covariance correlation coefficient analysis of pesticide

Rank	Value (%)	Leader	Quadrant
GDP	50	PNG	4 medium
Life	48	PNG	4 medium
Nobel	27	PNG	4 medium
Pet	25	PNG	4 medium
Toll	22	PNG	4 medium
Forest	3	Russia	2 weak

Table 2 Upper-covariance correlation coefficient analysis of pesticide

Rank	Value (%)	Leader	Quadrant
Life	52	China	1 medium
GDP	37	Korea	1 medium
Toll	32	China	1 medium
Forest	25	Korea	1 medium
Pet	19	PNG	3 weak
Nobel	12	PNG	3 weak

Table 3 Down-covariance correlation coefficient analysis of precipitation

Rank	Value (%)	Leader	Quadrant
Forest	38	China	4 medium
Pet	30	China	4 medium
GDP	24	China	4 medium
Nobel	15	China	4 weak
Toll	5	Korea	4 weak
Life	4	China	4 weak

Table 4 Upper-covariance correlation coefficient analysis of precipitation

Rank	Value (%)	Leader	Quadrant
Forest	73	PNG	3 strong
Pet	34	Malaysia	1 medium
Toll	21	Saudi Arabia	3 medium
GDP	17	Saudi Arabia	3 weak
Life	17	South Africa	3 weak
Nobel	11	Saudi Arabia	3 weak

coverage is strongly positive related to the precipitation, and (Table 4) weakly negative related to rain rate. The rank of each country is different for different factor (Tables 5 and 6).

Table 5 Forest etc. ranks of each countries

Forest	GDP	Toll	Life
Japan	USA	Mexico	Japan
Korea	Australia	Peru	Australia
PNG	Germany	China	Italy
Malaysia	Canada	South Africa	France
Brazil	UK	Indonesia	Korea
Peru	Japan	Italy	Canada
Indonesia	France	Australia	UK

Table 6 Nobel etc. ranks of each countries

Nobel	Pet	Pesticide	Rain
USA	Peru	China	PNG
UK	Argentina	Korea	Malaysia
Germany	Mexico	Japan	Indonesia
France	Brazil	Italy	Brazil
Canada	USA	Argentina	Peru
Japan	France	France	Japan
Italy	Italy	Germany	Korea

Clearly, more than half of the world's forests are found in only five countries: Brazil, Canada, China, Russian Federation and United States of America. The largest part of the forest is found in the tropical domain, followed by the boreal, temperate and subtropical domains. These domains are further divided into terrestrial global ecological zones that is important to human health.

2.2 *COVID-19 Cases and Deaths*

Covid-19 profoundly changed world and the heritage development agenda. Our hypothesis is that the pesticide residual has a negative impact on the pandemic situation. In semi-covariance analysis (Table 1), the death toll has positive movement associated with pesticide. And almost no negative association with pesticide (Table 2). Indeed, (Table 3) and (Table 4) show not many relationships between the rain and the Covid-19 cases. This virus works everywhere. Although some pesticides have been restricted or banned because they pose risks of cancer, birth defects, or neurological damage, little attention has so far been given to what may be their greatest risk: impairment of human and animal immune systems against the Covid-19 virus. According to this new study, there is considerable evidence that widely used pesticides may suppress immune responses to bacteria, viruses, making people significantly more vulnerable to disease.

2.3 *GDP*

Gross domestic product (GDP) is the standard measure of the value added created through the production of goods and services in a country during a certain period. As such, it also measures the income earned from that production, or the total amount spent on final goods and services.

GDP counts all final private and government spending as additions to income and output for society, regardless of whether they are actual productive, profitable or environment friendly. This means that obviously unproductive or even destructive activities are routinely counted as economic output and contribute to growth in GDP. For example, this includes spending directed toward extracting or transferring wealth between members of society rather than producing wealth, spending on investment projects for which the necessary complementary goods and labor are not available or for which actual consumer demand does not exist such as the construction of empty ghost cities or bridges to nowhere, unconnected to any road network; and spending on goods and services that are either themselves destructive or only necessary to offset other destructive activities, rather than to create new wealth, such as the production of weapons of war or spending on anti-terrorist measures.

In semi-covariance analysis (Table 1) and correlation analysis (Table 2), we can easily get the conclusion that GDP is positive related to pesticide, the higher the GDP the worse the pollution is. In (Table 3) and (Table 4) we can see, the higher nature rain region has lower GDP. As such a balance is needed while we seek for GDP grows.

2.4 Life Span

Life expectancy represents the average number of years that a group of persons, all born at the same time, might be expected to live, and it is based on the changing death rate over many past years. The concept of life span implies that there is an individual whose existence has a definite beginning and end. Until the middle of the twentieth century, infant mortality was approximately 50% of the total mortality of the population. If we do not take it into account for child mortality in total mortality, then the average life expectancy in the 12–19 centuries was approximately 55 years. If a medieval person was able to survive childhood, then he had about 50% chance of living up to 55 years. That is in reality, people did not die when they lived to be 25–40 years old, instead continued to live about twice as long today. There are great variations in life expectancy between different parts of the world, mostly caused by differences in public health, medical care, and diet.

In semi-covariance analysis (Table 1) and correlation analysis (Table 2), we can easily get the conclusion that life span is positive related to pesticide, that could be because the country using pesticide is good at mastering the chemicals and medical drug is also advanced, that prevents people from dying on sickness. In (Table 3) and (Table 4) we can see, the higher nature rain region has longer life. As such a balance is important while we resort for pesticide usage.

2.5 Nobel Prize

Nobel Prizes are awarded in the fields of Physics, Chemistry, Physiology or Medicine, Literature, and Peace. In 1968, Sweden's central bank Sveriges Riksbank established the Prize in Economic Sciences in Memory of Alfred Nobel, founder of the Nobel Prize. Nobel Prizes are widely regarded as the most prestigious awards available in their respective fields [8]. Alfred Nobel was a Swedish chemist, engineer, and industrialist most famously known for the invention of dynamite. He died in 1896. In his will, he bequeathed all of his "remaining realizable assets" to be used to establish five prizes which became known as "Nobel Prizes". Nobel Prizes were first awarded in 1901. The country won more Nobel Prize would be considered advanced in science.

In semi-covariance analysis (Table 1) and correlation analysis (Table 2), we can easily get the conclusion that Nobel prize is not related to pesticide usage. In (Table 3) and (Table 4) we can see, the higher heritage rain region has less Nobel winner. As such a balance is important between modern life and traditional heritage life.

2.6 Pet Ownership

As a dependent, a pet will add to your living expenses. When getting a pet, there are the initial costs like a bed, a crate, grooming items, a collar, a leash, a litter box, a scratching post, and other miscellaneous things needed immediately upon adoption. Then there are recurring expenses, such as the cost of food, treats, and toys. However, pets love us unconditionally, and that's priceless. There's nothing like coming home every day to a four-legged family member who's thrilled to see you. It's also difficult to put a price tag on their constant companionship [9]. A pet is someone to snuggle with, take walks with, accompany you on car rides, or hang with on the couch. Not to mention, the companionship of a pet can improve your mental health and overall well-being.

In semi-covariance analysis (Table 1) and correlation analysis (Table 2), we can easily get the conclusion that pet is negatively related to pesticide, the lower the pesticide the more cozy cats or dogs are around you. In (Table 3) and (Table 4) we can see, the higher nature rain region, more related life with pets. As such a balance is needed while we look for pets.

3 Methods

3.1 Collect, Analyze and Adjust Polls

According to Johns Hopkins University, national public health agencies, The US, India and Brazil have seen the highest number of confirmed cases, followed by France, Russia, Turkey and the UK. Confirmed cases have been rising steeply since the middle of last year, but the true extent of the first outbreaks in 2020 is unclear because testing was not then widely available. The 100 millionth Covid cases was recorded at the end of January—about a year after the first officially diagnosed case of the virus. Here is the map for the world pandemic situation as of July 8, 2021 (Fig. 1).

According to Professor Federico Maggi and Dr. Fiona Tang, University of Sydney, 64% of global agricultural land at risk of pesticide pollution (Fig. 2). Asia and Europe revealed as having regions at high-risk of pesticide pollution. 92 chemicals commonly used in agricultural pesticides in 168 countries, while boosting productivity have potential implications for human and animal health, particularly immune system becomes weak. As we can see the map here for pesticide residual coincide partially with the pandemic crisis.

Figure 3, 4 and 5 show the relationship between death toll, pesticide residual rain and forest. Figure 3 indicates some bilinear relationship between toll and pesticide.

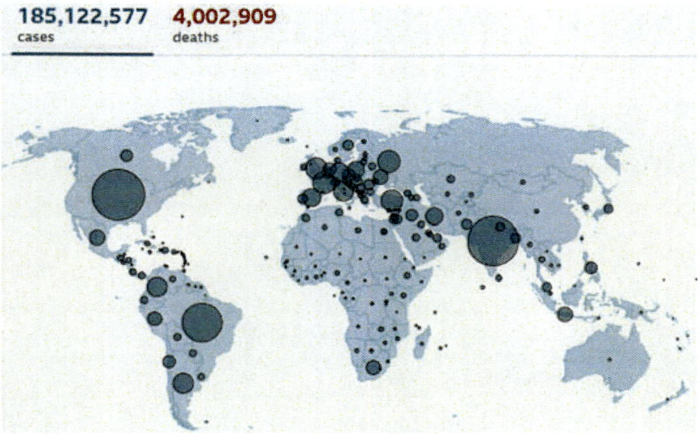

Fig. 1 COVID-19 pandemic map

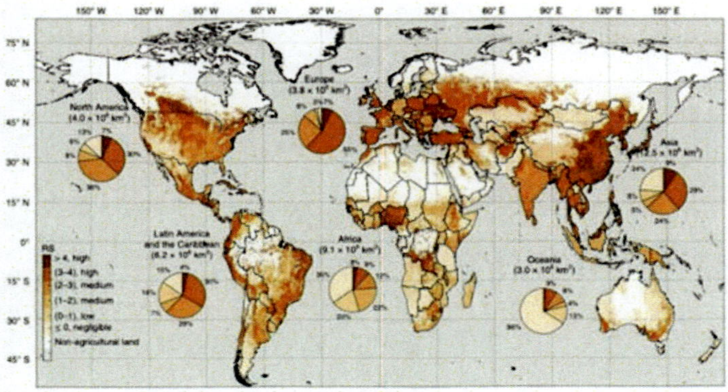

Fig. 2 Pesticide residual map

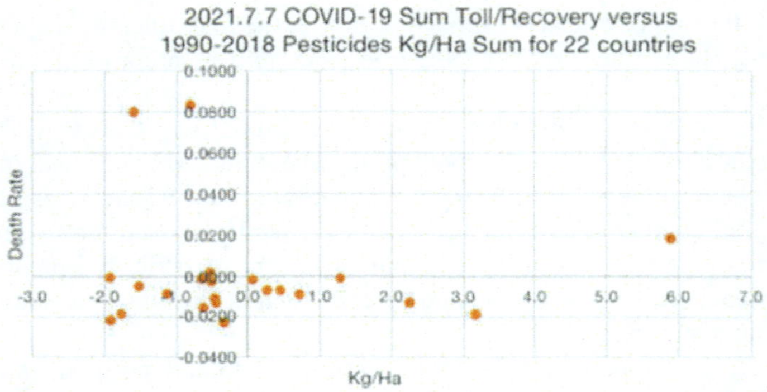

Fig. 3 COVID-19 death ratio versus pesticide residual normalized by precipitation

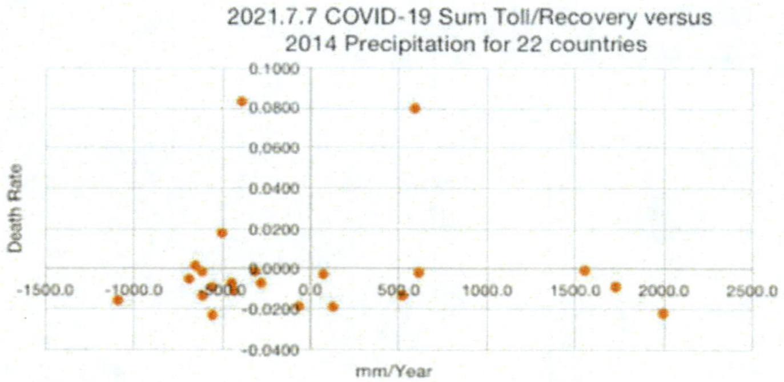

Fig. 4 COVID-19 death ratio versus precipitation

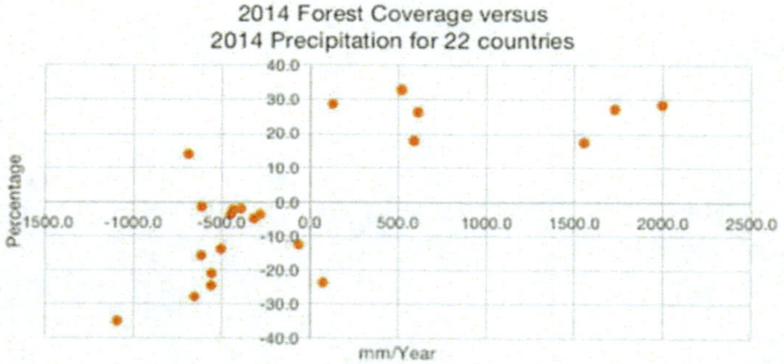

Fig. 5 Precipitation versus forest coverage

One factor due to immune system damage, one factor due to improved drug technology. Figure 4 indicating the toll is independent of rain or not. Figure 5 indicating more rain more forest.

3.2 Calculate Semi-covariance Coefficient

Unlike the Pearson coefficient, which only provides two directions, the semi-covariance approach provides a four-direction measurement between the target and variables. In semi-covariance, upper side covariance and downside covariance are calculated. If two variables move in the same direction above or below mean, such as an upward trend, the upper side covariance should be positive while the downside covariance should be 0. If two variables move in different directions above and below mean, such as downward trend, the upper side covariance should be 0 while the downside covariance should be negative. The value of Pearson coefficient will be dominated by the large covariance of moving pattern. Regarding math derivation, please see the following section.

$$Upperside\, correlation\, coefficient = \frac{E(ReLU((X - EX)(Y - EY)))}{\sqrt{E(X^2) - EX^2}\sqrt{E(Y^2) - EY^2}}.$$

$$Downside\, correlation\, coefficient = \frac{-E(ReLU(-(X - EX)(Y - EY)))}{\sqrt{E(X^2) - EX^2}\sqrt{E(Y^2) - EY^2}}.$$

4 Mathematical Derivation

The following simple derivation proves the relationship between our semi-covariance and Pearson correlation, which also demonstrates its fundamental relationship to the core of the latest deep learning method. It can be considered as an unsupervised mirrored pre-learned fixed-parameter pooling neuron network pairs on the complex plane.

For vectors X, Y, and Z, define the basic Rectified Linear Unit as $ReLU(X) = \max(0,\ X)$, then

$$Z = ReLU(0, Z) - ReLU(0, -Z).$$

Therefore, after taking the basic Pooling of the Expectation

$$E(Z) = E(ReLU(Z) - ReLU(-Z)).$$

It may be noted that

$$Z = (X - EX)(Y - EY).$$

Then, we have the semi-covariance

$$\begin{aligned} &E((X - EX)(Y - EY)) \\ &\quad = E(ReLU((X - EX)(Y - EY)) - ReLU(-(X - EX)(Y - EY))). \end{aligned}$$

Both sides are divided by the standard deviation

$$\begin{aligned} &\frac{E((X - EX)(Y - EY)}{\sqrt{E(X^2) - EX^2}\sqrt{E(Y^2) - EY^2}} \\ &\quad = \frac{E(ReLU((X - EX)(Y - EY)))}{\sqrt{E(X^2) - EX^2}\sqrt{E(Y^2) - EY^2}} \\ &\quad = \frac{E(ReLU(-(X - EX)(Y - EY)))}{\sqrt{E(X^2) - EX^2}\sqrt{E(Y^2) - EY^2}} \end{aligned}$$

We get

$$\begin{aligned} &Pearson\, correlation\, coefficient \\ &\quad = Upper\, correlation\, coefficient \; - \; Down\, correlation\, coefficient \end{aligned}$$

As upper and down correlation are calculated separately, each of them can be positive or negative. We group the positive or negative factors into two groups, and divide each for three subgroups i.e. strong (more than 2 out of 3 chances) medium and weak (less than 1 out of 5 chances) correlation (−1, −0.66, −0.2, 0, +0.2, + 0.66, +1). This way, we can better understand which factor goes with which factor, and which factor goes against which factor.

5 Discussion Conclusion

Here we present semi-covariance, an accurate way of measuring the non-linear correlation between variables. Our framework can successfully analyze the association between alternative factors and the response. The result of our analyses of the 2021 pandemic situation suggest that pesticide, precipitation, GDP, longevity, pet, scientist, and forest data [10–16] have substantial impacts each other in different directions.

While pesticide residual is proportional to pandemic death and the rest environmental and social economic factors are not independent either.

Due to each continental or neighborhood countries have similar environment or social data, we will not go through every country on the planet. We pick most of G20 and the country suffers more COVID-19 pandemic as the example for analysis.

In this project, our goal is to investigate whether the semi-covariance framework can accurately help us conduct association analysis between the alternative factors and the nonlinear factors related to pandemic, which is essentially a nonlinear data mining problem. Therefore, it is not surprising that semi-covariance gives more than Pearson.

Finally, by splitting the Pearson Coefficient into the semi-covariance, it reflects the fluctuating asymmetry, fitting the nature of things more accurately, and can be used for multiple factors in more aspects of the fields. It not only provides the interaction between the control and response variables, but also reveals the hidden pattern in the response variable. In future studies, we will study more about the forest coverage for the longevity and setup a complete digital twins prediction engine based on this algorithm.

Acknowledgements Research supported by the Jiangsu Computer Society of China with *Grant **KJFWRMJK(2021)*** "Research and implementation of intelligent IoT rehabilitation assistant platforms".

References

1. Lee, G.-H., Choi, K.-C.: Adverse effects of pesticides on the functions of immune system. Comp. Biochem. Physiol. Part C: Toxicol. Pharmacol. **235** (2020)
2. Markowitz, H.M.: Portfolio Selection: Efficient Diversification of Investments. Wiley, New York (1959)
3. Engle, R., Sheppard, K.: Evaluating the specification of covariance models for large portfolios. Mathematics (2007)
4. Haddad, V., Moreira, A., Muir, T.: When selling becomes viral: disruptions in debt markets in the COVID-19 crisis and the fed's response. Rev. Financ. Stud. (2021)
5. de Cabezón Irigaray, S., Javier, F., Carvajal, M., Dary, L., Grijalba, M.: Fernando, pesticides and longevity. In: Bentely, J.V., Keller, M.A. (eds.) Handbook on Longevity: Genetics, Diet & Disease. Nova Science Publishers (2009)
6. Bennett, B.M., Barton, G.A.: The enduring link between forest cover and rainfall: a historical perspective on science and policy discussions. For. Ecosyst. **5**, 5 (2018)
7. Woelffel, N., Silva-Filho, G., de Campos, R.H., Santos, H.G.: Influence of pesticide use on gross domestic product in Santa Maria de Jetibá-ES. Int. J. Adv. Eng. Res. Sci. **6**(8) (2019)
8. Pakes, A., Sokoloff, K.L.: Science, technology, and economic growth. Proc. Natl. Acad. Sci. **93**(23), 12655–12657 (1996)
9. Morgan, L., Protopopova, A., Birkler, R.I.D., et al.: Human–dog relationships during the COVID-19 pandemic: booming dog adoption during social isolation. Humanit. Soc. Sci. Commun. **7**, 155 (2020)
10. GDP ranking 2020. http://statisticstimes.com/economy/projected-world-gdp-ranking.php
11. CDC deaths 2021. https://covid.cdc.gov/covid-data-tracker/#cases_casesper100klast7days
12. Life expectancy 2015. https://en.wikipedia.org/wiki/List_of_countries_by_life_expectancy

13. Forest coverage 2014. https://en.wikipedia.org/wiki/List_of_countries_by_forest_area_(percentage)
14. Average precipitation in depth (mm per year) 2014. https://www.indexmundi.com/facts/indicators/AG.LND.PRCP.MM/rankings
15. United Nations Food and Agriculture Organization Pesticides indicators 2018. http://www.fao.org/faostat/en/#data/EP/visualize
16. Pets ownership 2020. https://www.petsecure.com.au/pet-care/a-guide-to-worldwide-pet-ownership/

State and Covariance Matrix Propagation for Continuous-Discrete Extended Kalman Filter Using Modified Chebyshev Picard Iteration Method

A. Imran, X. Wang, and X. Yue

Abstract In this paper, we propose a new method for the extended Kalman Filter state estimation for nonlinear systems with no closed-form solutions, given noisy state measurements are available with known uncertainties. The system is defined by a couple of sets of equations called the "moment equations." In the CD-EKF discrete, noisy state estimations are available at known time stamps. Propagation of the state estimation requires the integration of the moment equations that can diverge if the underlying system is stiff. We are employing the MCPI method at this stage, thus significantly improving the propagation accuracy compared to traditional methods. The proposed CD-EKF is applied to two problems (1) the famous Duffing Oscillator, a known stiff system, (2) to the Xu-Wang equations of relative orbital propagation, which define the relative motion of two satellites under the J2 perturbation of Earth.

Keywords CD-EKF · Relative orbital propagation · Modified chebyshev picard iteration · Non-linear systems

1 Introduction

Kalman Filter was proposed by Rudolf E. Kalman in 1960 [1]. However, Kalman Filter is an optimal estimation filter for linear systems and, the estimation accuracy deteriorates if the underlying systems are non-linear. In these conditions, the Extended Kalman Filter behaves better [2] and may outperform even UKF under some circumstances [3]. CD-EKF is used when the underlying system is based on non-linear differential equations, and noisy state samples are available at discrete time intervals. The prediction step of the CD-EKF requires integrating the moment equations between the time steps, which can be a very complex and resource-consuming process given the complexity of underlying systems. If the system is stiff and the time step is large, the filter can diverge. This has attracted a lot of research; Frogerais et al.

A. Imran (✉) · X. Wang · X. Yue
School of Astronautics, Northwestern Polytechnical University, Xi'an PRC, China
e-mail: aliimran900@mail.nwpu.edu.cn

H. Dai (ed.), *Computational and Experimental Simulations in Engineering*, Mechanisms and Machine Science 119,
https://doi.org/10.1007/978-3-031-02097-1_11

compared the results of various techniques for CD-EKF propagation and proposed a method based on RK-4 method [4]. Another major issue with the integration of the covariance equation is that the covariance matrix must be positive semi-definite. However, most integration techniques cannot guarantee that there are ways to ensure the positive semi-definiteness of the matrix though [5]. In this paper, we propose a new technique where the state is propagated using the MCPI method and the covariance matrix propagation is done through the method proposed by Mazzoni in conjunction with the MCPI method. The results are compared with other techniques. For the test case, two scenarios are tested, the CD-EKF is applied to the famous yet stiff Duffing Oscillator and the more complex Xu-Wang equations of relative orbital motion under the J2 perturbation [6]. In [7] the authors propose a variable time step system, whereas we are proposing a fixed time step system. The variable time step system is more efficient however, it comes with additional computations where the trust-region of the next integration step should be assessed.

2 Mathematical Model

2.1 Extended Kalman Filter

The underlying system is defined by Itô-type stochastic differential equation (SDE) given by:

$$\dot{\mathbf{x}}(t) = \mathbf{f}(\mathbf{x})dt + \mathbf{G}(t)d\omega(t) \tag{1}$$

where $\boldsymbol{x}$ is an n-dimensional state vector, $\boldsymbol{f}(\boldsymbol{x})$ is the underlying non-linear vector function $n \to n$. $\boldsymbol{G}(\boldsymbol{t}) \in n \times q$, and the Wiener-process $\boldsymbol{\omega}(\boldsymbol{t}) \sim \mathcal{N}(0, Q(t))$ and $\boldsymbol{Q}(\boldsymbol{t})$ is the $[q \times q]$ sized diffusion matrix known as the process noise.

The system states could be measured through uncertain measurements defined by

$$z_k = \boldsymbol{h}(\boldsymbol{x_k}) + \boldsymbol{v_k} \tag{2}$$

where z_k is the kth m-dimensional measurement at time $t = t_k = kt_s$ and t_s is the time step. $\boldsymbol{v_k} \sim \mathcal{N}(0, R)$ where $\boldsymbol{R}$ is the $m \times m$ square diffusion matrix which defines the uncertainty in the measurements, h is the translation function that converts current states to the measurement vector $m \to n$.

Once the system is defined the associated coupled moment differential equations can be defined by,

$$\dot{\boldsymbol{x}}(t) = \boldsymbol{f}(\boldsymbol{x}, t) \tag{3}$$

$$\dot{\boldsymbol{P}} = \boldsymbol{F}(\boldsymbol{x}, t)\boldsymbol{P} + \boldsymbol{P}\boldsymbol{F}(\boldsymbol{x}, t) + \boldsymbol{G}(t)\mathbf{Q}(t)\boldsymbol{G}(t)^T \tag{4}$$

where $\mathbf{F}$ is the Jacobian defined by

$$F(\mathbf{x}, t) = \frac{\partial f(\boldsymbol{x}, t)}{\partial \boldsymbol{x}} \tag{5}$$

First step in the solution of the EKF is the integration of the above defined moment equations to get the apriori state estimation $\hat{\mathbf{x}}_{k|k-1}$ and $\hat{\mathbf{P}}_{k|k-1}$. The apriori state covariance matrix can be estimated using the equations [5]

$$\mathbf{P}_{k|k-1} \approx \mathbf{M}_{\tau}\mathbf{P}_{k-1}\boldsymbol{M}_{\tau}^{T} + \mathbf{N}_{\tau}\mathbf{P}_{k-1}\boldsymbol{N}_{\tau}^{T} \tag{6}$$

where $\mathbf{P} = vec(P)$, and $vec()$ is the vectorization function that converts an $[n \times n]$ matrix to $[n^2 \times 1]$ vector. $\hat{\tau} = t_{k-1} + \frac{t_s}{2}$, and $\mathbf{M}$ and $\mathbf{N}$ are defined by

$$\mathbf{N}_{\tau} = \left[I - \boldsymbol{F}_{\tau}\frac{t_s}{2}\right]^{-1}, \mathbf{M}_{\tau} = \boldsymbol{N}_{\tau}\left[I + \boldsymbol{F}_{\tau}\frac{t_s}{2}\right]^{-1} \tag{7}$$

This requires the calculation of the Jacobian at half the time step, which implies that the state should be accurately estimated at both at $t_{k-1} + t_s/2$ and t_k. This problem can be solved by carefully setting up the MCPI integration method so that we do not have to evaluate both steps separately and can provide the apriori estimate of both $\hat{x_{half}}$ and $\hat{x_k}$ in a single step.

2.1.1 State Propagation Using the MCPI Method

Let us define an MCPI integrator $\mathcal{M}_{i \times u}$. Where i is the number of Chebyshev Gauss Lobatto (CGL) nodes and u is the number of Chebyshev Polynomials used. CGL nodes are defined by:

$$\hat{\tau}_j = cos(j\pi/(i-1)), j = 0, 1, \ldots (i-1) \tag{8}$$

where the first node corresponds to $t_{(k-1)}$, and the last node corresponds to t_k, MCPI only works over the independent variable range of $\hat{\tau} = -1 \leq \hat{\tau} \leq 1$ Thus the time domain has to be translated, For any system defined by (3) over a time span $t \in [t_{k-1}, t_k]$ The intermediary nodes can by defined at time nodes:

$$t = \frac{t_k - t_{k-1}}{2}\hat{\tau} + \frac{t_k + t_{k-1}}{2} \tag{9}$$

Once apriori information is available, we can propagate the Kalman filter and move to the prediction step; the prediction is given by

$$\boldsymbol{K}_k = \boldsymbol{P}_{k|k-1}\boldsymbol{H}_k^T(\boldsymbol{H}\boldsymbol{P}_{k|k-1}\boldsymbol{H}_k^T + \boldsymbol{R}_k)^{-1} \tag{10}$$

$$\hat{\boldsymbol{x}}_{k|k} = \hat{\boldsymbol{x}}_{k|k-1} + \boldsymbol{K}_k(z_k - \boldsymbol{h}(\hat{\boldsymbol{x}}_{k|k-1})) \tag{11}$$

$$\boldsymbol{P}_{k|k} = (I - \boldsymbol{K}_k\boldsymbol{H}_k)\boldsymbol{P}_{k|k-1} \tag{12}$$

$$\boldsymbol{H}_k = \left.\frac{\partial \boldsymbol{h}}{\partial \boldsymbol{x}}\right|_{x_{k|k-1}} \tag{13}$$

2.2 Modified Chebyshev Picard Iteration Method

Modified Chebyshev Picard Iteration Method (MCPI) was proposed by Dr. Xiaoli Bai [8], and has been used extensively in the field of orbital dynamics [9, 10].

with changes in (8) and (9), (3) becomes:

$$\frac{dx}{\hat{\tau}} = \frac{t_k - t_{k-1}}{2} f(\frac{t_k - t_{k-1}}{2}\tau + \frac{t_k + t_{k-1}}{2}, x) \tag{14}$$

The transformed function will be referred to as $\boldsymbol{F}(\hat{\tau}, x)$. MCPI works by estimating the integral and nodes, and for this we will be using the Chebychev-Guass-Lobatto(CGL)nodes defined by:

$$\hat{\tau}_j = cos(j\pi/n), j = 0, 1, \ldots, n \tag{15}$$

As we stated above by setting up the MCPI carefully we can evaluate the integral for both x and x_{half} in a single go, by setting N as an odd number the node at $\frac{N+1}{2}$ will lie exactly at $\frac{t_{k-1}+t_k}{2}$. The next step is to evaluate the weights based on an initial guess of x^0.

$$\frac{dx}{d\tau} = \hat{F}(\hat{\tau}, x^0) \approx \sum_{k=0}^{k=N'} A_k T_k(\hat{\tau}) \tag{16}$$

T_k is the Chebyshev polynomials of order k defined at time $\hat{\tau}$. The coefficients A_k $(k = 0, 1, \ldots, N)$ are calculated using

$$A_k = \frac{2}{N}\sum_{j=0}^{N}{}'' \hat{F}(\hat{\tau}, x^0(\tau_j))T_k(\hat{\tau}_j) \tag{17}$$

The notations can be further understood by going through Dr. Xiaoli Bai's work. A_k is calculated by summing the product of the force function F and the Chebyshev polynomials T_k. Next we formulate an equation that is suitable for Picard Iteration.

$$x^1(\tau) = \sum_{k=0}^{k=N}{}' \beta_k T_k(\hat{\tau}) \tag{18}$$

$$x^1(\hat{\tau}) = x_0 + \int_{-1}^{\tau} F(s, x^0(s).ds \tag{19}$$

Once we have a suitable equation we can evaluate the integrals using the properties of Chebyshev Polynomials. Once the integral is calculated the estimated state can be used as the guess to iteratively improve the estimate until we reach our tolerance levels.

3 Simulation and Analysis

3.1 *Duffing Oscillator*

Duffing Oscillator is a non-linear damped and driven oscillator given by the equation:

$$\ddot{x} = \delta\dot{x} + \alpha x + \beta x^3 = \gamma cos(\omega t) \tag{20}$$

where δ is the damping, α controls the linear stiffness, β is the non-linearity in the damping force, γ is the driving force amplitude and ω is the angular frequency of the driving force.

We applied the MCPI-EKF to the Duffing Oscillator problem and the results are shown in Fig. 1.

The associated covariance plot is for the MCPI-EKF error is shown in Fig. 2. A summary of various techniques applied for EKF state propagation is given in Table 1. And it can be seen how MCPI-EKF is superior to traditional methods.

3.2 *Relative Orbital Propagation*

The Xu-Wang model is the satellite relative orbital propagation model under J2 perturbation of earth and is given by [6].

$$\dot{v_x} = 2v_y\omega_z - x_r(\eta_r^2 - \omega_z^2) + y_r\alpha_z - z_r\omega_x\omega_z - (\varsigma_r - \varsigma)s_i s_\theta - r(\eta_r^2 - \eta^2) \tag{21}$$

$$\dot{v_y} = -2v_x\omega_z + 2v_z\omega_x - x_r\alpha_z + y_r(\eta_r^2 - \omega_z^2 - \omega_x^2) + z_r\alpha_x - (\varsigma_r - \varsigma)s_i c_\theta \tag{22}$$

$$\dot{v_z} = -2v_y\omega_x - x_r\omega_x\omega_z - y_r\alpha x + z_r(\eta_r^2 - \omega_x^2 - (\varsigma_r - \varsigma)c_i \tag{23}$$

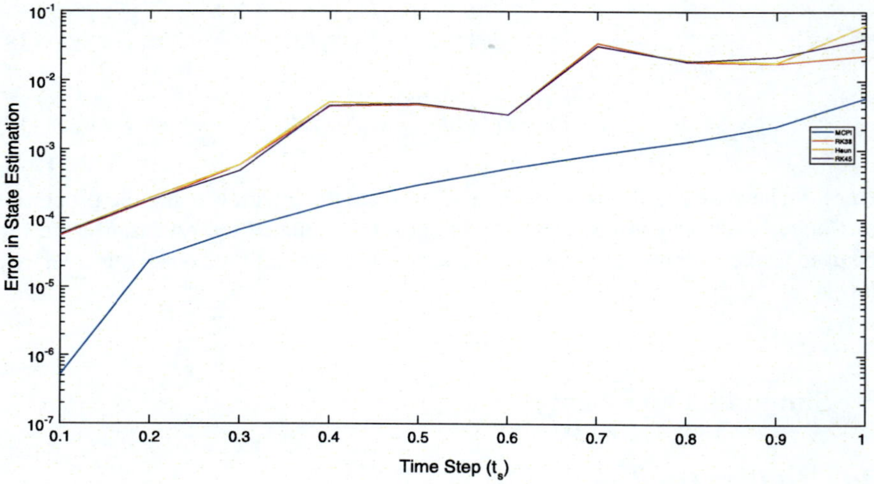

Fig. 1 Error for varying time steps with a constant measurement noise of 0.1

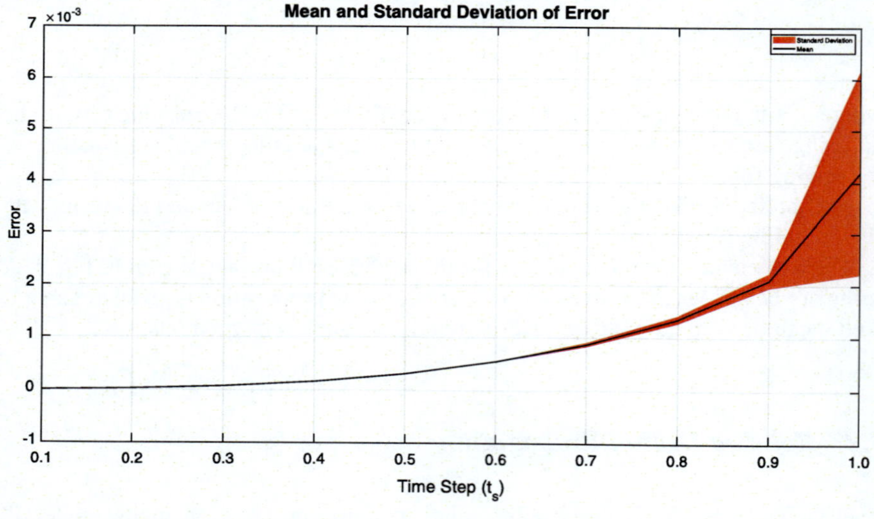

Fig. 2 Mean and Covariance plot for a constant measurement noise of 0.1

Table 1 Summary of the various integration tools for EKF state propagation

Duffing

	MCPI-EKF		RK3/8		Heun		RK45	
t_s	Pos	Vel	Pos	Vel	Pos	Vel	Pos	Vel
0.1	5.e-07	1.e-07	6.e-05	2.e-04	6.e-05	2.e-04	6.e-05	2.e-04
0.2	2.e-05	2.e-06	2.e-04	6.e-04	2.e-04	7.e-04	2.e-04	7.e-04
0.3	7.e-05	1.e-05	6.e-04	2.e-03	6.e-04	2.e-03	5.e-04	2.e-03
0.4	2.e-04	3.e-05	4.e-03	1.e-02	5.e-03	2.e-02	4.e-03	2.e-02
0.5	3.e-04	8.e-05	4.e-03	2.e-02	5.e-03	2.e-02	5.e-03	2.e-02
0.6	5.e-04	2.e-04	3.e-03	1.e-02	3.e-03	1.e-02	3.e-03	1.e-02
0.7	8.e-04	3.e-04	3.e-02	1.e-01	3.e-02	1.e-01	3.e-02	1.e-01
0.8	1.e-03	5.e-04	2.e-02	7.e-02	2.e-02	9.e-02	2.e-02	7.e-02
0.9	2.e-03	8.e-04	2.e-02	7.e-02	2.e-02	7.e-02	2.e-02	1.e-01
1.0	6.e-03	2.e-03	2.e-02	2.e-02	6.e-02	3.e-01	4.e-02	2.e-01

$$\dot{r} = v_x \tag{24}$$

$$\dot{v_x} = -\frac{\mu}{r^2} + \frac{h^2}{r^3} - \frac{k_{J2}}{r^4}(1 - 3s_i^2 s_\theta^2) \tag{25}$$

$$\dot{h} = -\frac{k_{J2}s_i^2 s_{2\theta}}{r^3} \tag{26}$$

$$\dot{\theta} = \frac{h}{r^2} + \frac{2k_{J2}c_i^2 s_\theta^2}{hr^3} \tag{27}$$

$$\dot{i} = -\frac{k_{J2}s_{2i}s_{2\theta}}{2hr^3} \tag{28}$$

So it's a set of 11 differential equations that need to be solved. We applied the MCPI-EKF to the Xu-Wang equations with a known measurement error of 0.1, and the results are shown in Table 2. The associated error and covariance plot is shown in Fig. 3.

4 Conclusion

In this paper, a new approach was proposed for solving the problems associated with the state propagation and covariance estimation of the Extended Kalman Filter. The filter behaves exceptionally well, even on very stiff systems. The approach ensures the stability of the state estimation and the covariance matrix estimation and extends Mazzoni's work. The MCPI-EKF was applied to the classical Duffing Oscillator problem and outperformed the traditional techniques. Similarly, it was also tested if the MCPI-EKF performs on the more complex Xu-Wang problem; the results also appear very promising.

Table 2 MCPI-EKF for the Xu-Wang equations and the error and associated covariance

Xu-Wang relative orbital motion model				
t_s	Pos Acc	Pos Cov	Vel Acc	Vel Cov
0.1	3.30E-02	1.50E-03	3.10E-02	0.11
0.2	4.40E-02	2.40E-03	4.50E-02	0.4
0.3	5.10E-02	3.00E-03	5.60E-02	0.86
0.4	5.50E-02	3.50E-03	5.70E-02	1.5
0.5	5.70E-02	3.80E-03	6.50E-02	2.2
0.6	6.20E-02	4.20E-03	6.70E-02	3.1
0.7	6.40E-02	4.60E-03	7.00E-02	4.1
0.8	6.40E-02	5.00E-03	6.80E-02	5.2
0.9	6.60E-02	5.40E-03	6.50E-01	6.5
1	7.00E-02	5.80E-03	6.20E-01	7.8

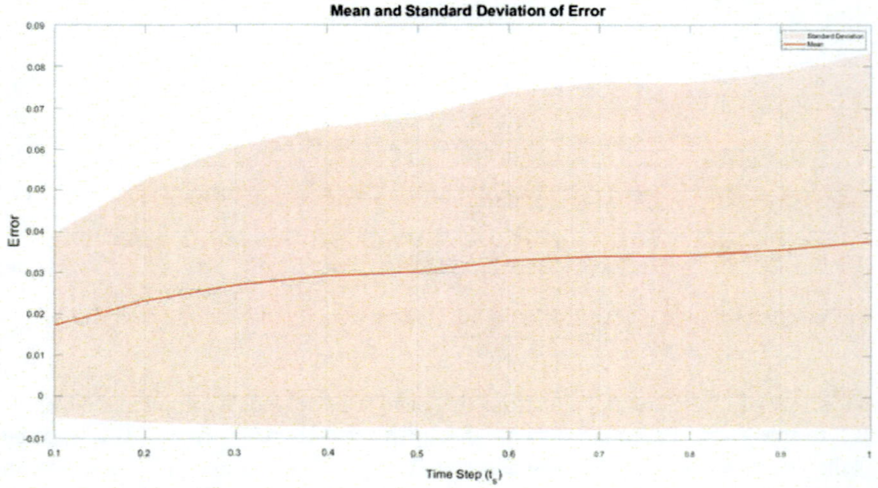

Fig. 3 Mean and Covariance plot for a constant measurement noise of 0.1

References

1. Kalman, R.E.: A new approach to linear filtering and prediction problems. J. Basic Eng. **82**(1), 35–45 (1960)
2. Jazwinski, Andrew: Stochastic processes and filtering theory. Academic Press, New York (1970)
3. LaViola, J.J.: A comparison of unscented and extended kalman filtering for estimating quaternion motion (2003)
4. Frogerais, P., Bellanger, J.-J., Senhadji, L.: Various ways to compute the continuous-discrete extended kalman filter. IEEE Trans. Autom. Control **57**(4), 1000–1004 (2012)

5. Mazzoni, Thomas: Computational aspects of continuous–discrete extended kalman-filtering. Comput. Stat. **23**(4), 519–539 (2007)
6. Guangyan, Xu., Wang, Danwei: Nonlinear dynamic equations of satellite relative motion around an oblate earth. J. Guid. Control Dyn. **31**(5), 1521–1524 (2008)
7. Kulikov, G.Y., Kulikova, M.V.: Accurate numerical implementation of the continuous-discrete extended kalman filter. IEEE Trans. Autom. Control **59**(1), 273–279 (2014)
8. Bai, X.: Modified Chebyshev–Picard iteration methods for solution of initial value and boundary problems. Ph.D. thesis, Texas A&M University (2010)
9. Bai, Xiaoli, Junkins, John L.: Modified Chebyshev–Picard iteration methods for solution of initial value problems. J. Astronaut. Sci. **59**(1–2), 327–351 (2012)
10. Imran, A., Wains, F.S., Wang, X., Xiaokui, Y.: Application of modified Chebyshev Picard iteration to the relative orbital dynamics problem. In: 2021 International Bhurban Conference on Applied Sciences and Technologies (IBCAST). IEEE (2021)

A Novel Model to Calculate the Fluctuating Pressure in Eccentric Annulus for Bingham Fluid

Jiangshuai Wang, Jun Li, Yanfeng He, Gonghui Liu, and Song Deng

Abstract The accurate calculation of fluctuating pressure is crucial to avoid the occurrence of kick and lost circulation in the process of tripping and casing running. For conventional oil-based (diesel) and gas-oil synthetic drilling fluids, Bingham model can better characterize their rheological properties. However, there is little research on the fluctuating pressure in eccentric annulus for Bingham fluid. In this paper, a novel model to calculate the fluctuating pressure in eccentric annulus for Bingham fluid was established. And, the self-adaptive Simpson integral method was used to solve the model. In addition, the model was verified by the classical Burkhardt model in concentric annulus and indoor experiment results in eccentric annulus. Moreover, the variation laws of fluctuating pressure under different factors were investigated. Here are some significant findings, (1) the maximum relative error between the established model and Burkhardt model is less than 8.9% in concentric annulus. And, the maximum relative error between the established model and indoor experiment results is less than 9.2% in eccentric annulus. (2) The fluctuating pressure gradient is negatively correlated with eccentricity, but positively correlated with size ratio of string to hole and dynamic shear stress. (3) In deviated wellbores and horizontal wellbores with large eccentricity, the tripping speed can be appropriately increased to reduce the non-production time. (4) In slim hole with narrow gap of annulus, the tripping speed should be strictly controlled to avoid excessive or too small wellbore pressure.

J. Wang (✉) · Y. He · S. Deng
School of Petroleum Engineering, Changzhou University, 21 Gehu Middle Road, Changzhou 213164, Jiangsu, China
e-mail: wjs125126@126.com

J. Li · G. Liu
College of Petroleum Engineering, China University of Petroleum-Beijing, 18 Fuxue Road, Changping 102249, Beijing, China

J. Li
College of Petroleum, China University of Petroleum-Beijing at Karamay, 355 Anding Road, Karamay 834000, Xinjiang, China

H. Dai (ed.), *Computational and Experimental Simulations in Engineering*, Mechanisms and Machine Science 119,
https://doi.org/10.1007/978-3-031-02097-1_12

Keywords Novel model · Fluctuating pressure · Eccentric annulus · Bingham fluid · Eccentricity

1 Introduction

With the application of drilling technologies such as extended reach drilling [1], horizontal well drilling [2, 3], managed pressure drilling [4] and deepwater drilling [5, 6], the accurate calculation of wellbore pressure has attracted extensive attention [7–9]. Especially, wellbore pressure will fluctuate in some special drilling operations. For example, when the drill string or casing moves up and down in the wellbore filled with drilling fluid, it will cause the fluctuation of wellbore pressure [10], that is, fluctuating pressure. The commonly used prediction models of fluctuating pressure in the field, such as Burkhardt model [11] and Schuh model [12], are established for concentric annulus. Generally speaking, in vertical wellbores the string is probably in the center of the wellbore, and the annulus is concentric. In this concentric annulus, the classical Burkhardt model and Schuh model for Bingham fluid have a high degree of agreement. However, in deviated wellbores and horizontal wellbores, the gravity action leads to the tendency of the string to be close to the bottom of the wellbore [13–15]. Therefore, the string is often in an eccentric position in the anulus. Neglecting the influence of eccentricity on fluctuating pressure will lead to overestimation of trip rate and an increase in non production time and drilling cost.

For the flow of Newtonian fluid, Power-Law fluid, Casson fluid, Herschel-Bulkley fluid and Robertson-Stiff fluid in eccentric annulus, scholars have conducted a lot of researches on corresponding calculation models of fluctuating pressure [16–20]. In details, in 1996, based on the model of fluctuating pressure in concentric annulus, Wang et al. [16] obtained the approximate solution of fluctuating pressure in eccentric annulus for Newtonian fluid, and established the empirical model of fluctuating pressure in eccentric annulus by using multiple parameter regression method. Then, in 1998, for the Power-Law fluid, Wang et al. [17] established the governing equation of the fluctuating pressure in the eccentric annulus, and obtained the fluctuating pressure and velocity distribution in eccentric annulus when running casing. In 2011, for the Casson fluid, Sun et al. [18] developed a new calculation method of fluctuating pressure. Using this new method can accurately predict fluctuating pressure in horizontal wells. In 2016, for the Herschel-Bulkley fluid, Li et al. [19] established the calculation model of fluctuating pressure in eccentric annulus and obtained the fluctuating pressure gradient. In addition, the influence of eccentricity on fluctuating pressure gradient was also discussed. In the same year, for Robertson-Stiff fluid, Li et al. [20] used narrow-slot flow model to simulate fluid flow in eccentric annulus, and established the calculation model of fluctuating pressure under steady laminar flow condition. Moreover, the influence of drilling fluid performance parameters on fluctuating pressure was discussed. In fact, compared with other fluid models, Bingham model can better characterize their rheological properties for conventional oil-based (diesel) and gas-oil synthetic drilling fluids [21–23]. In detail, when using Bingham mode, many good features are obtained at the same time, such as high fitting degree, simple rheological model and high hydraulic calculation efficiency. However, there is little research on the fluctuating pressure of Bingham fluid in eccentric annulus.

In view of the inadequacy of the current researches, in this paper, a novel model to calculate the fluctuating pressure in eccentric annulus for Bingham fluid was established. And, the self-adaptive Simpson integral method was used to solve the model. In addition, the model was verified by the classical Burkhardt model in concentric annulus and indoor experiment results in eccentric annulus. Moreover, the variation laws of fluctuating pressure under different factors were investigated. All these works are conducted to accurately predict fluctuating pressure and clearly understand its variation law in eccentric annulus.

2 Fluctuating Pressure Model Based on Bingham Fluid in Eccentric Annulus

2.1 Basic Assumptions

In this paper, since the calculation model of fluctuating pressure is established under the steady-state condition, the following assumptions are made:

(1) The hole is a regular circular well with known diameter.
(2) The parameters in the flow field do not change with time, that is, the flow is stable.
(3) The flow in the eccentric annulus is simplified as the flow in a narrow slot with different heights.

2.2 Eccentric Annulus Model

Generally, in deviated wellbores and horizontal wellbores, fluid flows in eccentric annulus. In this paper, the flow section in eccentric annulus is simplified as the following physical model (as shown in Fig. 1).

According to the Sine Rule, the gap of annulus at different eccentric angles are obtained as follows:

$$h = (r_2^2 - e^2 \sin^2 \theta)^{\frac{1}{2}} - r_1 + e \cos \theta \tag{1}$$

where, r_1 is the outer radius of string, m. r_2 is the inner radius of wellbore, m. e is the eccentric distance, m. θ is the eccentricity angle, °. h is the gap of annulus, m.

In addition, according to Fig. 1, a key parameter is defined to describe the eccentric degree of position relationship between the string and hole, namely eccentricity (ε). The calculation formula of eccentricity is as follows:

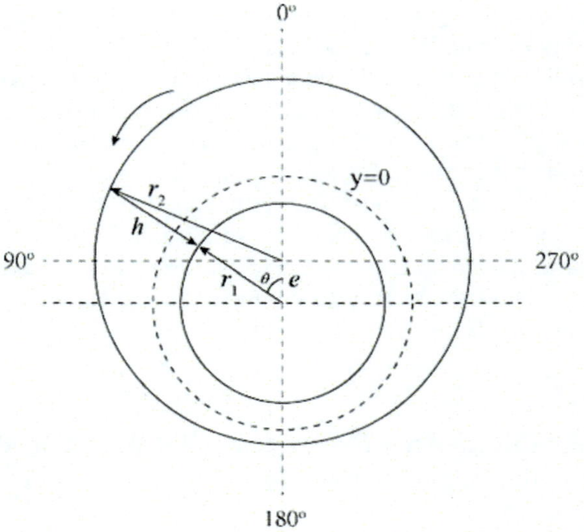

Fig. 1 Schematic diagram of the flow section in eccentric annulus

$$\varepsilon = e/(r_2 - r_1) \tag{2}$$

2.3 *Calculation of Flow Rate in Eccentric Annulus in the Process of String Movement*

When the flow is stable, the external force on the fluid element is zero. In this case, the relationship between shear stress and fluctuating pressure gradient [24] can be expressed as follows:

$$\tau = \frac{\Delta p_s}{L} y \tag{3}$$

where, y is the distance from the center of the annulus ($y = 0$), m. τ is the shear stress between flow layers, Pa. L is the length of microelements, m. $\Delta p_s/L$ is the fluctuating pressure gradient, Pa/m.

Formula (3) is the governing equation of the axial uniform flow in the narrow slot (eccentric annulus).

In the process of string movement, due to the adhesion of drilling fluid, drilling fluid will flow with the movement of string. The maximum velocity of drilling fluid in annulus is:

$$\upsilon_{\max} = \frac{3}{2}\left(\frac{r_1^2}{r_2^2 - r_1^2} + K_c\right)\upsilon_p \tag{4}$$

where, υ_p is the movement speed of string, m/s. K_c is the adhesion coefficient of fluid, dimensionless. Under laminar flow condition, the calculation formula of adhesion coefficient is as follows:

$$K_c = -\frac{1 - k^2 + 2k^2 \ln k}{2(1 - k^2) \ln k} \tag{5}$$

where, $k = r_1/r_2$, is the size ration of string to hole, dimensionless.

So, the flow rate (Q) caused by string movement in eccentric annulus is as follows:

$$Q = \pi (r_2^2 - r_1^2)\upsilon_{max} \tag{6}$$

2.4 Model of Fluctuating Pressure

For Bingham fluid, the rheological equation is as follows:

$$\begin{cases} \tau = \tau_0 + \mu_p\left(-\frac{du}{dy}\right), \tau > \tau_0 \\ \frac{du}{dy} = 0, \tau \le \tau_0 \end{cases} \tag{7}$$

where, τ_0 is the dynamic shear stress, Pa. μ_p is the plastic viscosity, Pa·s. And du/dy is the shear rate, s^{-1}.

Combined with (3) and (7), the velocity distribution of Bingham fluid in eccentric annulus can be obtained as follows:

$$\begin{cases} u = \frac{\frac{\Delta p_s}{L}\frac{1}{2}\left(\frac{h}{2}-y\right)\left(\frac{h}{2}+y-2r_0\right)}{\mu_p}, y > r_0 \\ u_0 = \frac{\frac{\Delta p_s}{L}\frac{1}{2}\left(\frac{h}{2}-2r_0\right)^2}{\mu_p}, y \le r_0 \end{cases} \tag{8}$$

There is a special area in which the velocity of fluid is the same. It's called the flow core. r_0 is the radius of the flow core, m. u_0 is the velocity of the fluid in the flow core, m/s.

By integrating the velocity over the whole flow area, the average velocity in eccentric annulus can be obtained as follows:

$$\overline{u}(\theta) = \left(2u_0 r_0 + 2\int_{r_0}^{\frac{h}{2}} u dy\right) \tag{9}$$

Substitute formula (8) into (9) and integrate, and ignore the higher order trace of $r_0/\frac{h}{2}$ under the condition of $\frac{h}{2} \ge r_0$. Then, the results are as follows:

$$\overline{u}(\theta)=\frac{\Delta p_{\mathrm{s}}}{L}\frac{1}{\mu_{\mathrm{p}}}\frac{h^{2}}{4}\left[\frac{1}{3}-\tau_{0}\left(\frac{\Delta p_{\mathrm{s}}}{L}h\right)^{-1}\right] \tag{10}$$

Taking the gap at position θ in Fig. 1 as the reference and taking the small increment $d\theta$. So, the arc length ds corresponding to $d\theta$ can be expressed as:

$$ds=(h+2r_{1})d\theta \tag{11}$$

Therefore, we can get the flow rate through the area of micro element in the eccentric annulus:

$$dQ=\frac{1}{2}h\overline{u}(\theta)ds \tag{12}$$

After sorting out and integrating the above formula, the flow rate of fluid through the whole eccentric annulus is as follows:

$$Q=\frac{1}{\mu_{\mathrm{p}}}\frac{1}{2}\int_{0}^{2\pi}\frac{h^{3}}{4}\left[\frac{1}{3}h\frac{\Delta p_{\mathrm{s}}}{L}-\tau_{0}\right]d\theta+\frac{1}{\mu_{\mathrm{p}}}r_{1}\int_{0}^{2\pi}\frac{h^{2}}{4}\left[\frac{1}{3}h\frac{\Delta p_{\mathrm{s}}}{L}-\tau_{0}\right]d\theta \tag{13}$$

Obviously, formula (6) and formula (13) have the same result. Therefore, combined with formula (6) and formula (13), the calculation model of fluctuating pressure gradient based on Bingham fluid in eccentric annulus can be obtained as follows:

$$\frac{\Delta p_{\mathrm{s}}}{L}=\frac{C+B}{A} \tag{14}$$

where,

$$\begin{cases} A=\frac{2r_{1}+1}{2\mu_{\mathrm{p}}}\int_{0}^{2\pi}\frac{h^{3}+h^{4}}{12}d\theta \\ B=\frac{2r_{1}+1}{2\mu_{\mathrm{p}}}\int_{0}^{2\pi}\frac{h^{2}+h^{3}}{4}\tau_{0}d\theta \\ C=\pi(r_{2}^{2}-r_{1}^{2})\upsilon_{\max} \end{cases} \tag{15}$$

It must be noticed that the integrand function in expressions A and B of formula (15) is complex and the original function is difficult to be solved, so the self-adaptive Simpson integral method is used to solve the model.

3 Model Validation

In order to verify the accuracy and reliability of the model established in this paper, the calculation results of the model are compared with those of the classical Burkhardt model in concentric annulus and indoor experiment results in eccentric annulus.

3.1 Model Validation Under the Condition of Concentric Annulus

Firstly, the results of the model are compared with those of Burkhardt model under the condition of concentric annulus. The inner radius of wellbore is 50 mm and the outer radius of string is 40 mm. In addition, other fluid parameters are shown in Table 1.

Figure 2 shows the comparison of the fluctuating pressure gradient calculated by the model established in this paper and Burkhardt model under the condition of

Table 1 Basic parameters of fluid

	τ_0 (Pa)	μ_p (Pa·s)
1# fluid	0.015	0.003
2# fluid	0.032	0.005
3# fluid	1.213	0.010

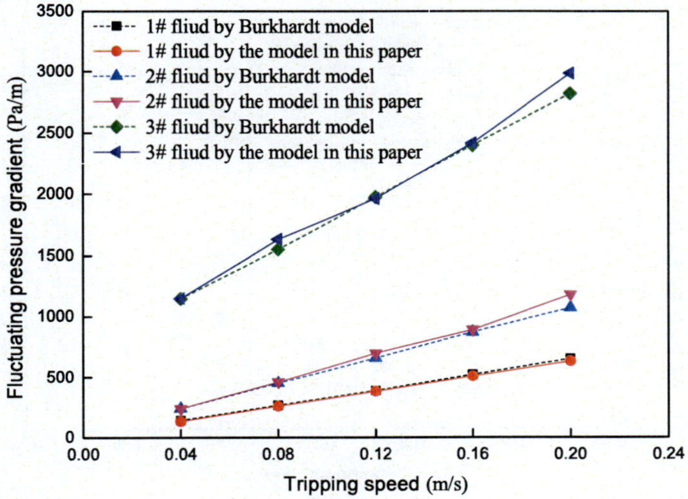

Fig. 2 Comparison of the fluctuating pressure gradient calculated by the model established in this paper and Burkhardt model under the condition of concentric annulus

concentric annulus. It can be seen that under three different fluid types, the calculation results of the model are in good agreement with those of Burkhardt model. And, the maximum relative error is less than 8.9%. Therefore, it is proved that the model can also accurately calculate the fluctuating pressure in the concentric annulus during tripping after properly simplifying the model. The proper simplification is to remove the eccentricity factor.

3.2 *Model Validation Under the Condition of Eccentric Annulus*

Furthermore, under the condition of eccentric annulus, the calculated results of the model are compared with the indoor experimental results [25]. The simulation parameters are the same as above.

Figure 3 shows the comparison of the fluctuating pressure gradient calculated by the model established in this paper and the indoor experimental results under the condition of eccentric annulus when the eccentricity is 0.5. It can be seen that under three different fluid types, the calculated results of the model are also in good agreement with the indoor experimental results. And, the maximum relative error is not more than 9.2%. Moreover, when the dynamic shear stress of drilling fluid is large (i.e. 3# fluid), there is a big difference between the calculated results of the model and the indoor experimental results. The reasons are as follows: except for the error of drilling fluid rheological parameter fitting, the error is related to the lateral vibration

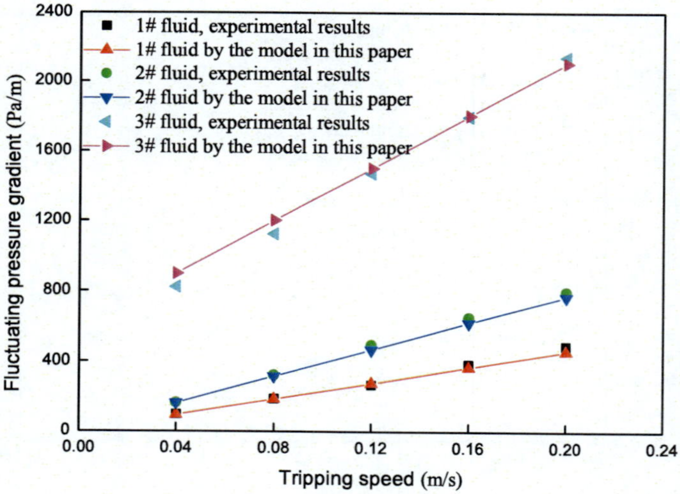

Fig. 3 Comparison of the fluctuating pressure gradient calculated by the model established in this paper and the indoor experimental results under the condition of eccentric annulus

of the string in the process of the experiment [25]. In addition, when the tripping speed is large, the calculated results deviate greatly from the indoor experimental results. This is for the reason that when the string is eccentric, the fluid mainly passes through the wide gap. In this case, the lateral vibration of the drill string has a great influence on the fluctuating pressure in the process of the experiment.

In conclusion, the calculated results of the model are in good agreement with those of the classical Burkhardt model and indoor experimental results, which proves the accuracy and reliability of the model. Therefore, the model established in this paper can not only predict the fluctuating pressure in eccentric annulus, but also can be used to predict the fluctuating pressure in concentric annulus after simplifying the model.

4 Influence of Different Factors on Fluctuating Pressure

Generally, the gas-oil synthetic drilling fluid, belongs to Bingham fluid. In this section, the influence of eccentricity, size ratio of string to hole and dynamic shear stress of drilling fluid on fluctuating pressure gradient were investigated. The inner radius of wellbore is 50 mm and the outer radius of string is 40 mm. In addition, dynamic shear stress of drilling fluid is 1.614 Pa, plastic viscosity of drilling fluid is 0.051 Pa·s.

4.1 Eccentricity

Figure 4 shows the variation of the fluctuating pressure gradient with the tripping speed under different eccentricity conditions. It can be seen that the fluctuating pressure decreases with the increase of eccentricity. In detail, when the eccentricity is 1, the fluctuating pressure gradient decreases to about 46% of the result in concentric annulus. Usually, the string is often eccentric in deviated wellbores and horizontal wellbores. Therefore, in the horizontal or inclined section with large eccentricity, the tripping speed can be appropriately increased to reduce the non-production time.

4.2 Size Ratio of String to Hole

Figure 5 shows the variation of fluctuating pressure gradient with tripping speed under different size ratio of string to hole conditions. It can be seen that the fluctuating pressure increases with the increase of size ratio of string to hole. In detail, when the size ratio of string to hole is greater than 0.7, the fluctuating pressure increases rapidly with the increase of tripping speed. Therefore, in slim hole with narrow gap of annulus, the tripping speed should be strictly controlled to avoid downhole

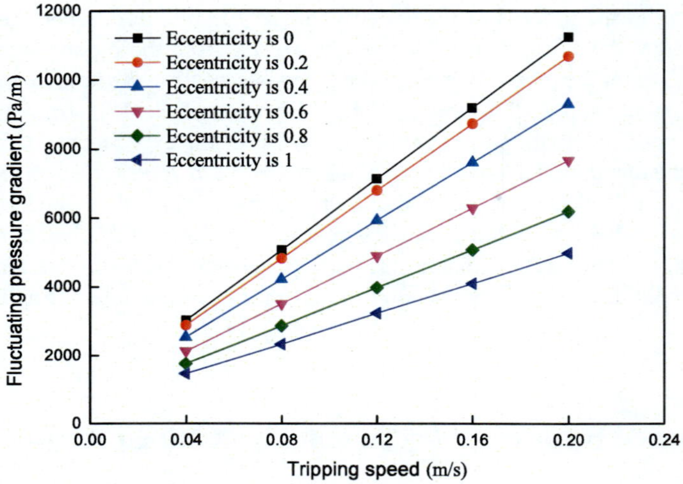

Fig. 4 The influence of eccentricity on fluctuating pressure gradient

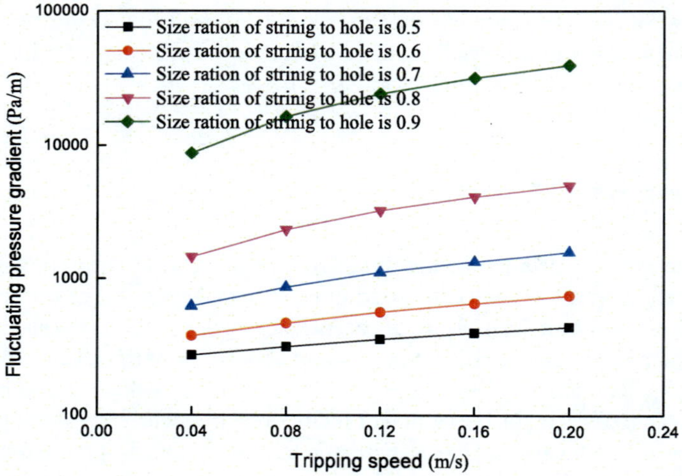

Fig. 5 The influence of size ratio of string to hole on fluctuating pressure gradient

complications caused by excessive or too small wellbore pressure, so as to ensure drilling safety.

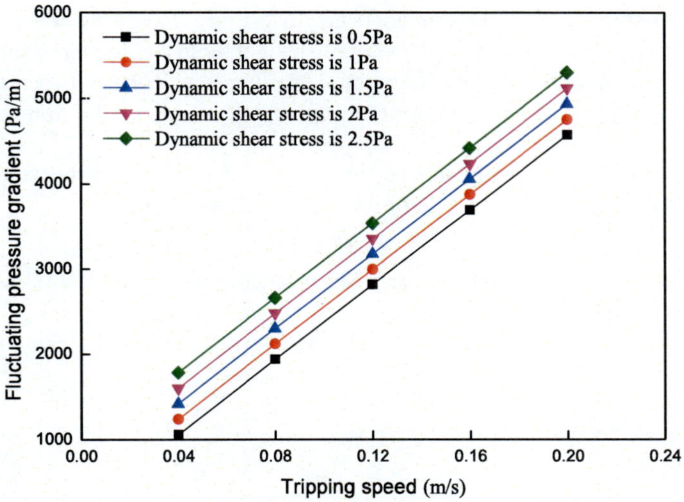

Fig. 6 The influence of dynamic shear stress on fluctuating pressure gradient

4.3 Dynamic Shear Stress

Figure 6 shows the variation of fluctuating pressure gradient with tripping speed under different dynamic shear stress of drilling fluid conditions. It can be seen that for Bingham fluid, the fluctuating pressure increases linearly with the increase of dynamic shear stress. This phenomenon can also be explained by the positive proportional relationship between fluctuating pressure gradient and dynamic shear stress in formulas (12) and (13).

To sum up, the fluctuating pressure gradient is negatively correlated with eccentricity, but positively correlated with size ratio of string to hole and dynamic shear stress. In addition, in deviated wellbores and horizontal wellbores with large eccentricity, the tripping speed can be appropriately increased to reduce the non-production time. Moreover, in slim hole with narrow gap of annulus, the tripping speed should be strictly controlled to avoid excessive or too small wellbore pressure.

5 Conclusions

(1) The model established for Bingham fluid is reliable. In details, the maximum relative error between the established model and the classical Burkhardt model is less than 8.9% in concentric annulus. And, the maximum relative error between the established model and indoor experiment results is less than 9.2% in eccentric annulus.

(2) The fluctuating pressure gradient is negatively correlated with eccentricity, but positively correlated with size ratio of string to hole and dynamic shear stress.
(3) In deviated wellbores and horizontal wellbores with large eccentricity, the tripping speed can be appropriately increased to reduce the non-production time.
(4) In slim hole with narrow gap of annulus, the tripping speed should be strictly controlled to avoid excessive or too small wellbore pressure.

Acknowledgements Project supported by the Key Program of National Natural Science Foundation of China (Project No. 51734010).

References

1. Zhu, N., Huang, W., Gao, D.: Dynamic wavy distribution of cuttings bed in extended reach drilling. J. Petrol. Sci. Eng. 108171 (2021)
2. Dong, X., Liu, H., Zhai, Y., et al.: Experimental investigation on the steam injection profile along horizontal wellbore. Energy Rep. **6**, 264–271 (2020)
3. Xu, Y., Sheng, G., Zhao, H., et al.: A new approach for gas-water flow simulation in multi-fractured horizontal wells of shale gas reservoirs. J. Petrol. Sci. Eng. 108292 (2021)
4. Kinik, K., Gumus, F., Osayande, N.: Automated dynamic well control with managed-pressure drilling: A case study and simulation analysis. SPE Drill. Complet. **30**(2), 110–118 (2015)
5. Zheng, S., Li, W., Cao, C., et al.: Prediction of the wellhead uplift caused by HT–HP oil and gas production in deep-water wells. Energy Rep. **7**, 740–749 (2021)
6. Wang, J., Li, J., Liu, G., et al.: Parameters optimization in deepwater dual-gradient drilling based on downhole separation. Pet. Explor. Dev. **46**(4), 819–825 (2019)
7. Zhang, H., Ni, H., Wang, Z., et al.: Optimization and application study on targeted formation ROP enhancement with impact drilling modes based on clustering characteristics of logging. Energy Rep. **6**, 2903–2912 (2020)
8. Wang, J., Li, J., Liu, G., et al.: Prediction of annulus pressure in variable pressure gradients drilling. Acta Petrolei Sinica **41**(4), 497–504 (2020)
9. Li, H., Wang, W., Liu, Y., et al.: An integrated drilling, protection and sealing technology for improving the gas drainage effect in soft coal seams. Energy Rep. **6**, 2030–2043 (2020)
10. Márcia, P.V., Gabrielle, F.M.O., Lindoval, D.F., et al.: Monitoring and control strategies to manage pressure fluctuations during oil well drilling. J. Petrol. Sci. Eng. **166**, 337–349 (2018)
11. Burkhardt, J.A.: Wellbore pressure surges produced by pipe movement. J. Petrol. Technol. **13**(6), 595–605 (1961)
12. Schuh, F.J.: Computer makes surge-pressure calculations useful. Oil Gas J. **62**(31), 96–104 (1964)
13. Tian, J., Yang, Y., Dai, L., et al.: Dynamics and anti-friction characteristics study of horizontal drill string based on new anti-friction tool. Int. J. Green Energy (2021). https://doi.org/10.1080/15435075.2021.1880909
14. Shwetank, K., Syahrir, R., Pandian, V., et al.: Explicit flow velocity modelling of yield power-law fluid in concentric annulus to predict surge and swab pressure gradient for petroleum drilling applications. J. Petrol. Sci. Eng. 107743 (2020)
15. Ali, E., Gursat, A.: Functional and practical analytical pressure surges model through herschel bulkley fluids. J. Petrol. Sci. Eng. **171**, 748–759 (2018)
16. Wang, H., Liu, X., Dong, J.: Approximate solution of stable fluctuation pressure of New-Tonian fluid in eccentric annular. Oil Drilling Prod. Technol. **18**(2), 18–21 (1996)

17. Wang, H., Su, Y., Liu, X.: Numerical analysis of stead surge pressure of Power-Law fluid in eccentric annuli. Acta Petrolei Sinica **19**(3), 104–109 (1998)
18. Li, Q., Wang, Z., Li, X., et al.: The computational model for surge pressure of Herschel-Bulkley fluid in eccentric annulus. Acta Petrolei Sinica **37**(09), 1187–1192 (2016)
19. Sun, Y., Li, Q., Kong, C., et al.: New prediction method on surge pressure in horizontal well basing on Casson fluid. Drilling Fluid Completion Fluid **28**(2), 29–31 (2011)
20. Li, Q., Wang, Z., Wang, Y., et al.: Computational model of fluctuating pressure of Robertson-stiff fluid in eccentric annulus. J. Xi'an Shiyou Univ. (Nat. Sci. Edition) **31**(03), 86–91 (2016)
21. Yan, J., Zhao, X.: Rheological properties of oil-based drilling fluids at high temperature and high pressure. Acta Petrolei Sinica **24**(03), 104–109 (2003)
22. Li, H., Wang, N., Tian, R., et al.: Study on rheological property and model of GTL based drilling fluids under deepwater condition. China Offshore Oil Gas **22**(06), 406–408 (2010)
23. Wang, Z.: Research and application progress of oil-based drilling fluid at home and abroad. Fault-Block Oil Gas Field **18**(04), 533–537 (2011)
24. Fan, H.: Practical Drilling Fluid Mechanics. Petroleum Industry Press, Beijing (2014)
25. Guo, Y.: Calculation Model and Experimental Study on Wellbore Pressure Fluctuation Under Tripping Condition. China University of Petroleum, Beijing (2014)

Interactive Restoration of Implicitly Defined Shapes

Jiayu Ren, Yoshihisa Fujita, and Susumu Nakata

Abstract Shape representation using implicit functions is one of the popular geometric modeling methods and used for variety of important applications in the area of computer graphics and simulations, such as surface reconstruction from unorganized point sets, shape modeling based on constructive solid geometry and fluid simulation based on particles. Although many different generation and rendering approaches have been proposed, the generated surfaces are likely to have irregular components such as isolations of long-narrow parts, unexpectedly attached spikes and noise-like ripples on smooth curves. Such problems are caused by a variety of reasons, including the limitations of surface generation algorithms, lack of robustness of scanning machine that cause defects over the input point set and the complexity of the desired shape to generate. It is necessary to propose algorithms that are able to give restorations to the irregular represented shapes as post-processing. In this paper, three methods have been proposed to apply user-interactive restorations of 2D implicitly represented shapes with irregular components, where users are expected to specify locations of irregular part as people often have good understanding of the shape that they create. Experiments have shown that the proposed methods are competent to make fast and effective restorations and corrections of irregular components on 2D implicitly represented shapes.

Keywords Computer graphics · Shape modeling · Implicit representation · Interactive shape restoration

J. Ren (✉)
Graduate School of Information Science Engineering, Ritsumeikan University, Shiga 525-8577, Japan
e-mail: gr0450ek@ed.ritsumei.ac.jp

Y. Fujita · S. Nakata
College of Information Science and Engineering, Ritsumeikan University, Shiga 525-8577, Japan

H. Dai (ed.), *Computational and Experimental Simulations in Engineering*, Mechanisms and Machine Science 119,
https://doi.org/10.1007/978-3-031-02097-1_13

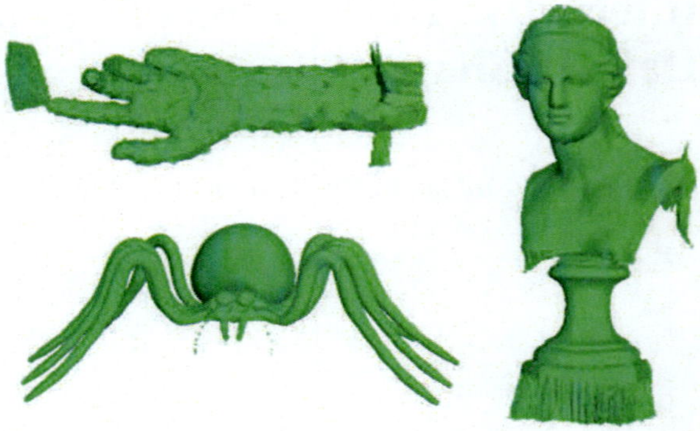

Fig. 1 Examples of irregular surfaces that generated from "Grid of Polynomials" [3]. Upper-left: noisy surface with redundant parts; lower-left: surface with isolate parts; right: surface with redundant parts

1 Introduction

Implicit 2-dimensional (2D) or 3-dimensional (3D) shapes generated based on scattered data are wildly used in different areas, such as computer graphics, fluid simulations and animations. They naturally ensure generation of smooth and closed manifolds based on scanned data points [1]. Many scanning techniques have been developed for point set acquisition, such as structured light or laser-based range scanners, multi-view stereo camera systems, and depth cameras [2]. When data points are obtained from different scanning systems, the quality of the data for surface reconstruction varies a lot, as well as the generated shapes. In addition, [2] says that point acquisition contains the issues of "missing data", "noise", "outliers" and "non-uniform sampling". The presence of such issues may lead irregular components at the generated shape, including isolated components, redundant components and noisy components, and they are shown in Fig. 1.

The contribution of this paper is to present a solution for interactive restoration of irregular implicitly defined 2D shape that gives different methods to make corrections of irregular parts on implicit defined curves within a user-specified region. The proposed technique is fast and able to make effective, smooth and continuous restorations. Evaluations have also been presented to test the effectiveness of the proposed techniques.

2 Implicit Representations

A typical way to define a 2D or 3D shape implicitly is to find an appropriate function $f(\boldsymbol{x})$ or $f(x, y)$ and solve the zero-value set of the function as an implicit function $f(\boldsymbol{x}) = 0$, for representation of the expected shape. One of the approaches to solve the zero-value set of the function is to introduce a set of constraint points $\boldsymbol{x}_1, \boldsymbol{x}_2, \ldots, \boldsymbol{x}_n$ as input, and at each point, the interpolation condition can be written as $f(\boldsymbol{x}_i) = 0$, and $f(\boldsymbol{x})$ can be acquired by interpolating of the constraint points

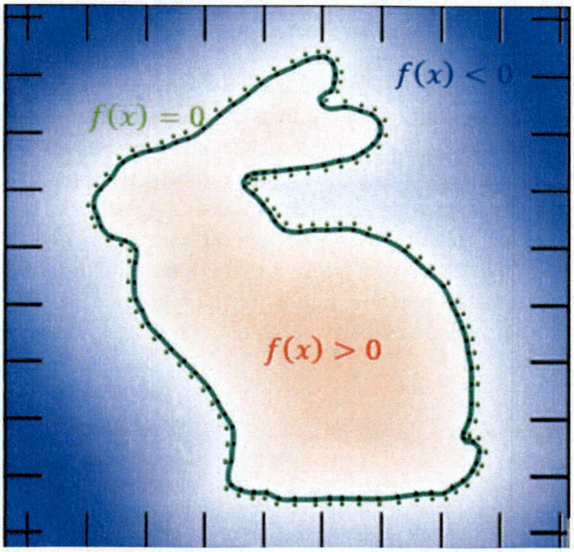

Fig. 2 A color map of function $f(\boldsymbol{x})$ and the 2D curve defined at the zero point set of $f(\boldsymbol{x})$, with constraint points and normal vectors. The generation method is discussed in [4]

via interpolation conditions. Figure 2 shows an example of an interpolated function $f(\boldsymbol{x})$ and an implicit curve defined by $f(\boldsymbol{x}) = 0$. The interior and exterior of the shape take opposite signs and the red part inside the curve shows locations where the function takes positive values and the blue part outside the curve shows locations where the function takes negative values, and the depth of the color is proportional to the absolute value of the function.

Different methods have been proposed to obtain the function $f(\boldsymbol{x})$, and using Radial Basis Functions (RBF) is one of the approaches. The idea is to interpolate scattered data points using RBF as a base function. Turk et al. [4] and Carr et al. [5] comprehensively discussed this approach. Moreover, Nakata et al. [3] and Itoh et al. [6] proposed a different method that is based on piecewise polynomials. In their methods, signed input points are converted into Spline coefficients and a scalar field in unit cube can be represented by means of a liner combination of B-Spline base functions, so that complex shapes with a great number of data points can be generated in a fast speed.

Although different methods to generate both 2D and 3D shapes have been proposed, and researchers have made their best effort to make fast and accurate representations, due to different reasons, the problem of irregular parts has always been a typical issue in computer graphics area, and relatively less work have been done to tackle this issue. This paper gives three methods to allow users interactively correct isolated, redundant and noisy parts.

3 Irregular Parts on Implicitly Represented Shapes

Despite different approaches have been proposed to generate implicit shapes, one of the common issues is that in some situations, irregular parts may be generated unexpectedly. This problem is mainly caused due to defects of input point sets. Fortunately, it is possible to make corrections as post-processing in order to get an expected result as people often have a good understanding to the desired shape and able to identify and mark irregular components so that to make corrections, which is known as user-interactive operations. The aim of this paper is to give respective solutions to perform interactive restorations of different types of irregular parts, including isolated parts, redundant parts and noisy parts.

3.1 Isolated Parts

Isolation means some parts from the surface or curve get separated apart with each other while ideal situation is such parts should connected together. In RBF based reconstruction techniques such as Turk's method [4], it is required to give input data points as well as corresponding normal vectors as input data, the reconstructed result gives good interpolation if densely sampled surface points are given. However, surface points can be separated due to the limitations of scanning and such sparsity often gives unexpected interpolation and sometime produces isolated parts. Another typical situation to get isolation problems is at narrow locations. If the expected curve is thin and long, the distance from a surface point to the opposite point will be shorter than that to its neighbor point. In this situation, the interpolation is not robust because the point will be more likely to connect to the opposite point instead of its neighboring points. Figure 3 gives an illustration to this situation.

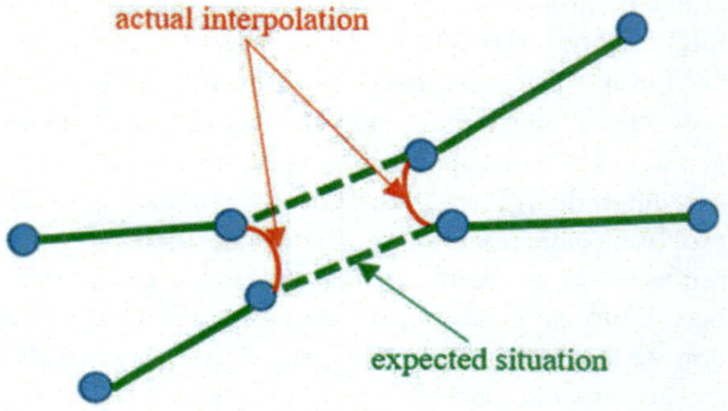

Fig. 3 The situation of isolated interpolation at narrow locations. Constraint points as given as the blue dots. Connections will be made as the red curves shown in the figure rather than the dashed green lines

3.2 Redundant Parts

Redundant parts includes redundant bulges (Fig. 4) and cylindrically extended components (Fig. 5). They often appear as protruding parts extern from the surface in the form of bulges or cylinder shapes. A correctly generated shape requires correct and evenly distributed directions of normal vectors equipped with input constraint points, and when the direction distributions of vectors become irregular unexpectedly, values of the function around corresponding region will be misguided and a larger area of positive or negative values will be produced, and finally get redundant components.

Another reason that cause redundant parts on the generated shape is blanks at input point cloud. Due to defects of scanning process, absence of constraint points become a common situation at locations where the scanning laser is difficult to reach. In this situation, constraint points at the two ends of a blank may not be interpolated straightly, but interpolated with curvature, and produce a redundant bulge or even a component that extends till the end of plot region.

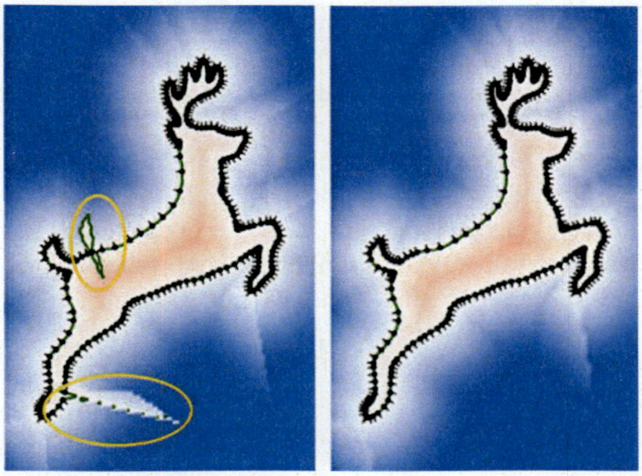

Fig. 4 An implicit curve with redundant parts caused by wrong direction of normal vectors (left) with ideal situation (right). Grid of Polynomial [3] is used for curve generation

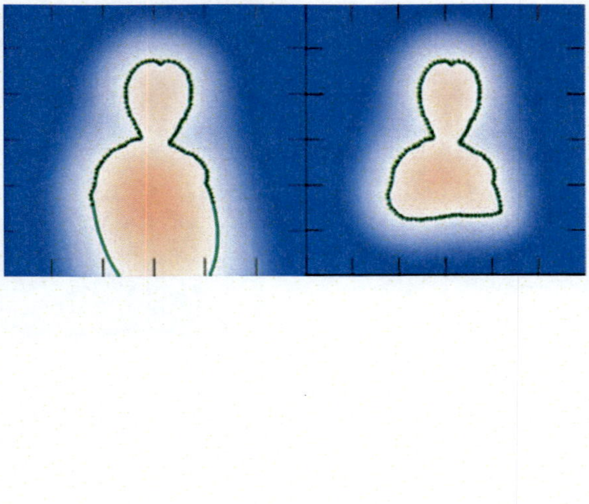

Fig. 5 An implicit curve with redundant parts that extends cylindrically caused by absence of constrain points (left) with ideal situation (right). Turk's method [4] is used for curve generation

3.3 Noisy Parts

A noisy part means a part of the generated shape contain high frequency components. To get a smooth shape, the input constraint points as well as directions of normal vectors have to be evenly distributed. However, due to limitations of scanning, such requirements may not be fully satisfied, and the deployment of points become scattered and fluctuated. In this case, noisy parts are most likely to happen especially among complex structures. Figure 6 (right) gives a typical example of an implicit curve with noisy part.

4 Correction of Irregular Parts

In this section, different methods to correct corresponding type of irregular part from two dimensional implicit curves have been discussed. The conceptual procedure of user-interactive restorations is the user can specify an area by mouse clicking and apply correction algorithms at the specified area. In the proposed methods, such area is defined by a set of points or a rectangular region. Figure 7 shows examples of specified regions defined via points and rectangle respectively.

In principle, the idea to make corrections of irregular parts is to modify the original function $f(\boldsymbol{x})$ and adjust or update some of its values. One of the ways to adjust function values is to add other functions to $f(\boldsymbol{x})$ in order to change function values at expected locations and get the desired shape, and this approach is especially useful to make corrections at scattered areas or change the sign of function values. The other way is to fully update all of the function values within a certain area so that they can be replaced with correct ones which can define a correct shape. Two techniques have been used in this approach, and they are Laplace interpolation applied to discretized functions and Spline interpolation. Laplace interpolation can be used to replace wrong

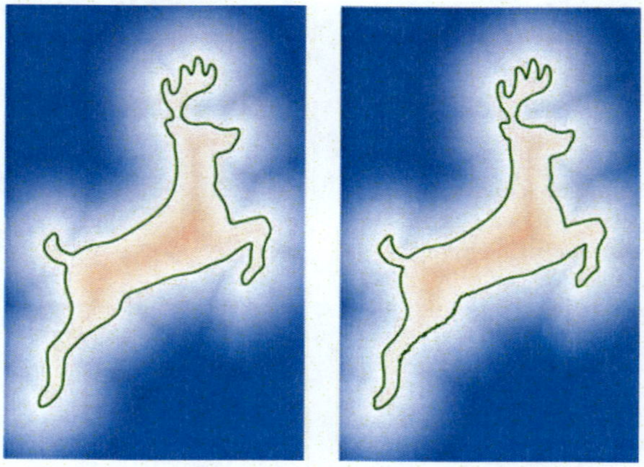

Fig. 6 An example of noisy implicit curve (right) with expected situation (left) generated via 'Grid of Polynomial' [3]

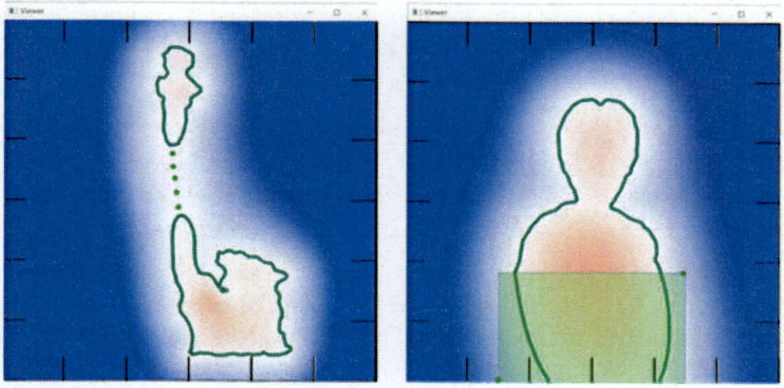

Fig. 7 An illustration of procedure of user-interactive restorations. Left: five points specified in-between two isolated parts. Right: two points specified at two end points of the counter-diagonal of the rectangular region

values after acquire a discrete form of $f(\boldsymbol{x})$ by sampling and Spline interpolation can be used to restore continuous function from updated discrete function. Details of the proposed methods are given in the following part.

4.1 Connecting Isolated Parts

In this section, an algorithm to connect isolated parts has been introduced. As shown in Fig. 8, for an isolated curve defined as $f(\boldsymbol{x}) = 0$, the values of $f(\boldsymbol{x})$ become negative instead of positive in between two isolated parts, and if such values is changed with correct ones (positive values that smooth adapt with surroundings), the isolated curve will be connected. One solution is to add positive-valued functions $g(\boldsymbol{x})$ to original function $f(\boldsymbol{x})$ in the middle of isolated parts, such as Gaussian

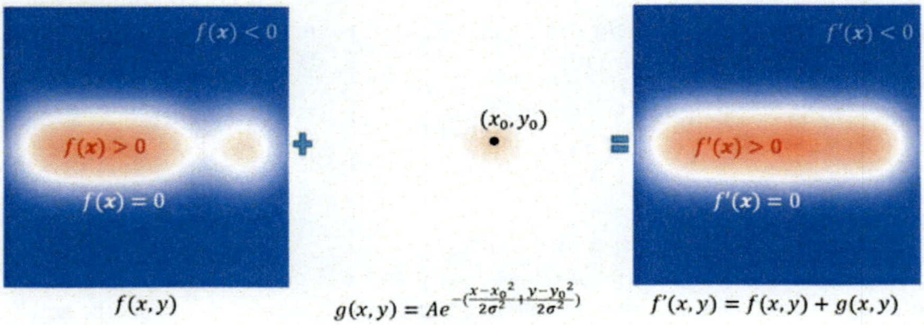

Fig. 8 The strategy of connecting isolated parts. In this case, straight connection is expected

functions, as they take larger values at their main beam and the value far away from the center will be much smaller and cause least damage to the original function. Thus, the output function is represented as

$$f'(\boldsymbol{x}) = f(\boldsymbol{x}) + \sum_{i=1}^{k} g_i(\boldsymbol{x}), \quad g_i(\boldsymbol{x}) = A_i e^{\frac{\|x-x_i\|^2}{2\sigma_i^2}} \tag{1}$$

where $f(\boldsymbol{x})$ is the original function, A_i and σ_i are the amplitude and standard deviation of each Gaussian function and $\boldsymbol{x}_i$ are the center of each Gaussian function as well as the location of each user-specified point. The procedure is shown in Fig. 8.

It is essential to determine appropriate amplitude and standard deviation of each Gaussian function. In the proposed method, they are determined based on users' manual specify. Users are able to decide the number and location of Gaussian functions, the amplitudes as well as the standard deviations based on the expected shape of restoration. In conclusion, Gaussian functions are used to fill the gap between isolated parts in the proposed method. To make a successful restoration, it is expected that the user can specify the locations based on the following rule:

- The locations, amplitudes and standard deviations are completely decided by users.
- The users can keep trying different parameters until obtain appropriate curves.
- If users need thin connection, they should put a larger number of points, and set amplitudes and standard deviations smaller.
- If users need thick connection, they should put a smaller number of points, and set amplitudes and standard deviations larger.

4.2 *Remove Redundant Parts*

In this section, an algorithm to remove redundant parts has been introduced. As shown in Fig. 9, the strategy to remove redundant parts is to mask function values around redundant parts, then replace with new functions that are continuously and smoothly connect to their neighbor values. Thus, the aim of restoration is to output the result as $f'(\boldsymbol{x})$ in the form of

$$f'(\boldsymbol{x}) = \begin{cases} g(\boldsymbol{x}) \ (x \in \Omega) \\ f(\boldsymbol{x}) \ (x \notin \Omega) \end{cases}, \tag{2}$$

where region Ω is the user-specified region.

An important requirement is continuously and smoothly adaption, one of the solutions is to solve a boundary problem of Laplace's equation. We expect the newly generated function satisfy the condition defined as Eq. (3) while taking neighbor values into consideration as boundary condition, so that it will be able to achieve

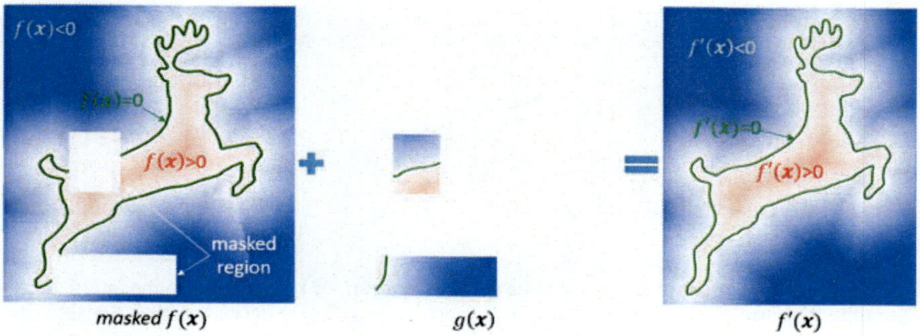

Fig. 9 Strategy to remove redundant parts

smooth adaption.

$$\begin{cases} \nabla^2 g(x, y) = \frac{\partial^2 g(x,y)}{\partial x^2} + \frac{\partial^2 g(x,y)}{\partial y^2} = 0 \ (x \in \Omega) \\ \nabla^2 g(x, y) = f(x, y) \qquad\qquad\qquad (x \in \partial\Omega) \end{cases}. \tag{3}$$

where $\partial\Omega$ is the boundary of the specified region and Ω is the specified region.

Therefore, In order to retrieve the function $g(\boldsymbol{x})$ within the specified region, the following procedures have been performed.

- The user specify a rectangular region as the specified region to apply correction within the region.
- To perform sampling of $f(\boldsymbol{x})$ at grid points (the number of grid points can be specified by the user) and obtain f_{ij} as a discrete form of $f(\boldsymbol{x})$.
- To determine new values g_{ij} within the specified region as the solutions of discrete Laplace's equation based on surrounding boundary values via Eq. (4) which is acquired as a discrete form of finite difference approximations of Eq. (3) $(x \in \Omega)$.

$$g_{i,j} = \frac{g_{i-1,j} + g_{i+1,j} + g_{i,j-1} + g_{i,j+1}}{4}. \tag{4}$$

- As the discrete function becomes f_{ij} outside the specified region and g_{ij} inside the specified region, the final step is to obtain the final function, $g(\boldsymbol{x})$ as the Spline interpolation of the combined discrete values, where Quadratic B-spline function has been used as base function of Spline interpolation.

Figure 10 illustrates the whole procedure.

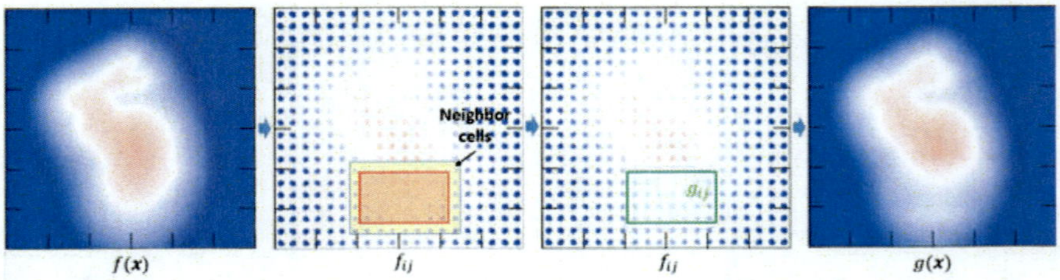

Fig. 10 Illustration of the procedure of remove redundant part via Laplace and Spline interpolation. The red and green rectangular region shows the specified correction region where red region indicates cells before correction and green region indicates cells after correction, and neighbor cells are considered the cells that adjacent to the specified region

4.3 Noise Reduction

In this section, an algorithm to remove noisy parts has been introduced. The idea is to perform low resolution sampling to original function $f(\boldsymbol{x})$ and get $f(\boldsymbol{x}_i)$, where $i \in [1, n_1]$ and n_1 is sampling resolution adjustable by the user. The output result is considered as the Spline interpolation of the discrete samples $f(\boldsymbol{x}_i)$. Neighbor values is also considered in order to give a final result that achieve continuously and smoothly adaption at the boundary of the specified region. Thus, the output function $g(\boldsymbol{x})$ can be written as Eq. (5).

$$g(\boldsymbol{x}) = \sum_{i=-1}^{n_1} \sum_{j=-1}^{n_2} c_{ij} B_0(x - i) B_0(y - j). \tag{5}$$

where n_1 and n_2 is the number of knots (discrete samples and neighbor cells) at x-axis and y-axis respectively; c_{ij} are the Spline coefficients and $B_0^Q(t)$ is the Quadratic Base Function of Spline interpolation, which is defined as Eq. (6).

$$B_0^Q(t) = \begin{cases} \frac{t^2}{2} + t + \frac{1}{2} & (-1 \le t < 0) \\ -t^2 + t + \frac{1}{2} & (0 \le t < 1) \\ \frac{t^2}{2} - 2t + 2 & (1 \le t < 2) \\ 0 & otherwise \end{cases} \tag{6}$$

The Spline coefficients can be solved by solving a liner equation defined by interpolation conditions as Eq. (7), where $\boldsymbol{x}_i$ are coordinates of knots.

$$g(\boldsymbol{x}_i) = f(\boldsymbol{x}_i) \tag{7}$$

In conclusion, the procedure to remove noise part is described as follow.

- The user specify a rectangular region.

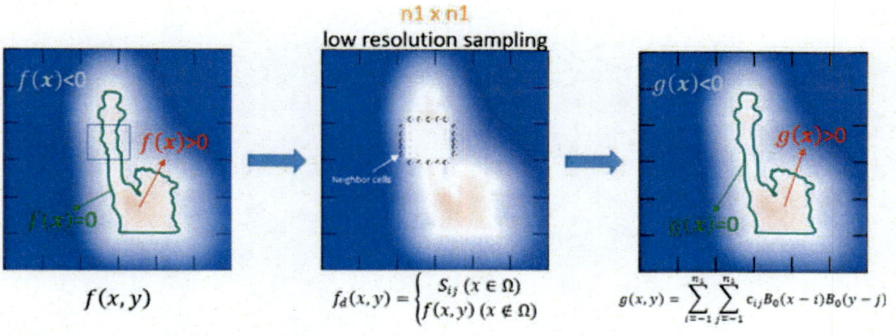

Fig. 11 Illustration of the procedure to remove noisy parts. Where the rectangular region shows the specified region, $f_d(x, y)$ is the set of discrete samples of $f(x, y)$ and $g(x, y)$ is the interpolated output, and n_1 denotes sampling resolution

- Low resolution sampling in performed inside the region.
- Find the neighbor cells around the region.
- The output is the Spline interpolation of the discrete samples and neighbor cells.

The full procedure is illustrated in Fig. 11.

5 Restoration Results

In this section, evaluation results of the proposed methods have been presented. We use three sets of points to generate test examples, and they are "megami" (163 data points), "ritsbunny" (109 data points) and "roman" (136 data points). We construct implicit curves using Turk's method that is presented in [4]. The proposed algorithms have shown that they are effective to address their respective issue without serious delay.

5.1 Evaluation of Isolate Connection

A point set that is separated apart has been made as test example shown in Fig. 12.

Restoration examples can be seen in Fig. 13. Based on the desired shape of connection, the user can adjust parameters (amplitudes and standard deviations) and specify locations to put Gaussian functions as desired, and in the test examples, locations of Gaussian function are shown at the green dots. Different shapes of restoration are shown in Fig. 13a–d refer to "thin connection", "straight connection", "bulky connection" and "bend connection", based on the rule that described in Sect. 4.1.

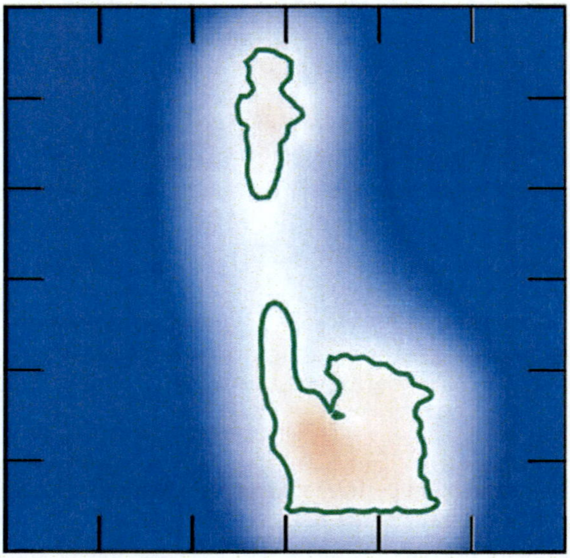

Fig. 12 Test example of isolated curve "megami". The shape of connection can be determined by the user

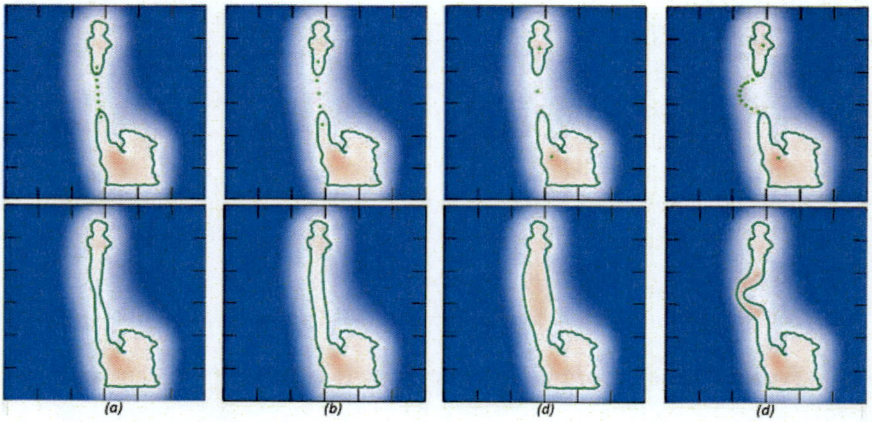

Fig. 13 Test results of connection of isolated curves with different shapes

5.2 Evaluation of Redundant Removing

Two examples have been created as shown in Fig. 14 upper and lower row as a redundant bulge and a cylindrically extended part respectively.

Test examples to remove different kinds of redundant parts have been summarized in Fig. 15, where the example to remove redundant bulge, cylindrically extended part,

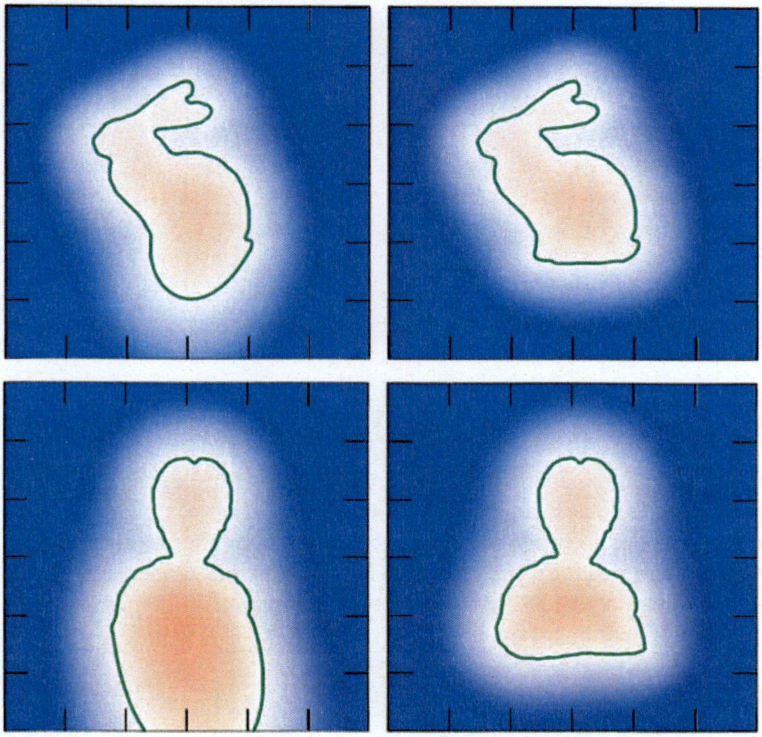

Fig. 14 Test example "ritsbunny" and "roman" shown at upper row and lower row respectively. The left figure show irregular situation while the right figure show expected situation

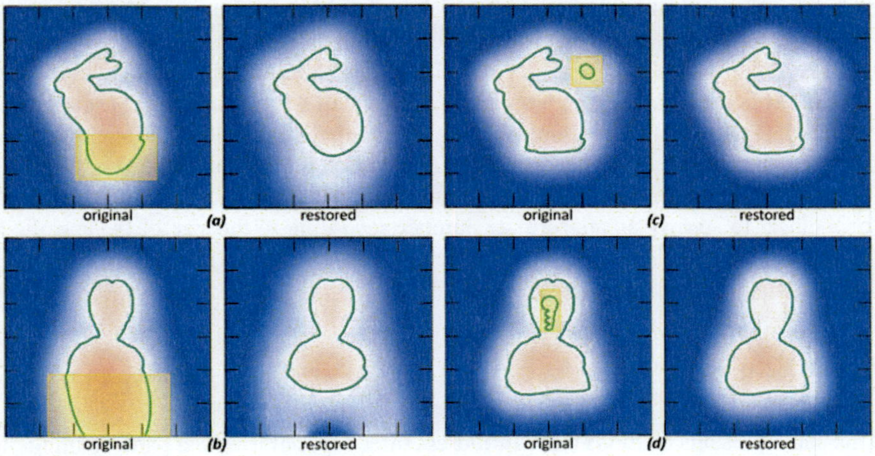

Fig. 15 Examples of removing redundant part. The rectangular region shows the user-specified correction region

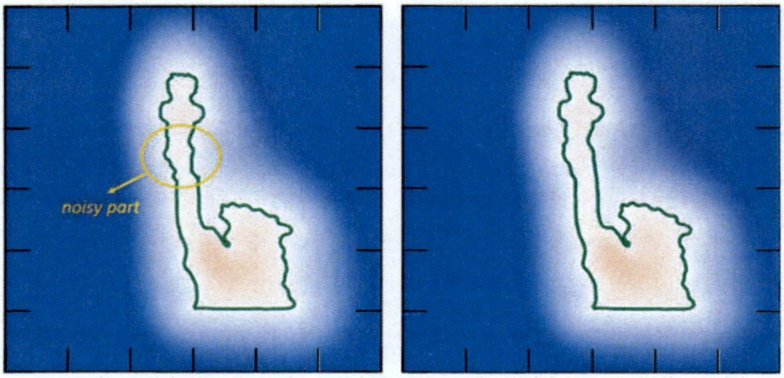

Fig. 16 A noisy curve "megami" (left) with expected result (right)

redundant part that outside the curve (isolated-redundant part) and redundant part that inside the curve has been shown in a–d respectively.

It is seen that the proposed method is able to remove different sizes of redundant component at any locations. However, due to the nature of Laplace interpolation, the redundant part may not be fully removed, and this situation is not a perfect restoration if the user wants a straight re-connection.

5.3 *Evaluation of Noise Reduction*

A point set that with random displacement of normal vectors has been created in order to make an example of implicit curve with noisy part shown in Fig. 16.

Results of restored curves restored from different numbers of discrete knot in low resolution samples can be seen in Fig. 17. This number can be adjusted by the user. It is seen that setting a smaller number will get a smoother restoration as expected. However, if this parameter is set too small, the restoration will fail because the discrete samples cannot carry enough information of the shape of the curve while if this parameter is set too large, smoothing effect will be weak. Thus, it is required for the user to specify an appropriate number according to the size of the specified region.

6 Conclusion

This paper presents methods that can address three types of irregular parts on implicit curves. A method that to add Gaussian functions has been used to connect isolated parts, Laplace solver has been used to replenish values of the function within a specified region that has been masked in order to remove redundant parts and low

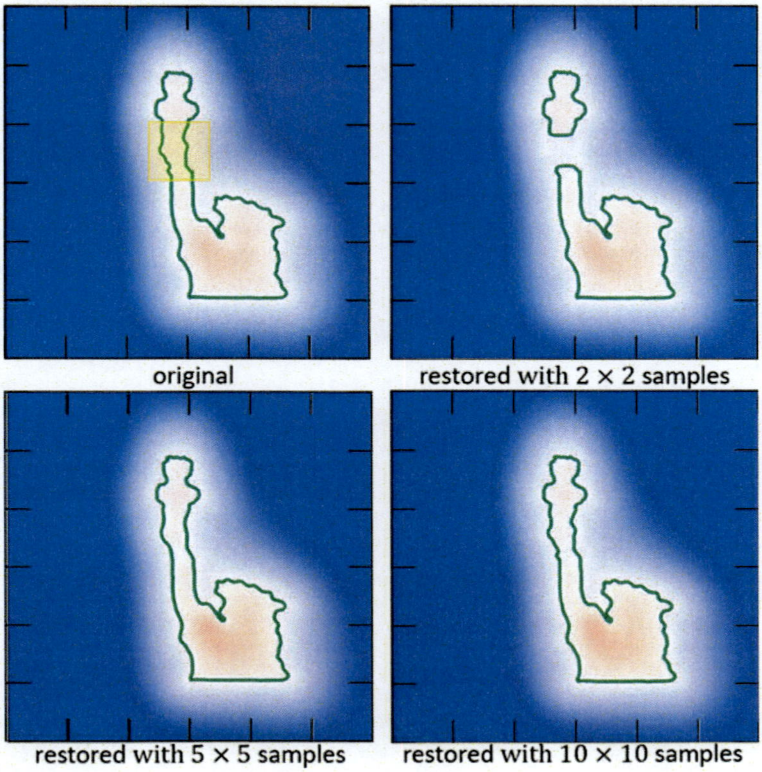

Fig. 17 Examples of removing noisy components with different sampling resolutions. The rectangular region shows the user-specified correction region

resolution sampling has been used for de-noise. Finally, Spline interpolation has been used to restore continuous function after obtain a discrete solution. Evaluations have been made in order to test the robustness of the proposed methods and results of evaluations have seen that they are capable to accomplish their respective objective. However, limitations of the proposed methods do exist as well, the common limitation of all proposed methods is that they are only tested in implicitly represented 2D shapes. In the future, it is expected that they will able to be implemented in correction of 3D shapes. In addition, flat restoration is difficult to achieve by the proposed method when removing redundant parts.

References

1. Huang, Z., Carr, N., Ju, T.: Variational implicit point set surfaces. ACM Trans. Graph. **38**(4), 13p, Article 124 (2019)

2. Liu, Y., Song, Y., Yang, Z., Deng, J.: Implicit surface reconstruction with total variation regularization. Comput. Aided Geom. Des. **52–53**, 135–153 (2017)
3. Nakata, S., Aoyama, S., Makino, R., et al.: Real-time isosurface rendering of smooth fields. J. Vis. **15**, 179–187 (2012)
4. Turk, G., O'Brien, J.F.: Modelling with implicit surfaces that interpolate. ACM Trans. Graph. **21**(4), 855–873 (2002)
5. Carr, J.C., Beatson, R.K., Cherrie, J.B., Mitchell, T.J., Fright, W.R., McCallum, B.C.: Reconstruction and representation of 3d objects with redial basis functions
6. Itoh, T., Nakata, S.: Fast generation of smooth implicit surface based on piecewise polynomial. CMES Comput. Model. Eng. Sci. **107**(3), 187–199 (2015)

Numerical Simulation Research on Seal Failure of Remedial Cement Sheath in Oil and Gas Wells

Jiwei Jiang, Zhixue Chen, Weiwei Hao, Jun Li, Yan Xi, Wenbao Zhai, Xuefeng Chen, and Bo Li

Abstract Due to insufficient cement return height during cementing operations, there may be a channel for oil and gas leakage between the wellbore and the formation, causing environmental problems such as groundwater pollution. Therefore, it is necessary to reinject the cement slurry into the original wellbore annulus and evaluate the sealing performance of the remedial cement sheath. However, the current theoretical and applied research on this aspect is relatively lacking. In response to this, a new model for calculating elastoplastic strain of the remedial cement sheath/original cement sheath based on damage mechanics was established, and the plastic deformation behavior of the interface area was studied. The seal failure mechanism of the wellbore remedial cement sheath is clarified, and the influence of the mechanical parameters of the remedial cement sheath on the cumulative plastic strain and the micro-annulus of the remedial cement sheath-casing interface is analyzed. The results show that under the combined action of the internal pressure of the casing and the non-uniform ground stress, the cumulative plastic deformation of the inner wall of the remedial cement sheath and the micro-annular gap at the interface are the main reasons for the sealing failure of the remedial cement sheath. Increasing the Poisson's ratio, cohesion strength and internal friction angle of the remedial cement sheath and reducing the elastic modulus of the remedial cement sheath, is helpful to reduce the accumulated plastic strain and improve the sealing performance of the remedial cement sheath.

Keywords Abandoned wellbore · Remedial cement sheath · Plastic deformation · Seal failure · Numerical simulation

---

J. Jiang (✉) · Z. Chen · W. Hao · W. Zhai · X. Chen · B. Li
CNPC Engineering Technology R&D Company Limited, Beijing 102206, China
e-mail: tianya0603@qq.com

J. Li
China University of Petroleum-Beijing, Beijing 102249, China

Y. Xi
Beijing University of Technology, Beijing 100124, China

H. Dai (ed.), *Computational and Experimental Simulations in Engineering*, Mechanisms and Machine Science 119,
https://doi.org/10.1007/978-3-031-02097-1_14

1 Introduction

Due to the problem of insufficient cement return height during cementing operations, there may be oil and gas leakage channels between the wellbore and the formation in the later development process, causing environmental problems such as groundwater pollution. Therefore, it is necessary to reinject the cement slurry into the original wellbore annulus and evaluate the sealing performance of the cement slurry. However, the current theoretical and applied research on this aspect is relatively lacking. After refilling the original wellbore with cement sheath, if it is developed and used again, it will undergo a series of integrity tests (such as pressure test, leakage test and integrity test, etc.), perforation operations and stimulation operations, etc. [1], higher pressure changes may be experienced in the wellbore. Related literature studies have shown that a large change in the internal pressure of the casing may cause plastic strain in the cement sheath, produce micro-annulus, debond from the outer wall of the casing, and lose its sealing performance [2–8]. Therefore, it is of great significance to study the elastoplastic changes of the remedial cement sheath and the original cement sheath under the change of internal pressure, and reveal the sealing failure mechanism of the remedial cement sheath.

Based on the basic theory of damage mechanics, this paper establishes the elastoplastic strain calculation model of the remedial cement sheath/original cement sheath, studies the sealing failure mechanism of the remedial cement sheath, and analyzes the effect of the mechanical parameters of the remedial cement sheath on the cumulative plastic strain and the remedial cement sheath/original cement sheath. The influence law of the micro-annular gap at the cement sheath-casing interface provides a reference for the study of the seal failure mechanism of the wellbore remedial cement sheath.

2 Elastic–Plastic Model of Wellbore Remedial Cement Sheath

2.1 Establishment of the Elastic–Plastic Model of the Remedial Cement Sheath

In order to minimize the influence of boundary effects on the simulation results and balance the calculation accuracy and model running time, the overall size of the model is set to 2 m × 2 m × 5 m, and the length of the remedial cement sheath and the original cement sheath are both 2.5 m (shown in Fig. 1) [9, 10]. The model assumes that the casing is a linear elastic material, the original cement sheath, the remedial cement sheath and the formation rock are elastoplastic materials, and the remedial cement sheath and the original cement sheath have different mechanical parameters, as shown in Table 1.

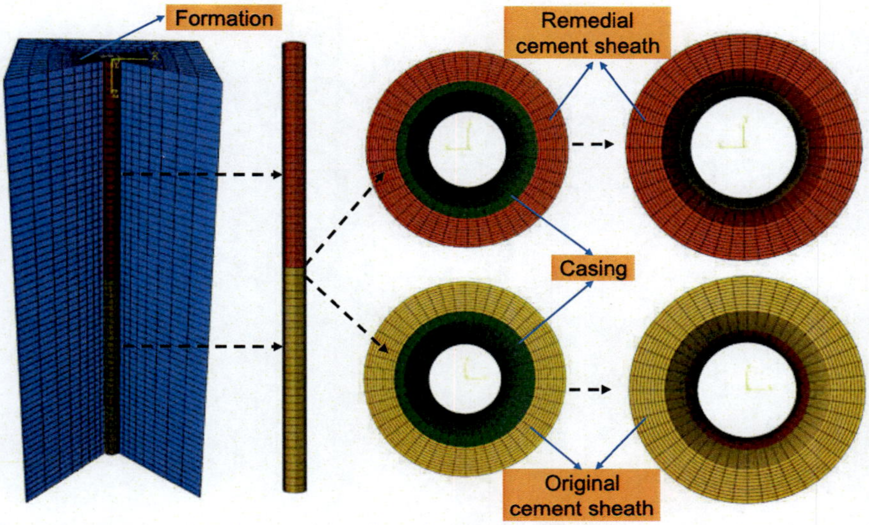

Fig. 1 Model of casing-remedial/original cement sheath-formation

The 3000–3005 m well section of a simulated well is selected as the research object. The overlying strata at the location of this well section is 45.8 MPa, the maximum horizontal stress is 52.4 MPa, and the minimum horizontal stress is 48.6 MPa.

2.2 *Criteria for Judging the Failure of the Remedial Cement Sheath*

The friction and vertical contact between the two contact surfaces are considered in the model building process. For the friction surface and the vertical contact surface, the Coulomb friction law and the hard contact based on the penalty function are used respectively. The interface setting between the remedial cement sheath, the original cement sheath and the casing adopts the cohesive failure mode to characterize the mechanical damage degree of the remedial cement sheath/original cement sheath-casing interface. The Mohr–Coulomb failure model is used as the criterion for judging the failure of the remedial cement sheath, the original cement sheath and the formation [11]:

$$\frac{1}{2}(\sigma_\theta - \sigma_r) + \frac{1}{2}(\sigma_\theta + \sigma_r)\sin\varphi = C\cos\varphi \tag{1}$$

Table 1 Model parameters of casing-remedial/original-cement sheath -formation

Name	Outer diameter/mm	Elastic modulus/GPa	Poisson's ratio	Internal friction Angle/°	Cohesive strength/MPa	Critical normal strength/MP	Critical shear strength/MPa	Critical fracture energy/(J m^{-2})
Casing	177.8	210	0.3	\	\	4.5	0.2	100
Remedial cement sheath	244.5	5.6	0.25	24	7.6			
Original cement sheath	244.5	7.8	0.17	27	8.0			
Formation	\	23	0.32	30	5.0			

where φ is the internal friction angle of the remedial cement sheath/original cement sheath, °; C is the cohesion strength of the remedial cement sheath/original cement sheath, MPa; σ_θ, σ_r are the circumferential stress and radial stress of the remedial cement sheath/original cement sheath, MPa.

3 Seal Failure Mechanism of Wellbore Remedial Cement Sheath

Use the prestress field function to apply ground stress to the entire model. The casing internal pressure is equal to the hydrostatic column pressure plus the construction pressure, minus the friction loss along the wellbore, and the casing internal pressure is applied to the inner surface of the casing. The model calculation process takes into account the variation of non-uniform ground stress, casing internal pressure and formation mechanical properties with well depth.

The internal pressure of the casing continues to increase with the continuous increase of the well depth. Therefore, in order to analyze the influence of the casing internal pressure on the accumulated plastic strain, the casing internal pressure was changed while keeping the ground stress constant. Figure 2 shows the cumulative plastic deformation changes under different internal pressures. It can be seen from the Fig. that as the internal pressure of the casing increases, the cumulative plastic strain of the remedial cement sheath and the original cement sheath both increase. The cumulative plastic strain of the remedial cement sheath is greater than the cumulative plastic strain of the original cement sheath, and the cumulative plastic strain of the remedial cement sheath increases faster than the cumulative plastic strain of the original cement sheath. The results show that the deeper the well, the larger the width of the micro-annulus, and the refill cement sheath is significantly affected by the casing internal pressure, and plastic failure is more likely to occur.

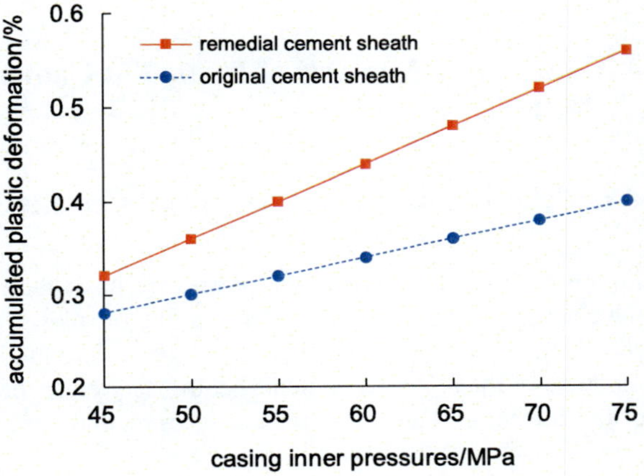

Fig. 2 Variation of accumulated plastic deformation under different casing inner pressures

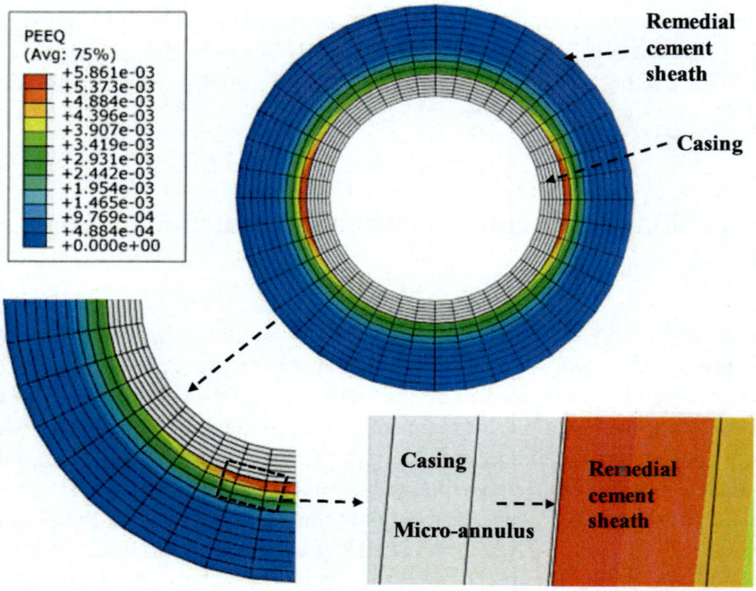

Fig. 3 Accumulated plastic deformation and micro-annulus at the interface between remedial cement sheath and casing

Figures 3 and 4 show the cumulative plastic deformation and micro-annulus of the remedial cement sheath-casing interface and the original cement sheath-casing interface. Due to the combined effect of casing internal pressure and non-uniform ground stress, there are accumulated plastic strains on the inner wall of the remedial cement sheath and the inner wall of the original cement sheath, and micro-annuluss are generated at the interface, which is more likely to form a gas channeling channel from the bottom of the well to the wellhead.

4 Sensitivity Analysis of Remedial Cement Sheath Seal Failure

4.1 Elastic Modulus of Remedial Cement Sheath

Figure 5 shows the cumulative plastic strain of the remedial cement sheath with the change of elastic modulus. The cumulative plastic strain of the remedial cement sheath increases with the increase of the elastic modulus. The greater the elastic modulus of the remedial cement sheath, the greater the possibility of micro-annulus at the interface. Therefore, the use of remedial cement sheath with a low elastic

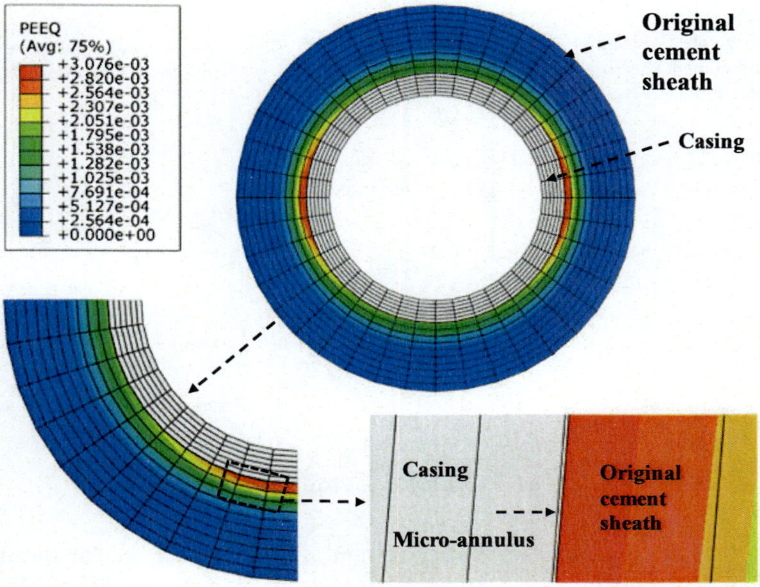

Fig. 4 Accumulated plastic deformation and micro-annulus at the interface between original cement sheath and casing

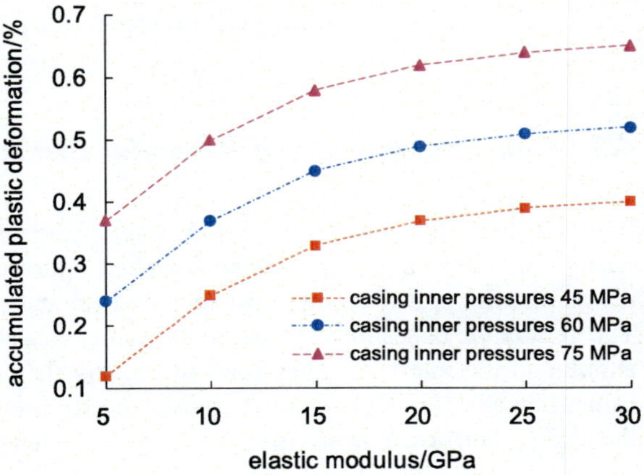

Fig. 5 Variation of cumulative plastic strain of remedial cement sheath with elastic modulus

modulus is beneficial to reduce the accumulated plastic strain and improve the sealing performance of the refill cement sheath.

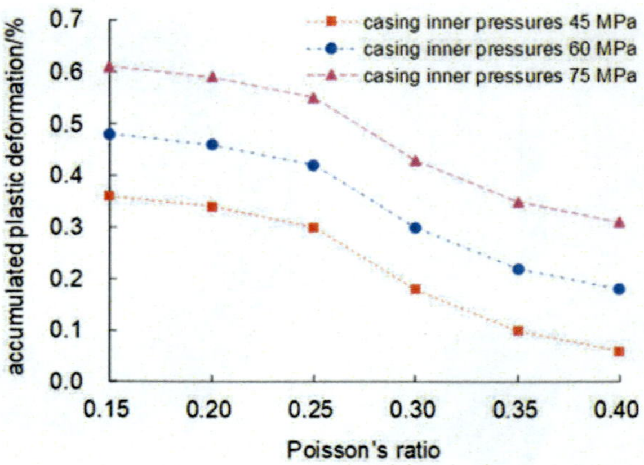

Fig. 6 Variation of cumulative plastic strain of remedial cement sheath with Poisson's ratio

4.2 *Poisson's Ratio of Remedial Cement Sheath*

Figure 6 shows the cumulative plastic strain of the remedial cement sheath with the change of Poisson's ratio. It can be seen from the figure that the cumulative plastic strain of the remedial cement sheath decreases with the increase of Poisson's ratio. The larger the Poisson's ratio of the remedial cement sheath, the smaller the possibility of micro-annulus at the interface. Therefore, the use of remedial cement sheath with a high Poisson's ratio is beneficial to reduce the accumulated plastic strain and maintain the sealing performance of the remedial cement sheath.

4.3 *Cohesive Strength of Remedial Cement Sheath*

Figure 7 shows the cumulative plastic strain of the remedial cement sheath with the change of cohesion. It can be seen from the figure that as the cohesion increases, the cumulative plastic strain of the remedial cement sheath is significantly reduced. The greater the cohesion of the remedial cement sheath, the lower the risk of micro-annulus at the interface. Therefore, increasing the cohesive force of the remedial cement sheath is an effective way to reduce the accumulated plastic strain and enhance the sealing performance of the remedial cement sheath.

4.4 *Internal Friction Angle of Remedial Cement Sheath*

In the Mohr-Columb criterion, as the internal friction angle increases, the friction coefficient increases, and the shear strength of cement increases with the increase in mechanical load. Figure 8 shows the cumulative plastic strain of the remedial

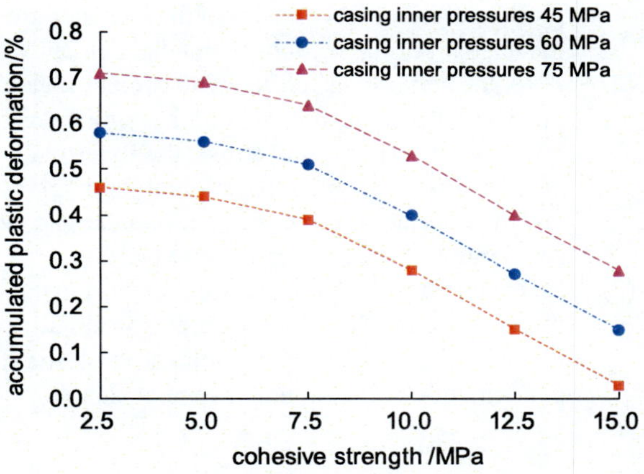

Fig. 7 Variation of cumulative plastic strain of remedial cement sheath with cohesive strength

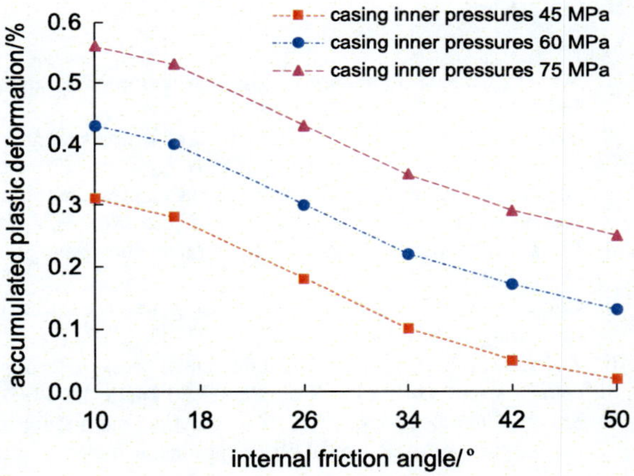

Fig. 8 Variation of cumulative plastic strain of remedial cement sheath with internal friction angle

cement sheath with the internal friction angle. With the increase of internal friction, the cumulative plastic strain of the remedial cement sheath is significantly reduced. The larger the internal friction angle of the remedial cement sheath, the lower the risk of micro-annulus at the interface.

5 Conclusions and Recommendations

(1) Based on the principle of damage mechanics, the finite element method is used to establish a new model for calculating elastoplastic strain of the remedial cement sheath/original cement sheath based on damage mechanics. The

plastic deformation behavior of the interface area is studied, and the seal failure mechanism of the wellbore remedial cement sheath is determined.

(2) After the wellbore is refilled with cement sheath, in the process of being developed and utilized again, under the combined action of casing internal pressure and non-uniform ground stress, the inner wall of the remedial cement sheath and the original cement sheath both undergo cumulative plastic deformation, and the interface produces micro-annular gap, which is the main reason for the seal failure of the remedial cement sheath.

(3) Increasing the Poisson's ratio, cohesion and internal friction angle of the remedial cement sheath, reducing the elastic modulus of the remedial cement sheath, is conducive to reducing the cumulative plastic strain and improving the sealing performance of the remedial cement sheath.

References

1. Wei, T.: Cementing quality logging evaluation of oil and gas wells, pp. 227–228. Petroleum Industry Press, Beijing (2010)
2. Yao, X., Zhou, B., Li, M., et al.: Effect of casing pressure testing on the sealing performance of cement sheath. J. Xi'an Shi you Univ. (Nat. Sci. Ed.) **24**(3), 31–34 (2009)
3. Xia, Y., Liu, A.: Research on effect of casing pressure test to cement-casing interface hydraulic sealing property. Drill. Fluid Complet. Fluid **28**(b11), 4–6 (2011)
4. Li, H., Wang, R.: Analyses of effect of factors on casing pressure test. Petrol. Drill. Tech. **22**(4), 44–46 (1994)
5. Chen, Z.: Analysis of influences of casing pressure testing on acoustic amplitude logging after cementing. Petrol. Drill. Tech. **31**(3), 36–37 (2003)
6. Zhou, B., Yao, X., Su, D.: The influence of pressure testing of casing string on the integrity of cement sheath Drill. Fluid Complet. Fluid **26**(1), 32–34 (2009)
7. Shi, Y., Guan, Z., Xi, C., et al.: An analytical method for the calculation of allowable internal casing pressure based on the cement sheath integrity analysis. Nat. Gas Ind. **37**(7), 89–93 (2017)
8. Chu, W., Shen, J., Yang, Y., et al.: Calculation of micro-annulus size in casing-cement sheath-formation system under continuous internal casing pressure change. Pet. Explor. Dev. **42**(3), 379–385 (2015)
9. Fan, M., Liu, G., Li, J., et al.: Effect of cementing quality on casing stress of shale gas well under heat-mechanical coupling. China Petrol. Mach. **44**(8), 1–5 (2016)
10. Guo, X., Li, J., Liu, G., et al.: Influence of cement sheath defect on casing stress under temperature-pressure effect. China Petrol. Mach. **46**(4), 112–118 (2018)
11. Xu, H., Zhang, Z., Shi, T., et al.: Influence of the WHCP on cement sheath stress and integrity in HTHP gas well. J. Petrol. Sci. Eng. **126**, 174–180 (2015)

Intelligent Recognition of Waterline Value Based on Neural Network

Kun Zhang, Chaoran Kong, Fuquan Sun, Chenglong Cong, Yue Shen, and Yushan Jiang

Abstract The accuracy of waterline recognition will affect the safety and shipping efficiency of the vessel. Due to weather conditions, observation experience, obstacles and other subjective and objective factors, the results of waterline recognition are inaccurate, which will lead to a series of problems. For intelligent recognition of waterline, a concise convolutional neural network with batch normalization transformation is proposed. The method uses drone to obtain image data of ship waterline. According to image features, gray level conversion, image enhancement and shape processing are carried out to reduce the influence of irrelevant background information. The network automatically extracts high-dimensional features with less convolution layer overlays. Parts of the network introduce batch normalization, and the training samples are normalized in small batch and then output through fully connection layer. An evaluation mechanism is introduced in the network to automatically recognize the ship waterline value by summarizing all the detected waterline data in the video. The experimental results show that the method can quickly identify the waterline image. Our proposed method is not affected by subjective factors. It has strong environmental adaptability and high recognition accuracy. The results are more objective.

Keywords Waterline · Drone video · Image enhancement · Neural network · Batch normalization

K. Zhang (✉) · C. Kong · F. Sun · C. Cong · Y. Jiang
School of Mathematics and Statistic Technologies, Northeastern University at Qinhuangdao, Hebei 066004, China
e-mail: zkhbqhd@neuq.edu.cn

F. Sun
e-mail: sunfuquan@neuq.edu.cn

Y. Jiang
e-mail: jiangyushan@neuq.edu.cn

Y. Shen
Company of YSUSOFT Information System at Qinhuangdao, Hebei 066004, China
e-mail: 1771566@stu.neu.edu.cn

H. Dai (ed.), *Computational and Experimental Simulations in Engineering*, Mechanisms and Machine Science 119,
https://doi.org/10.1007/978-3-031-02097-1_15

1 Introduction

Waterline recognition is an important basis for safety assessment of ship weight and ship stowage. There are several water gauge calibration lines on the hull indicating the ship draft [1, 2]. The weight of the cargo loaded on the ship can be obtained by measuring the draft value of the ship. There are many objective factors affecting the accuracy of the ship's weighing results. The recognition accuracy of the ship's waterline is one of the most important factors. Especially for large ships, per centimeter of the ship's water gauge scale represents tens or even hundreds of tons of weight. The accuracy of waterline recognition will affect the safety and shipping efficiency of the vessel. It is necessary to provide a more scientific, reasonable and accurate measurement method of ship draught value to overcome the influence of natural environment and other factors.

At present, it mainly relies on experienced observer to obtain the ship draft value, which means it mainly relies on visual observation to get the approximate value. In order to get a more accurate value, the observer needs multiple observations and averaging, however, this method still affected by subjective factors and has limitations. For example, it is difficult to read the value in the windy waters. What's more, different viewing angles will cause errors too. At the same time, observers carrying out outboard operations, working at heights, etc. may result in the risk of casualties. To overcome the shortcomings of manual detection, ultrasonic measurement [3–6] calculates the distance from the main deck to the water surface by measuring the echo return time of the ultrasonic, but practice as shown that the speed of ultrasonic is easily affected by sound velocity error, air density, humidity, temperature, real time changes in waves and so on. Those factors will lead to measurement accuracy errors. The laser ranging method [7–9] is similar to the ultrasonic measurement, the round trip time of laser is measured to calculate the actual draft of the ship, but they are susceptible to floating objects and water waves. The pressure sensing method [10, 11] is based on the relationship between water pressure and water depth to measure the ship draft, due to the density of water quality in different sea areas, the measurement of ship draft will be greatly disturbed, resulting in certain errors.

Compared with the above methods, the image detection method [12] has certain advantages. It is what you see is what you get measurement method. The experiential random error can be avoided by image recognition. Traditional image processing methods such as edge detection, morphological enhancement and image segmentation, etc. can assist image recognition. Especially, with the development of deep learning technology, intelligent recognition of ship waterline images [13–15] becomes possible. The rapid development and wide application of image processing technology provide a new idea and method for ship waterline recognition.

1.1 Deep Learning

Deep learning [16] has been widely applied to computer vision, target recognition and natural language processing, etc. Convolution Neural Network (CNN) is applied in computer vision and works well in object recognition. Convolutional neural networks do not require manual extraction of features that are easy to train and can achieve good results [17, 18]. Classic convolutional neural networks are ALexNet [19], VGGNet [20], Google Inception Net [21], and ResNet [22]. These four networks have appeared one after another, the depth and complexity are also progressive. They won the classification project competition championship of the ILSVRC (Imagenet Large Scale Visual Recognition Challenge) respectively. In recent years, the main breakthroughs are in deep learning and convolutional neural networks. The dramatic increase in performance is almost accompanied by a deepening of the number of layers of convolutional neural networks.

Researchers and developers can design neural network structures for research, testing, deployment and even utility. The rapid development of deep learning and the wide application in image processing provide a new idea and method for ship waterline recognition. There are several successful experiences of deep learning network in classification and recognition tasks [23, 24] illustrated that deep learning technique can improve the objectivity and accuracy of ship waterline measurement.

1.2 Contribution of This Paper

The main contributions of this paper are summarized as follows:

(1) The training data is collected from the actual port environment, and the ship waterline data is collected by drone. The data is labeled according to the standard of the waterline reading of the ship. In order to enhance the generalization performance and accuracy of the model, image data are deformed and enhanced. Finally, the model is trained with labeled data after pretreatment.

(2) A scale classification convolutional neural network is proposed, which uses less convolution layer stacking to eliminate the impact of a large number of computations on the overall performance of the network. The network can not only capture the feature information of different types of images to enhance the discriminability of samples but also improve the computing speed of the network.

The rest of this paper is organized as follows: in Sect. 2, data collection and preprocessing are described. This section includes video frame selection and image deformation and enhancement. Section 3 introduces the standard for ship waterline data labeling and describes the construction of the proposed waterline intelligent recognition model and the calculation of the final recognition results. In Sect. 4 the accuracy of the method is verified by experiments, and the performance of the method is analyzed from several angles. The fifth section summarizes the paper and puts forward suggestions for application and extension.

2 Data Acquisition and Preprocess

In order to realize the intelligent recognition of waterline by using neural network method, we use a supervised training method [25] that is using the labeled data to train the neural network to achieve the waterline identification. The quality and quantity of training data is very important. So we need to do the following data acquisition and preprocess work.

2.1 Training Set Acquisition

The deep learning approach requires massive data to train its neural networks, and the amount of data impacts greatly on the quality of the classifiers [26]. For the ship waterline recognition task, thousands of images are required for each sample, so sample collection is a hard working. In order to collect data safely and quickly, we use drone aerial photography to obtain a large number of waterline video. Considering the actual scene conditions, we use drones to collect the scales and adapt different ships and different positions of the hull corresponding to avoid class imbalance problem [27].

These video frame images cannot be directly used as training images, because different shooting light, distance, angle, color of the ship, scale shape of the waterline, etc. will influence recognition results. Due to the drone operator control errors, windy weather, obstacles appear and other factors, there will be many invalid images captured in the aerial video too. Other parts of the ship images are shown in Fig. 1a. Incomplete waterline images are shown in Fig. 1b, target area is too small, as show in Fig. 1c. Other invalid situations are large unrelated areas that appear on the image, as show in Fig. 1d.

The image contains the complete position information of waterline can be recognized as the effective image. For images with valid information of the ship waterline, there will be different situations such as different hull structures, weather conditions and environmental conditions. What's more, different hull scale may have different shapes, different paint colors, as show in Fig. 2a. The weather and environment conditions result in different corrosive levels of water lines, as show in Fig. 2b. And even the effect of night darkness, as shown in Fig. 2c.

(a) Other parts of the ship

(b) Incomplete waterline

(c) Too small waterline target

(d) Other invalid situations

Fig. 1 Examples of invalid waterline images in different situations

2.2 *Image Enhancement*

Poor performance may be happened owing to lacking of highly effective image preprocessing approaches, which are typically required before the feature extraction [28]. We have preprocessed the collected ship waterline data sets. Since the waterline of the ship is a scale drawn at the position of the ship's head, tail and the ship's middle side of the ship. The head and tail of the hull named bow and quarter. They are curved boards on both sides of the stem and stern. The scale lines at both ends of the ship are drawn on the curved hull, as show in Fig. 3. So the data is expanded by the concave-convex deformation of the image and the sample of the water ruler scale is added. The data is adapted to the different rules of the different hull position scale shapes [29, 30].

The color of the hull and the brightness of the image are interference information for the waterline identification, so we performed image enhancement preprocessing to eliminate the impact of these interference information on model training. Firstly, convert the color image to a gray scale image, then perform image enhancement based on Logarithmic and Fourier transform. For an image, it can be expressed as the product of the illumination component and the reflection component, that is:

$$m(x,\ y) = i(x,\ y)r(x,\ y) \tag{1}$$

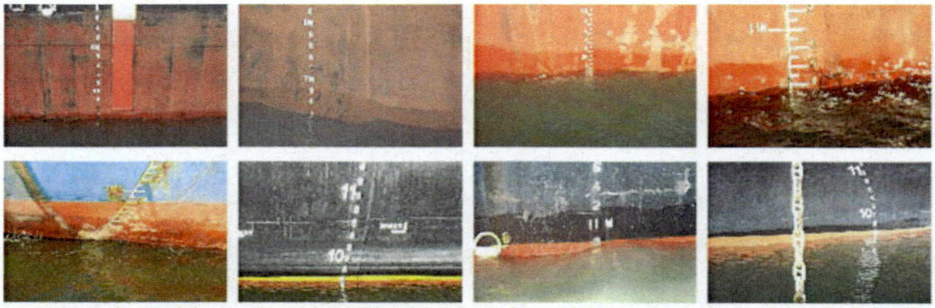

(a) Different paint color of the ship

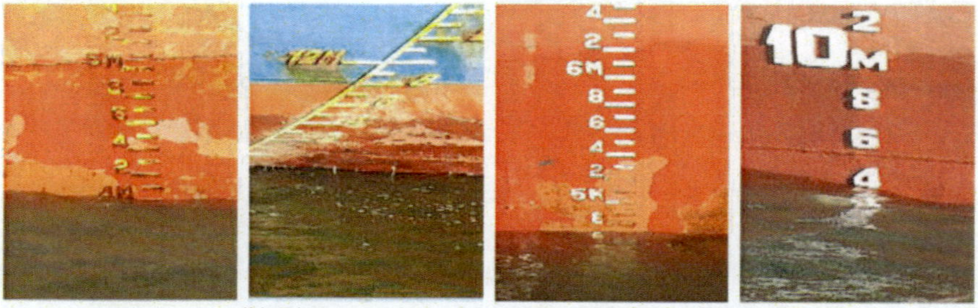

(b) The corroded white paint of the waterline

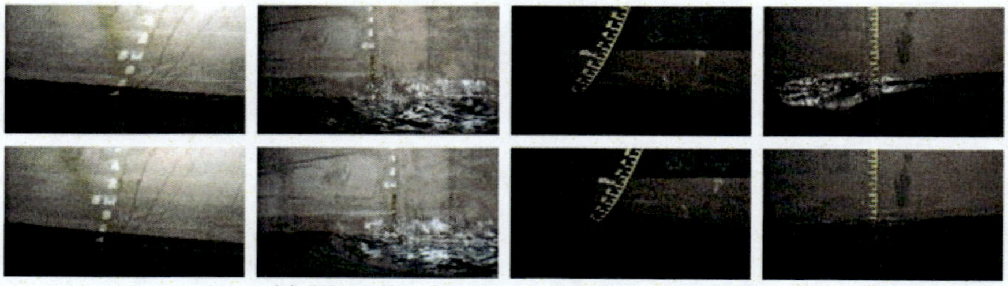

(c) Waterline image photography in night

Fig. 2 Examples of valid waterline images in different situations

where m(x, y) is the image, i(x, y) is the illuminant component, and r(x, y) is the reflection component. We take the logarithm on both sides of the above formula to get the formula (2):

$$ln(m(x, y)) = ln(i(x, y)) + ln(r(x, y)) \quad (2)$$

In order to use a high-pass filter in the frequency domain, we perform a Fourier transform as formula (3):

$$F\{\ln(m(x, y))\} = F\{\ln(i(x, y))\} + F\{\ln(r(x, y))\} \quad (3)$$

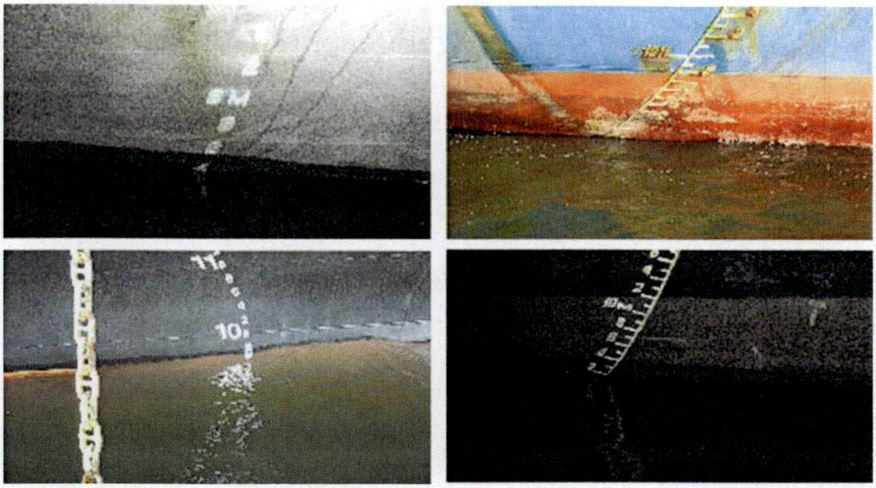

Fig. 3 Schematic diagram of draft marks drawn on the curved hull

Define F {ln(m(x, y))}as M(u, v), then the image is high-pass filtered so that the high-frequency components are increased and the low-frequency components are reduced as formula (4):

$$N(u, v) = H(u, v)M(u, v) = (k_1(e^{c(-D(u,v)/(d0^2))}) + k_2)M(u, v) \tag{4}$$

In order to adjust the degree of sharpening, two variables k_1 and k_2 are introduced. k_1 can adjust the amplification of the high-frequency component, and k_2 can adjust the attenuation of the DC component. Here, $D(u, v) = \sqrt{(u - M/2)^2 + (v - N/2)^2}$, M and N represent the size of the spectral image, (M/2, N/2) is the center, d0 is the cutoff frequency of the Gaussian high-pass filter, c is an adjustable parameter. We do the inverse Fourier transform on N (u, v) and use the exponential function to restore the logarithm finally as formula (5).

$$m'(x, y) = e^{F^{-1}\{N(u,v)\}} \tag{5}$$

3 Waterline Intelligent Recognition Model

In this section, we propose a concise convolutional neural network to obtain a larger receptive field in the stacked convolutional layer, which is a feature extraction network with repeated superimposed convolution layers. Classification features are learned by the model itself, so intelligent recognition is possible. The network can solve multi-classification tasks. This paper aims to enhance the recognition performance of complex images on the network by continuously deepening the network. We first introduce how to label the acquired sample image, after expanding the labeled data we transformed the image gray level on the expanded data set and enhanced it. Secondly, we introduce our scale classification convolutional neural network combined with batch normalization and use the preprocessed data to train the network and finally achieve the detection of the waterline image of the ship.

3.1 Sample Labeling

The quality of training data depends on the accuracy of the label. The accuracy of the label must reach the application level on the basis of the actual detection value. According to the following principles, there are two marking methods for drawing marks: one is metric system, which is represented by Arabic numerals, and the height is set to 10 cm, and the distance between upper and lower words is 10 cm, as shown in Fig. 4. The other is the British system, which is represented by Arabic and Roman numbers. Each number is six inches high and six inches apart. Lines and numbers are used to mark the hull plates on both sides of the bow and stern. This paper uses the first method to unify the sample labeling. When the water surface reaches the lower edge of a number on the water line mark, the number represents the current water line value; when the water line has just submerged the number, the actual value of the point is the number plus the corresponding word height; when the water line is half of the word height (or other proportion), the number represents the current water line value. Represents the height of the corresponding number plus (or minus) half (or other proportion); when the water surface is fluctuating, the actual water line position should be determined according to the average value of several observed values.

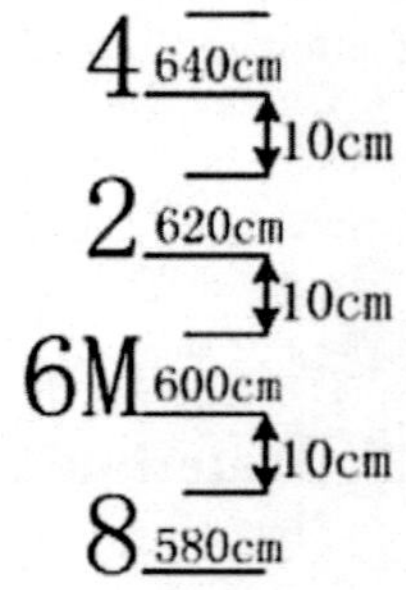

Fig. 4 Metric system drawing marks

3.2 Neural Network Structure

We added batch normalization after every full connection layer except the classification layer. The neural network structure diagram is shown in Fig. 6. The upper part of Fig. 6 is the size of the feature map and the bottom is the operational layer of the network. We propose a scale classification convolutional neural network. There are four convolution modules in the network, and each convolution module contains multiple stacked convolution layers. Each module contains a maximum pool level to reduce the size of the feature map. The fully connected layer is initialized by Xavier and uses dropout to limit the work of certain neurons with a small probability to make the model more versatile. After each convolutional layer, there is a nonlinear activation function to enhance the nonlinearity of the network. After the fully connected layer, the small batch data is normalized by adding a BN layer. The final fully connected layer directly outputs the judgment result of the category.

According to the waterline information marked in the training data, the label information is transformed into a unique heat vector form, and then the cross entropy loss function is used to calculate the training error. In the back- propagation phase, a stochastic gradient optimizer is used to optimize by setting a certain rate of descent. The weights and offsets in the network are adjusted based on the calculated errors. Use softmax classifier to process the output of the last fully connected layer and get the probability of the result. Compare the highest probability serial number to the

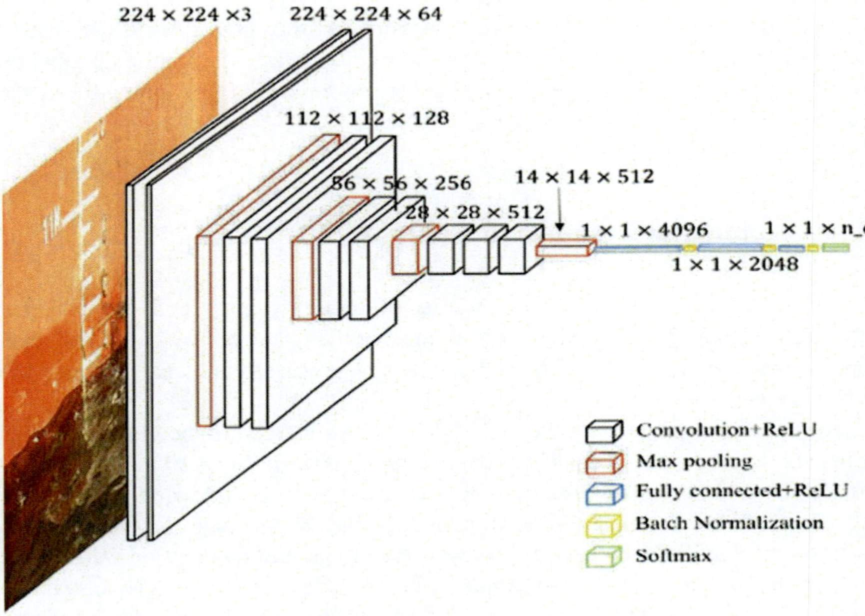

Fig. 6 Neural network structure diagram

Table 1 The overall structure of the model

Layer number	Type	Kernel number/size/step
1 layer	Conv1_1 + Relu	64/3*3/1
2 layer	Conv1_2 + Relu	64/3*3/1
pool_1	Max pooling	/2*2/
3 layer	Conv2_1 + Relu	128/3*3/1
4 layer	Conv2_2 + Relu	128/3*3/1
pool_2	Max pooling	/2*2/
5 layer	Conv3_1 + Relu	256/3*3/1
6 layer	Conv3_2 + Relu	256/3*3/1
pool_3	Max pooling	/2*2/
7 layer	Conv4_1 + Relu	512/3*3/1
8 layer	Conv4_2 + Relu	512/3*3/1
9 layer	Conv4_3 + Relu	512/3*3/1
pool_4	Max pooling	/2*2/2
10 layer	FC1 + BN + Relu	4096
11 layer	FC2 + BN + Relu	2048
12 layer	FC3 + Softmax	n_cls

actual waterline label of the sample and return to the correct category. The overall structure of the model is shown in Table 1.

The scale classification convolutional network performs high dimensional feature extraction with fewer convolutional layer stacks, normalizes the feature distribution by batch normalization. The concise nature network improves the training speed and the accuracy of network, so we named it C-Net.

4 Experimental Results and Analysis

In this section, we first show the results of image warping and enhancement processing. Then, the recognition performance of the C-Net network was analyzed. Finally, the ship draft video collected on site was processed and analyzed, and the analysis results were given.

The experimental data consists of 14,409 images marked in the video collected at the ship's port with a frequency concentration between 7 and 10 m. We classify the unrelated areas separately. According to the actual reading value, the pictures are classified. For example, the label of 9 m is classified as one category, and 9.5 or 10 m are classified as others. In this way, the problem of recognition waterline is transformed into a classification problem. The data set contains images of various complex situations such as night, rust, wind and waves, and different lighting. Some original training set images without data expansion are shown in Fig. 7. We use the

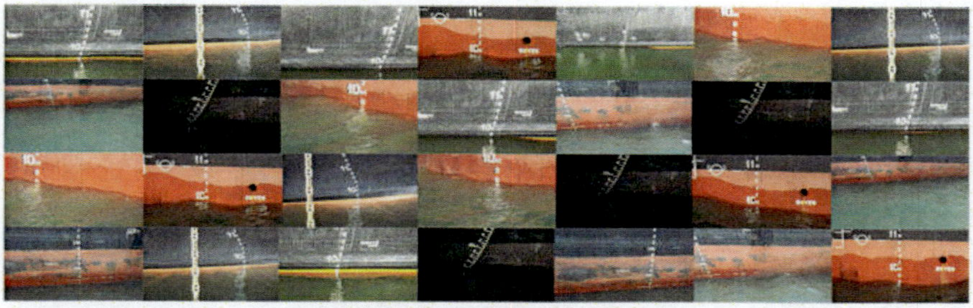

(a) Train set image in 9m

(b) Train set image in 10m

Fig. 7 Some original training set images without data expansion

fivefold cross-validation technique. The data set was divided into 5 groups that did not overlap with each other. One of the groups was taken as a test set and the other four were used as a training set. The final result is the average of the five sets test results. In addition, we analyzed the improved network qualitatively for different evaluation indicators. Finally, we discussed the robustness of the model.

4.1 Data Expansion and Image Enhancement

In the real environment, the waterline will float up and down within a certain scale, and the image frames extracted from the video occupy a large proportion in the middle value. In order to make the data more balanced, we performed multi-scale sparse sampling on the original video data to equalize the sparse data distribution problem with dense edges in a video. We extend the original training data with concave and convex deformations to accommodate the scale lines on bow and quarter hulls at both ends of the ship. The images of ship waterline after concave-convex transformation are shown in Fig. 8. Data augmentation can guarantee the validity of the model.

The sample data contains images of different hulls and different environmental conditions. All images were deformed and enhanced after gradation transformation.

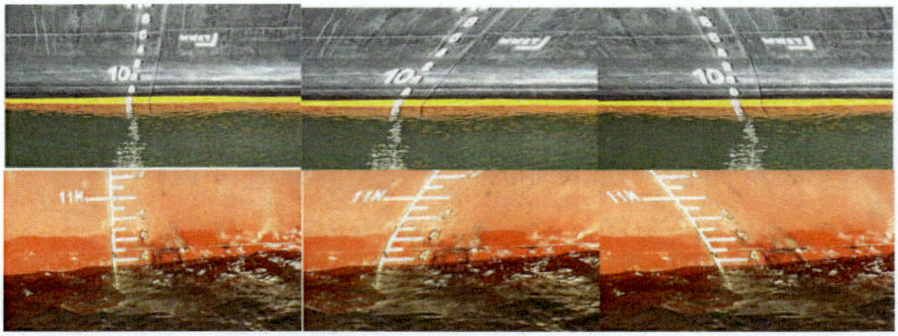

(a) Original image (b) Convex image (c) Concave image

Fig. 8 The images of ship waterline after concave-convex transformation

Image enhancement not only improves image quality but also effectively prevent the influence of water surface artifacts on recognition, as shown in Fig. 9.

4.2 Comparison and Analysis of Waterline Classification Results

In this paper, we simplify the network model to reduce the amount of network computation, so as to improve the training speed and accuracy of the network. We evaluated the proposed methods for waterline images. The network is initialized with pre-training weights, and then the C-Net network is trained using the tagged ship waterline image.

The training results of the C-Net with BN layer are shown in Fig. 10. From network training accuracy and loss curve, it can be found that when the BN layer is added to the network, the curve tends to be smooth and the training accuracy of the model reaches about 0.97 and when iterating to 1850 steps. We compared the proposed methods with the C-Net without the BN layer, VGG16 and Alex networks. Figure 11 shows the accuracy curves of different models. During the training process, the accuracy of VGG16 is continuously improved, but the convergence speed is slow. The Alex network has been oscillating and not converging during the training process. Joining the C-Net network of the BN layer, the convergence speed is faster and the training precision is higher.

Figure 12 shows the time statistics when the network training accuracy reaches 0.9. We found that the C-Net network training time that we proposed to join the BN layer was six times shorter than the VGG16 network.

Figure 13 shows the mAP and accuracy of the network in the final test results. On the identification of the water gauge image of the ship, our network achieved a test accuracy of 91.34%, which is 3.03% higher than that of the VGG16 network. It can

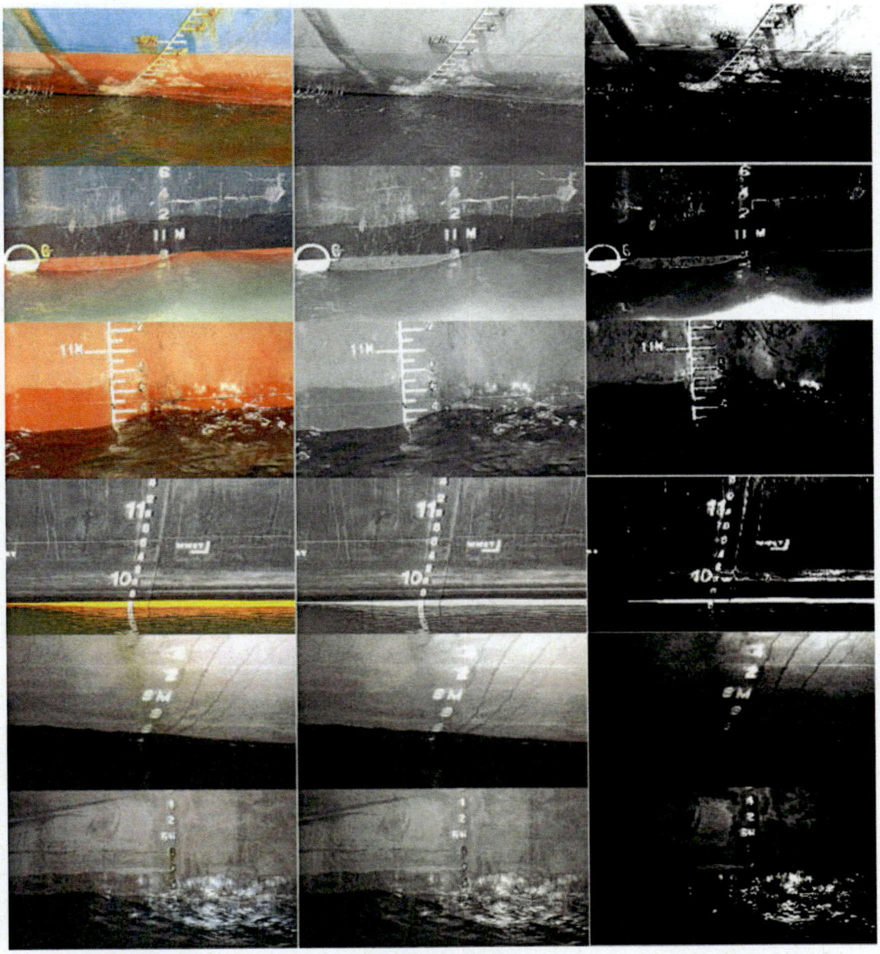

(a) Original image (b) Grayscale image (c) Enhanced image

Fig. 9 Image enhancement results

be seen that the method of this paper achieves better results on the identification of the water scale image of the ship. The C-Net network combined with BN is effective in solving such problems as the unobvious gap between sample classes and the small size of data sets in waterline images.

In order to narrow the gap with the actual measurement data of workers, we use the competitive fairness scoring method. The predicted values that are too large and too small in the identified image data will be eliminated and the number of invalid samples will be eliminated. Finally, the effective image is used to average the prediction results of the C-Net network to obtain the watermark scale prediction

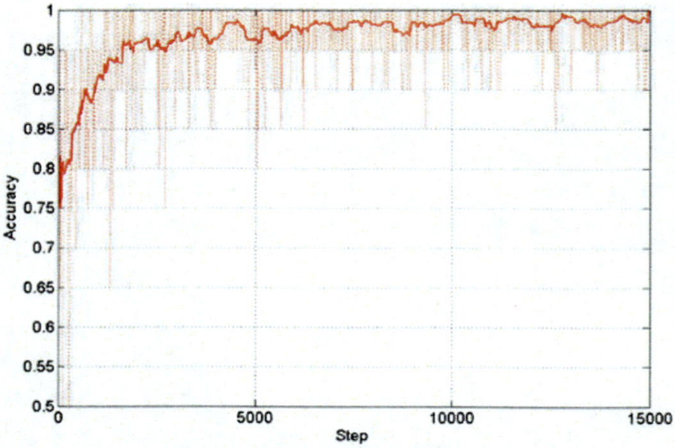

(a) Network training accuracy curve

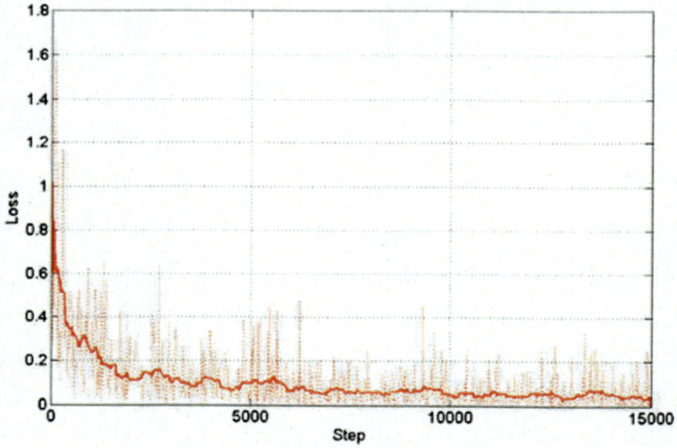

(b) Network training loss function curve

Fig. 10 C-Net training accuracy and loss curve

of the model. The change curve of water line coordinate and the average value are shown in Fig. 14.

From the video, sample 36 (or more) video frame images, identify each waterline scale, and then calculate the average, through the competitive fairness scoring method, the final recognition results of the waterline can be obtained. When an invalid image appears, it is judged as error, and can be removed or calculated by zero in average calculation. The average value can reduce the influence of wave fluctuation and improve the accuracy. Figure 15 shows some results of the C-Net waterline recognition.

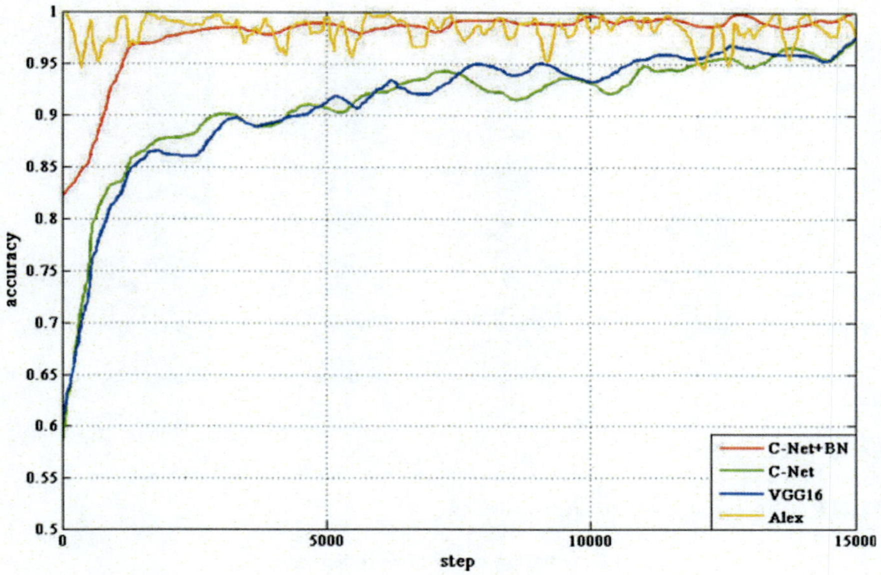

Fig. 11 The accuracy curves of different models

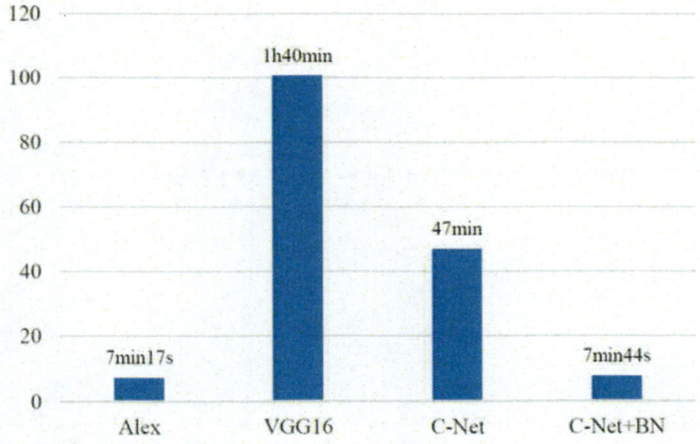

Fig. 12 Time spent in network training of different models

5 Conclusions and Future Work

This paper presents an intelligent method for waterline recognition based on neural network. The drone is used to obtain the waterline image data and image processing

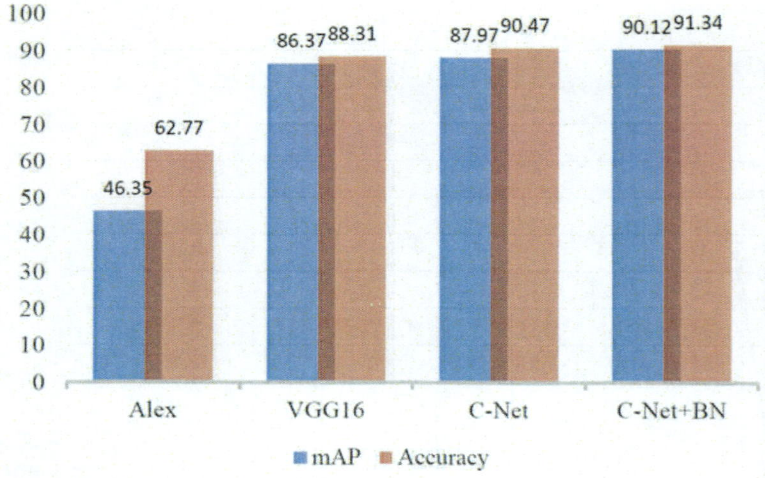

Fig. 13 Testing accuracy of different models (%)

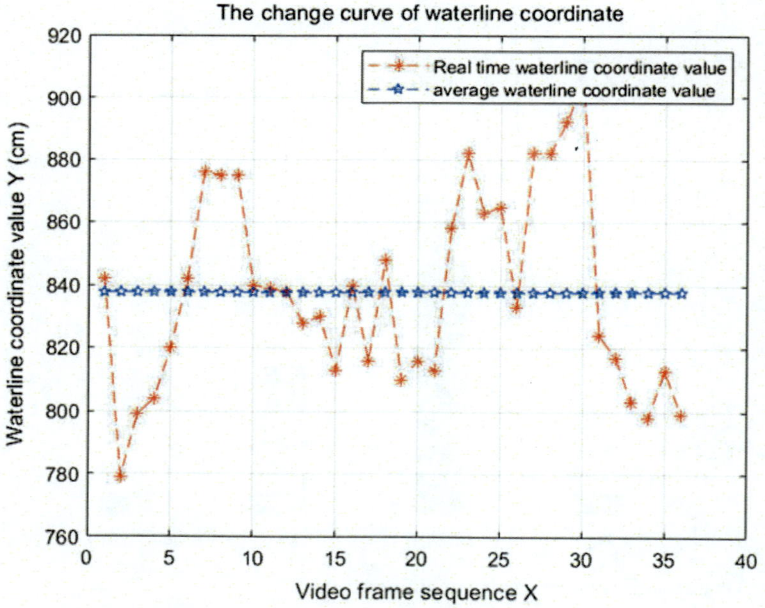

Fig. 14 The change curve of water line coordinate and the average value

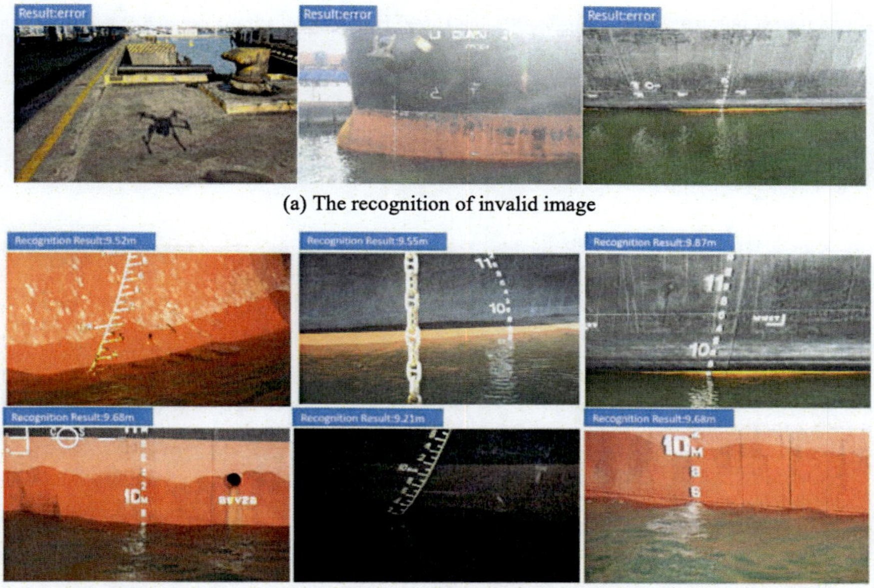

(a) The recognition of invalid image

(b) The recognition of valid image

Fig. 15 The C-Net waterline recognition

method is used to preprocess the training image in order to remove interference information. A scale classification convolutional neural network (named C-net) combined with batch normalization method is used to train for waterline recognition. The image labeling work conforms to the actual application standard, which lays a foundation for obtaining a high-quality training set, and finally realizes the automatic detection of the ship waterline. Compared with the artificial method, the method we proposed is not limited to the operator's visual and experience level, and the operation is convenient, which greatly saves labor. Compared with the electronic device detection method, the method is not limited to an extremely harsh weather environment, and the obtained detection result is more objective and accurate.

It is worth mentioning that the accuracy of sample labeling plays an important role in model recognition, the labeling of samples is completed by experienced port staff. It took us about one year to collect samples of different ship types, and the work of sample collection is still going on.

The limitation of our work is that the correlation between pre-training data set and waterline image data set is low. Pre-training weight does not provide a good initialization parameter for the network, and it does not help the network performance much. A suitable transfer learning approach will help improve the performance of the network. We would like to treat it as our future work.

Acknowledgements This article does not contain any studies with human participants or animals performed by any of the authors. This study was funded by the National Key R&D Program of

China (Grant number 2018YFB1402800), the Hebei Higher Education Research Practice Project (Grant number 2018GJJG422).

References

1. Kim, J., Han, Y., Hahn, H.: Embedded implementation of image-based water-level measurement system. IET Comput. Vision **5**(2), 125–133 (2011)
2. Park, B.S., Kang, D., Kang, I.K., Kim, H.M.: The analysis of the ship's maneuverability according to the ship's trim and draft. J. Fisheries Mar. Sci. Educ. **27**(6), 1865–1871 (2015)
3. Eggers, F., Kaatze, U.: Broad-band ultrasonic measurement techniques for liquids. Meas. Sci. Technol. **7**(1), 1 (1996)
4. Xinli, L., Xianqiao, C., Huaihan, L., Xiumin, C.: Research on data processing method of detection for dynamic ship draft based on multi-beam sonar system. In: International Conference on Transportation Information and Safety (ICTIS), June, pp. 633–636 (2015)
5. Xiaobo, S., Xinli, L., Huaihan, L., Xiumin, C., Xianqiao, C.: The design research of dynamic measurement system of inland ship draft. In: International Conference on Transportation Information and Safety (ICTIS), pp. 637–640 (2015)
6. Mudi, X., Siyin, Z., Lu, L., Peng, N.: Research on data processing method of real-time detection system for dynamic ship draft. Chinese J. Sci. Instrum. **33**(1), 173–180 (2012)
7. Xiuhua, L., Wenfu, W., Junrong G.: The method and development trend of laser ranging. In: 5th international conference on intelligent human-machine systems and cybernetics, vol. 2, pp. 7–10 (2013)
8. Yunfei, Z., Yilu, G., He, W., Lixuan, L.: Influence of Water on Underwater Distance Measurement by a Laser Range Finder, pp. 1–5. Oceans IEEE, Aberdeen (2017)
9. Wenwei, C., Ji, Y., Jie, X., Canhong, J., Lian, C.: A new measurement system of ship draft. Shipbuild. China **54**(1), 166–171 (2013)
10. Buick, J.M., Cosgrove, J.A., Douissard, P.A., Greated, C.A., Gilabert, B.: Application of the acousto-optic effect to pressure measurements in ultrasound fields in water using a laser vibrometer. Rev. Sci. Instrum. **75**(10), 3203–3207 (2004)
11. Zarnik, M.S., Belavic, D.: Study of LTCC-based pressure sensors in water. Sens. Actuators, A **220**, 45–52 (2014)
12. Gu, H.W., Zhang, W., Xu, W.H., Li, Y.: Digital measurement system for ship draft survey. Appl. Mech. Mater. **333–335**(1), 312–316 (2013)
13. Xin, R., Jianghui, P.: Ship draft mark recognition based on image processing. J. Shanghai Maritime Univ. **2** (2012)
14. Jintao, Y., Haitao, G., Chuanguang, L., Jun, L.: Coast dock extraction method based on waterline and perceptual organization. In: IEEE International Geoscience and Remote Sensing Symposium (IGARSS), pp.6201–6204 (2016)
15. Zhengwei, H., Yang, F., Zhong, L.: Mining channel water depth information from IoT-based big automated identification system data for safe waterway navigation. IEEE Access **6**, 75598–75608 (2018)
16. LeCun, Y., Bengio, Y., Hinton, G.: Deep learning. Nature **521**(7553), 436 (2015)
17. Gu, S., Ding, L.: A complex-valued VGG network based deep learning algorithm for image recognition. In: Ninth International Conference on Intelligent Control and Information Processing (ICICIP), pp. 340–343 (2018)
18. Zhong, G., Zhang, K., Wei, H., Zheng, Y., Dong, J.G.: Marginal deep architecture: stacking feature learning modules to build deep learning models. IEEE Access **7**, 30220–30233 (2019)
19. Alex, K., Ilya, S., Geoffrey, E.H.: Imagenet classification with deep convolutional neural networks. Neural Inform. Process. Syst. **25**(1) (2012)

20. Simonyan, K., Zisserman, A.: Very deep convolutional networks for large-scale image recognition. arXiv preprint arXiv:1409.1556 (2014)
21. Christian, S., Wei, L., Yangqing, J., et al.: Going deeper with convolutions. In: 2015 IEEE Conference on Computer Vision and Pattern Recognition (CVPR), Boston, MA, pp. 1–9 (2015)
22. Kaiming, H., Xiangyu, Z., Shaoqing, R., Jian, S.: Deep residual learning for image recognition. In: IEEE Conference on Computer Vision and Pattern Recognition (CVPR), pp. 770–778 (2016)
23. Zhong, G., Jiao, W., Gao, W., et al.: Automatic design of deep networks with neural blocks. Cogn. Comput. **12**, 1–12 (2020)
24. Yue, Z., Gao, F., Xiong, Q., et al.: A novel semi-supervised convolutional neural network method for synthetic aperture radar image recognition. Cognitive Comput. (2019)
25. Censor, Y., Zenios, S.A.: Parallel Optimization: Theory, Algorithms and Applications. Oxford University Press (1998)
26. Zhang, Z., Duan, F., Sole-Casals, J., Dinares-Ferran, J., Cichocki, A., Yang, Z., et al.: A novel deep learning approach with data augmentation to classify motor imagery signals. IEEE Access, vol. 7, pp. 15945–15954 (2019)
27. Weiss, G.M.: Mining with rarity: a unifying framework. ACM SIGKDD Explorations Newsl **6**(1), 7–19 (2004)
28. Oloyede, M.O., Hancke, G.P., Myburgh, H.C.: Improving face recognition systems using a new image enhancement technique, hybrid features and the convolutional neural network. IEEE Access **6**, 75181–75191 (2018)
29. Pan, S.J., Yang, Q.: A survey on transfer learning. IEEE Trans. Knowl. Data Eng. **22**(10), 1345–1359 (2010)
30. Xiaofeng, Z., Shize, G., Hong, S., Liang, G., Di, X., Nan, Z.: Feature-based transfer learning based on distribution similarity. IEEE Access **6**, 35551–35557 (2018)

Compact Ultra-Wideband Antenna for Microwave Imaging Applications

Lulu Wang and Sachin Kumar

Abstract This paper presents a new low-cost patch antenna design for microwave imaging (MI) systems. The antenna configuration consists of a microstrip feed line, a rectangular-shaped ground plane, and a circular-shaped radiating patch modified with first- and second-order Koch fractal shapes, which provides a wide usable fractional bandwidth of more than 118%. The antenna is fed by a 50 Ω SMA connector positioned at the starting of the microstrip transmission line. A sequence of first- and second-order Koch fractal structures is also used on the ground plane side edges on a continuous basis, which decreases the length of the ground and thereby miniaturises the length of the antenna. In order to improve impedance matching in the ultra-wideband (UWB) range, a hexagonal-shaped slot is cut from the ground surface of the antenna. The proposed antenna has a symmetrical structure. It exhibits an excellent omnidirectional radiation pattern even at higher frequencies. The impedance bandwidth ($S_{11} \leq -10$ dB) of the proposed antenna is 3.1–11.8 GHz. The antenna offers a satisfactory gain level, and its radiation efficiency is greater than 70% across the entire resonating band. The proposed antenna design is straightforward, compact, and it can be easily printed on the circuit board of the MI devices/systems. The designed UWB antenna could be effective in addressing the problems of low resolution.

Keywords Biomedical engineering · Microwave antenna · Microwave imaging system · RF biosensors

L. Wang
Biomedical Device Innovation Center, Shenzhen Technology University, Shenzhen 518118, China
e-mail: wanglulu@sztu.edu.cn

S. Kumar (✉)
Department of Electronics and Communication Engineering, SRM Institute of Science and Technology, Kattankulathur 603203, India
e-mail: sachinkr@srmist.edu.in

H. Dai (ed.), *Computational and Experimental Simulations in Engineering*, Mechanisms and Machine Science 119,
https://doi.org/10.1007/978-3-031-02097-1_16

1 Introduction

Microwave imaging (MI) has been proposed as a potential technique to overcome the limitations of X-ray mammography, which has received attention for breast cancer detection. Over the last three decades, several research groups have investigated MI for scanning the passive dielectric properties (conductivity and permittivity) of biological tissues using microwave frequency spectrum [1–3]. MI has been used to diagnose various diseases, including skin cancer, heart disease, breast cancer, lung cancer, pulmonary perfusion, and brain stroke [4–6]. MI has been extensively studied for monitoring breast cancer. The applicability of MI for monitoring breast cancer tissues is based on a significant change in the endogenous plastic somatic cells of breast tissues when a cancer tissue occurs [7, 8].

Holographic microwave imaging (HMI) has been developed for industrial and biomedical applications. HMI technique is a non-destructive imaging method for obtaining the shape of objects from the external scattering field of objects, and the harm of microwave signals to the human body is minimal. HMI technology as a means of medical detection has piqued the interest of many people. Several HMI-based research findings have been published in the last decade, including imaging algorithms, scanning methods, and antennas that facilitate the use of HMI methods [9–11].

The performance of image resolution is directly affected by the antenna, which is an essential component in the MI system. To obtain accurate imaging information, the microwave antenna must have ultra-wideband (UWB), low-profile, high gain, small size, and good time-domain characteristics [12, 13]. The existing UWB antenna has the disadvantages of low resolution, complex structure, high cost, and large volume when applied to microwave biological imaging. Many researchers have proposed various UWB antennas for breast cancer detection. Patch antennas, compared to horn antennas, are easier to use in liquid to achieve a better impedance match with an irradiated body [14–17]. However, the existing patch antennas for biomedical applications have limitations, including low radiation efficiency, large size, and high cost. Furthermore, the reduced dynamic range adversely affects the HMI system's capability to detect small tumors.

This paper presents a compact modified circular-shaped UWB patch antenna for MI applications to address the problems of low resolution, complex structure, high cost, and large size. The UWB antenna ground plane and radiator edges are customized using first- and second-order Koch fractal shapes to achieve miniaturized geometry. The antenna design operates in the frequency range of 3.1–11.8 GHz. The proposed antenna is assumed to work in a high dielectric medium to reduce impedance mismatch to an irradiated human body.

2 Antenna Configuration

Figure 1 shows the configuration of the proposed antenna, which consists of the modified circular-shaped radiating patch, a microstrip feeding line, and a rectangular-shaped ground plane. The radiating patch and ground plane are printed on the FR-4 substrate of a thickness of 1.6 mm, the dielectric constant of 4.4, and a loss tangent of 0.016. The proposed antenna exhibits a compact size of 24 mm × 30 mm.

The radiating patch is connected with a microstrip line of width W_f and length L_f (shown in Fig. 1a). The width of the microstrip feed line is fixed as 3 mm. On the

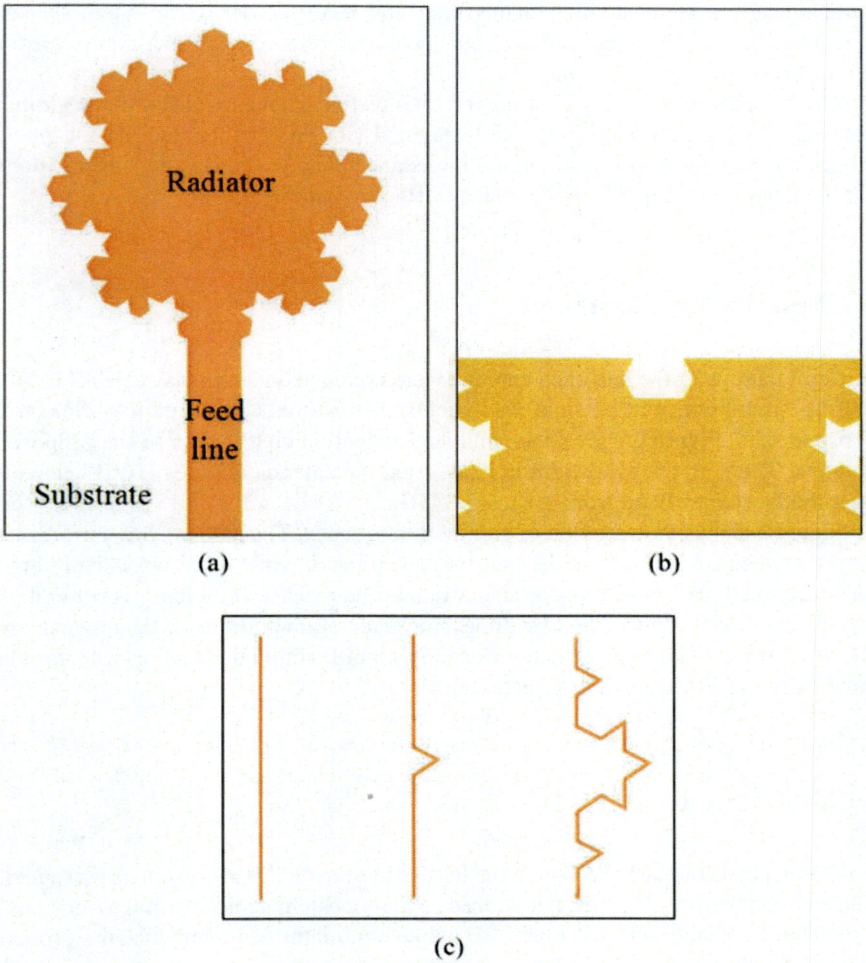

Fig. 1 Geometry of the proposed antenna: **a** patch, **b** ground plane, **c** fractal shape iterations (zeroth-, first-, and second-order, respectively)

Table 1 Design specifications of the antenna

Features	Value (mm)
Substrate length	30
Substrate width (W_{gnd})	24
Feed line length (L_f)	11.5
Feed line width (W_f)	3
Ground plane length (L_{gnd})	11.25

other side of the substrate, a ground plane of width W_{sub} and length L_{gnd} is present, as shown in Fig. 1b. To achieve miniaturized geometry, the antenna ground plane and resonator edges are customized using first- and second-order Koch fractal shapes, shown in Fig. 1c. The proposed antenna is connected to a 50 Ω SMA connector for microwave signal transmission.

The impedance matching is improved by inserting a hexagonal slot in the ground plane of the antenna. Changing the hexagonal slot dimensions and ground plane length has a significant impact on the impedance bandwidth. Design specifications for the proposed single layer antenna are listed in Table 1.

3 Results and Discussion

The simulations of the designed antenna were conducted using Ansys HFSS 15.0®. During simulation, the antenna was enclosed in a box with a relative dielectric constant of 1. Figure 2 shows the simulated reflection coefficients of the proposed antenna. The −10 dB bandwidth indicates that the antenna features UWB behavior over the frequency band from 3.1 to 11.8 GHz.

Figure 3 shows the variation of gain with frequency. The antenna has a moderate gain (between 2 and 4 dB) at the low-frequency band, while it has a relatively high gain (around 5 dB) at the high-frequency band. The radiation efficiency is better than 70% in the UWB range. The two-dimensional radiation patterns of the antenna are shown in Fig. 4. The H-plane displays nearly omnidirectional behavior, whereas the patterns in the E-plane are bi-directional.

4 Conclusion

A low-cost and compact UWB antenna for use in a far-field HMI system is presented. The antenna consists of a circular-shaped radiating patch, a microstrip feed line, and a rectangular-shaped ground plane. To achieve miniaturized geometry, the ground plane and radiator edges of the proposed antenna are modified using first- and second-order Koch fractal shapes. Also, a hexagonal-shaped slot is cut into the ground surface to improve impedance matching in the resonant frequency band. The results have

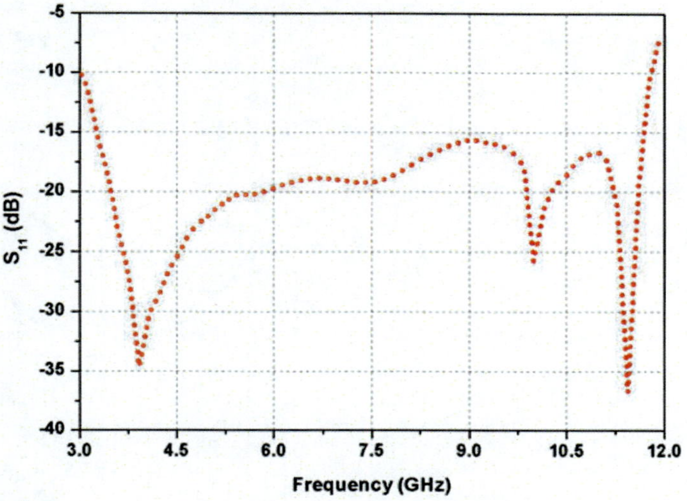

Fig. 2 Reflection coefficients of the UWB antenna

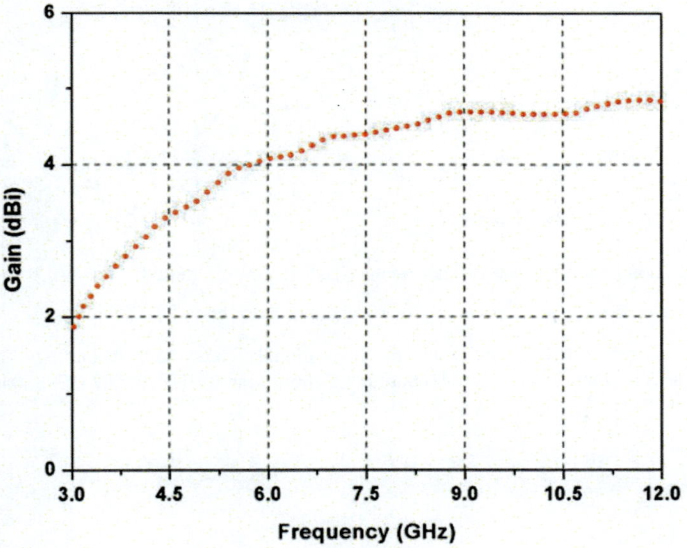

Fig. 3 Gain of the UWB antenna

shown that the antenna covers the UWB with moderate gain and offers good radiation performance. The proposed UWB antenna could be very much helpful in addressing the problems of low resolution.

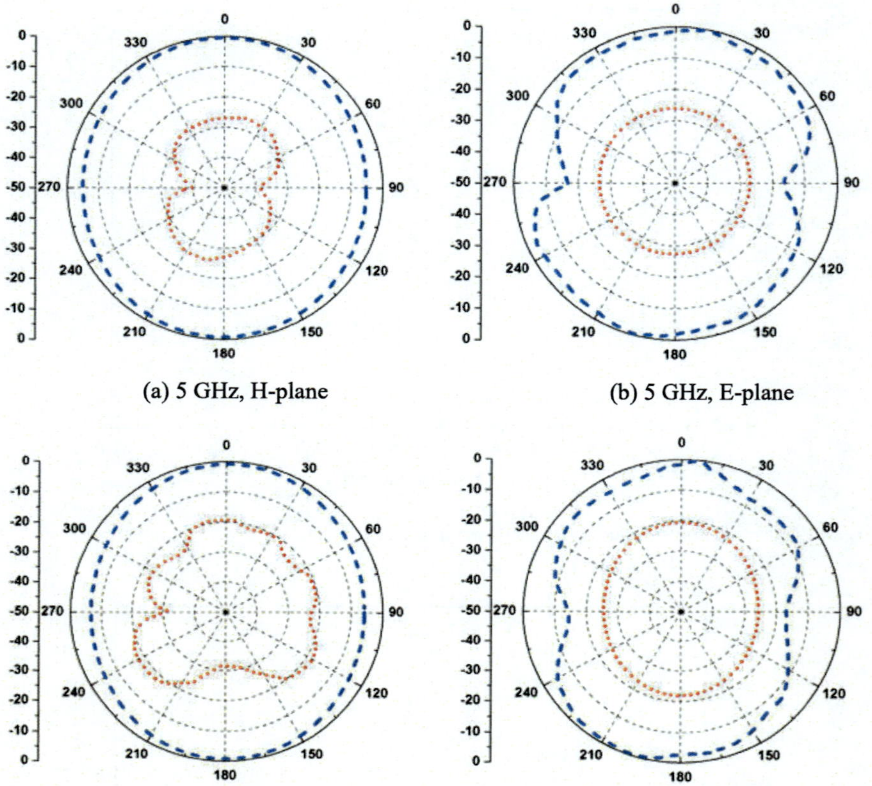

(a) 5 GHz, H-plane (b) 5 GHz, E-plane

(c) 10 GHz, H-plane (d) 10 GHz, E-plane

Fig. 4 Radiation patterns of the UWB antenna (dashed: co-pol.; dotted: cross-pol.)

Funding This research was funded by the International Science and Technology Cooperation Project of the Shenzhen Science and Technology Commission (GJHZ20200731095804014).

References

1. Benny, R., Anjit, T.A., Mythili, P.: An overview of microwave imaging for breast tumor detection. Prog. Electromagn. Res. B **87**, 61–91 (2020)
2. Kwon, S., Lee, S.: Recent advances in microwave imaging for breast cancer detection. Int. J. Biomed. Imaging **2016**, 5054912 (2016)
3. O'Loughlin, D., O'Halloran, M., Moloney, B.M., Glavin, M., Jones, E.: Microwave breast imaging: clinical advances and remaining challenges. IEEE Trans. Biomed. Eng. **65**(11), 2580–2590 (2018)
4. Zamani, A., Rezaeieh, S.A., Abbosh, A.M.: Lung cancer detection using frequency domain microwave imaging. Electron. Lett. **51**(10), 740–741 (2015)

5. Amin, B., Elahi, M.A., Shahzad, A., Porter, E., McDermott, B., O'Halloran, M.: Dielectric properties of bones for the monitoring of osteoporosis. Med. Biol. Eng. Compu. **57**, 1–13 (2019)
6. Salvador, S.M., Fear, S.M., Okoniewski, E.C., Matyas, M., John, R.: Exploring joint tissues with microwave imaging. IEEE Trans. Microw. Theory Tech. **58**(8), 2307–2313 (2010)
7. Meaney, P.M., Fanning, M.W., Li, D., Poplack, S.P., Paulsen, K.D.: A clinical prototype for active microwave imaging of the breast. IEEE Trans. Microw. Theory Tech. **48**(11), 1841–1853 (2000)
8. Osumi, N., Ueno, K.: Microwave holographic imaging method with improved resolution. IEEE Trans. Antennas Propag. **32**(10), 1018–1026 (1984)
9. Wu, H., Ravan, M., Amineh, R.K.: Holographic near-field microwave imaging with antenna arrays in a cylindrical setup. IEEE Trans. Microw. Theory Tech. **69**(1), 418–430 (2021)
10. Smith, D., Leach, M., Elsdon, M., Foti, S.J.: Indirect holographic techniques for determining antenna radiation characteristics and imaging aperture fields. IEEE Antennas Propag. Mag. **49**, 54–67 (2007)
11. Lazaro, A., Girbau, D., Villarino, R.: Simulated and experimental investigation of microwave imaging using UWB. Prog. Electromagn. Res. **94**, 263–280 (2009)
12. Islam, M.T., Mahmud, M.Z., Islam, M.T., Samsuzzaman, M.: A low cost and portable microwave imaging system for breast tumor detection using UWB directional antenna array. Sci. Rep. **9**, 15491 (2019)
13. Sugitani, T., Kubota, S., Toya, A.X.X., Kikkawa, T.: A compact 4 × 4 planar UWB antenna array for 3-D breast cancer detection. IEEE Antennas Wirel. Propag. Lett. **12**, 733–736 (2013)
14. Alkurt, F.Ö., Karaaslan, M., Furat, M., Ünal, E., Akgöl, O.: Monopole antenna integrated cavity resonator for microwave imaging. Opt. Eng. **60**(1), 013106 (2021)
15. Borja, B., Tirado-Méndez, J.A., Jardon-Aguilar, H.: An overview of UWB antennas for microwave imaging systems for cancer detection purposes. Prog. Electromag. Res. B **80**, 173–198 (2018)
16. Amineh, R.K., Ravan, M., Trehan, A., Nikolova, N.K.: Near-field microwave imaging based on aperture raster scanning with TEM horn antennas. IEEE Trans. Antennas Propag. **59**(3), 928–940 (2011)
17. Mahmud, M.Z., Islam, M.T., Misran, N., Almutairi, A.F., Cho, M.: Ultra-wideband (UWB) antenna sensor based microwave breast imaging: a review. Sensors **18**(9), 2951 (2018)

Medical Compound Figure Detection Using Inductive Transfer and Ensemble Learning

Mehdi Mehtarizadeh and Mohammad Reza Zare

Abstract Information retrieval in medical publications can be assisted with content-based image retrieval. Figures have become an inseparable part of publications. Around 40% of figures that appear in medical publications are compound and composed of two or more individual figures. Using medical image processing models, valuable metadata can be extracted from these figures which can assist image retrieval. However, a model that was originally designed to analyse a single figure is not able to analyse a compound figure. Thus, Compound Figure Detection (CFD) methods are used to first determine whether a figure is compound or not. Then, relative models and algorithms can be applied to individual figures. In this work, a model based on deep learning and inductive transfer is proposed to realise CFD. The model relies on pre-trained Convolutional Neural Networks as feature extractors. The model uses feature vectors as visual features (or visual words) and makes a dictionary of available visual features within the dataset. Every image is described based on the frequency of visual words it contains. Various versions of the model then form an ensemble of models to better generalise on unseen data. It is shown that the ensembled final model achieved 97.27% accuracy on the ImageCLEF2016 medical task test dataset.

Keywords Compound figure detection · Medical image analysis · Transfer learning · Ensemble learning

M. Mehtarizadeh (✉) · M. Reza Zare
University of Leicester, Leicester LE1 7RH, Leicester, UK
e-mail: mm917@leicester.ac.uk

M. Reza Zare
e-mail: mrz3@leicester.ac.uk

H. Dai (ed.), *Computational and Experimental Simulations in Engineering*,
Mechanisms and Machine Science 119,
https://doi.org/10.1007/978-3-031-02097-1_17

1 Introduction and Literature Review

Information Retrieval (IR) has achieved an indispensable role in the context of biomedical research literature. Online repositories such as PubMed Central contain millions of research papers that are made up of text and visual data such as charts, diagrams, and medical images. Most IR systems rely solely on text-based data and accept text queries. The efficiency of these systems can be enhanced by enabling them to process non-text data. Images comprise a significant volume of non-text data in digital published research and contain useful data that can play a vital role in increasing precision and recall in IR. In practice a notable portion of figures published in papers are placed in close proximity, e.g. in groups of two or more in order to save space in printing. Such figures are referred to as compound figures. Ability to determine whether an arbitrary figure is compound or not is a pre-requisite to analysing figures. Compound figure detection (CFD) is defined as the task of recognising whether a (medical) figure is compound, i.e. made up of more than one figure, or is only a single figure [20, 24]. It is crucial to differentiate between CFD and compound figure separation (CFS). In typical medical image processing pipelines, CFS is a next step to CFD and examines methods to correctly separate a compound figure into single figures that have formed it. After realisation of CFS, individual figures can be classified using relevant techniques. These classes can be considered as early metadata extracted from medical publications. In this paper, a supervised learning model based on transfer learning is suggested for CFD. Previous research on CFD has proposed various models which can be classified according to the underlying approach used in their design, into handcrafted and deep learning models [11]. Models based on handcrafted features rely on customised feature extractors to locate and describe points of interest in images. A combination of low and high-level features have been used in [13, 20, 21]. Border profile feature extractor suggested in [13] target visual features such as horizontal and vertical lines, that exist in compound images. In [24] difference of Gaussians has been used to locate visual features or local key points and SIFT descriptors have been used to represent these points of interest. To reduce the size of feature space, visual features are clustered using K-means clustering algorithm and a visual codebook is formed to represent the entire training dataset. A support vector machine (SVM) is trained as the classifier [24].

In [21] two methods have been suggested for CFD, one model that detects horizontal and vertical separating lines, and another model that detects connected regions assuming that separate connected regions form separate subfigures in a compound figure. A model is suggested that makes use of text captions accompanying figures in publications in [12] which looks for certain delimiter charactes in text captions and then assigns a label to each figure. In an image based model suggested in [12], the areas in the form of bounding boxes are detected and if there is more than one such area, the figure is labeled as compound. Final model is a hybrid of both.

Table 1 List of previous models suggested for CFD along with model accuracy

Model proposed by	Accuracy (%)	Visual feature types
Pelka and Friedreich [13]	85.39	Handcrafted
Wang et al. [21]	82.82	Handcrafted
Taschwer and Marques [20]	76.9	Handcrafted
Zare and Müller [24]	92.5	Handcrafted
Li et al. [12]	90.74	Handcrafted
Lee and Zare [11]	96.93	Learned

In [20] a combination of low and high-level features has been used. Features such as straight lines are found by Hough transform and statistical operators like mean and variance have been used on pixel values to detect horizontal and verical lines.

A model based on transer learning and late feature fusion has been suggested in [11]. This is the only past work that deploys learned features rather than handcrafted features. The model uses pre-trained CNNs as feature extractors and feeds extracted features into an SVM for final training. To improve the overall accuracy of the model, "score-based fusion operators" as proposed in [3] have been used [7].

Table 1 shows accuracy values achieved by the models reviewed earlier. As seen, the model proposed in [11] which utilises learned features, outperforms other models. Section 2 provides a detailed explanation of the methodology.

2 Methodology

In this section the suggested model for CFD is set out. CFD is a classification problem. The model suggested in this paper is based on supervised learning, inductive transfer, and bag of visual words (BoVW).

2.1 Bag of Visual Words

One of the dominant methods to model corpora of images is BoVW which originates from text retrieval [8]. The foundation of BoVW is feature extraction from images. These features form visual words. In most previous models suggested for CFD, handcrafted feature extractors were used. However, with the advances in deep learning and CNNs coming to prominence, the concept of learned features has received a significant attention. CNNs are made up of two different components, a convolutional component that extracts features from input images and a multi-layer perceptron (MLP) that maps features to output classes [15]. CNN architectures like VGG have millions of trainable parameters and need large datasets of hundreds of thousands of training images to converge well [17]. One such dataset is ImageNet. It has over a million natural images with around 1000 classes [6]. Utilizing knowledge learned by a CNN in one domain to a different domain is referred to as transfer learning [1, 2, 16, 19, 22, 23, 25]. Several research experiments have attempted to suggest models for image classification under transfer learning. Some of the models like in [2, 22] and [1, 16, 25] are proposed to realise medical image classification and analysis tasks whereas some others come from satellite image processing context [10, 22]. The common approach to transfer learning in CNNs is to replace the existing trained MLP with a new MLP and train it while keeping the weights of the convolutional component frozen. In this work, however, CNNs trained on ImageNet have been used as feature extractors. Figure 1 shows the pipeline used to train the proposed model. Training images are fed into a CNN which does not have a classifier component. The output from the last convolution layer before pooling, is a tensor of shape $w \times h \times d$. This tensor, which is referred to as a set of feature maps, can be imagined as d matrices of $w \times h$. Every element in these matrices is a pixel of d dimensions and can be considered a visual key-point or a visual word. The parameters w, h, and d depend on CNN architecture. Before training the classifier, feature maps are clustered in an unsupervised manner using the K-Means clustering method where $K < 10,000$. As a result, K clusters represent the training set as K visual words. These visual words replace the whole feature maps of the training set and reduce model size significantly. Finally, training images need to be described using the K visual words by creating a visual word-image matrix. The latter matrix is analogous to a term-document matrix in text retrieval and classification. Each element of the matrix is a measure of importance of the corresponding visual word in the corresponding image with regards to the whole dataset, and is calculated by multiplying term frequency (TF) by inverse

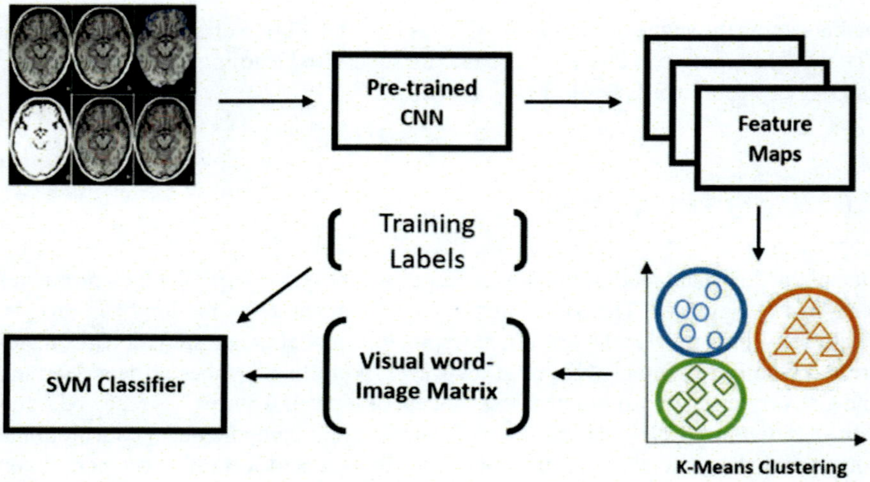

Fig. 1 Bag of visual words with features from a convolutional neural network

document frequency (IDF). The final training stage is to train an SVM that accepts the visual word-image matrix and training labels (compound, or non-compound) as input and learns the mapping from training images to the labels.

2.2 *Feature Map Selection*

When using pre-trained CNNs as feature extractors, it is possible to use all produced feature maps from the final convolution layer before pooling. However, this increases the size of the visual descriptor vectors. For instance, in VGG16, the final convolution layer before pooling produces 512 feature maps of 14×14 dimensions [17]. Therefore, using all feature maps will result in 196 vectors each having 512 elements. At this point, if more valuable feature maps can be extracted, the size of visual descriptor vectors remains manageable. Moreover, clustering gains a significant speed-up. It is crucial to select those feature maps that address the visual differences between compound and non-compound figures. Therefore, measuring how strongly a feature map activates on a region of interest, is an appropriate way of selecting the most differentiating feature maps. In Sect. 3.3, more explanation is provided.

2.3 *Ensemble Learning*

The base model suggested in this work aims to solve a real world problem. Therefore the ability to generalise on unseen data is important. The final models for each CNN were a group of 5 individually trained models that achieved the highest accuracy with regards to the number of clusters.

3 Results

This section provides more information on how the model explained in Sect. 2 is implemented.

3.1 *Dataset*

The dataset used for training, validation, and testing of the proposed model is provided by ImageCLEF 2016 medical task. The 2016 dataset has over 24,000 images in total [9]. Table 2 shows more details of the ImageCLEF 2016 medical task dataset. Also, a similar dataset provided by ImageCLEF 2015 medical task was used to create augmented data. 10,000 augmented training images were created for each class using

Table 2 ImageCLEF 2016 CFD subtask dataset

Class	Real training images	Augmented training images	Test images	Total
Compound	12,350	10,000	1806	24,156
Non-compound	8,650	10,000	1,650	20,300
Total	21,000	20,000	3,456	44,456

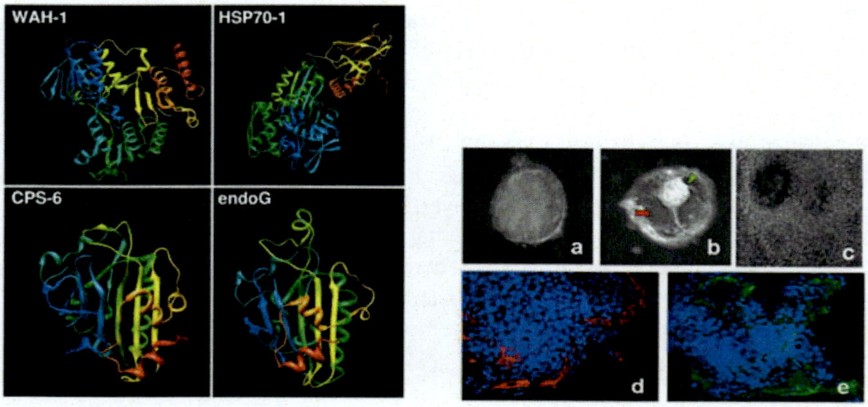

(a) Homogenous compound figure (b) Heterogenous compound figure

Fig. 2 Examples of homogenous and heterogenous compound figures

techniques such as rotation, flipping, and cropping. Trained models are evaluated by accuracy. Accuracy of class C in model M is defined by the following fraction:

$$accuracy(M, C) = \frac{Number\ of\ correct\ predictions\ of\ C}{Total\ number\ of\ test\ items\ of\ C} \tag{1}$$

One of the challenges of CFD, which also exists in the dataset, is lack of a clear horizontal or vertical separating line in compound images. Moreover, not every compound figure is a homogenous figure. A homogeneous compound figure is one that is made up of sub-figures of the same modality, whereas a heterogenous compound figure is one that is made up of sub-figures of different modalities. Figure 2 shows a homogenous compound figure and a heterogenous compound figure.

3.2 Bag of Visual Words

Selecting an optimised number for clusters is a rather blind search. There are many parameters involved such as number of training images, selected CNN, and problem type. In this work, the following constraints were set for K:

$$500 \leqslant K \leqslant 10,000 \tag{2}$$

Starting at $K = 500$ and incrementing K by 500 in every iteration, the model was trained several times to achieve the best accuracy. Figure 3a shows how the accuracy of a model using VGG 16 changes with different amounts of K. In the defined boundary for K, K = 7500, has the highest accuracy. VGG16, VGG19, MobileNet V2, XCeption, and Inception V3 are the CNNs that have been used in the models

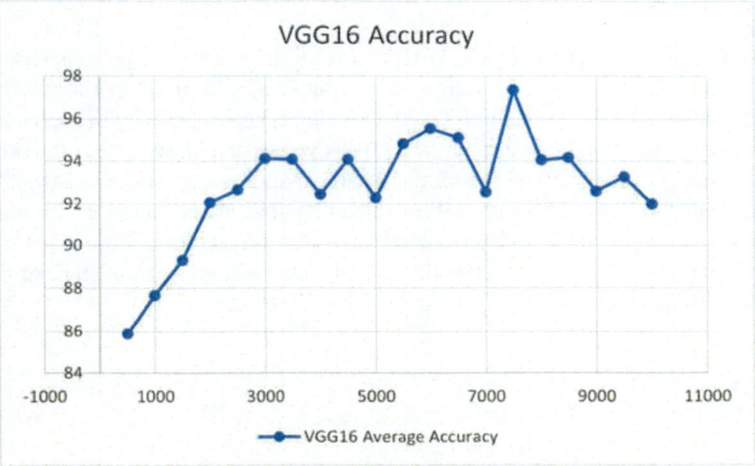

(a) VGG 16 Accuracy with Different Values of K

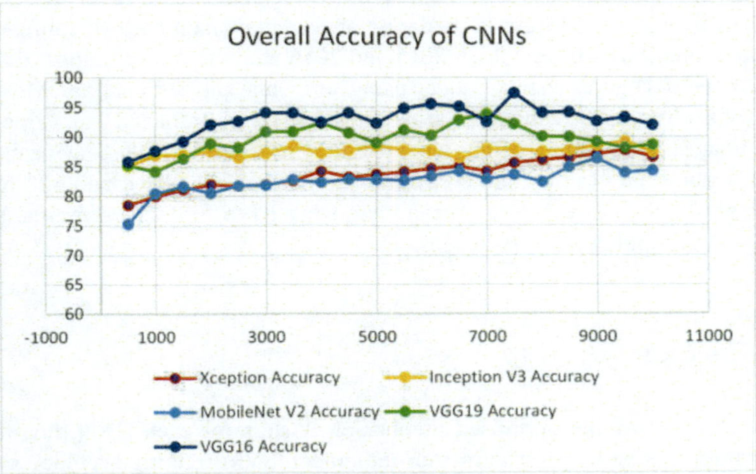

(b) Overall Accuracy of CNNs with Regards to K

Fig. 3 Accuracy and number of clusters (K)

Table 3 Accuracy of ensembled models

Feature extractor CNN	Accuracy
MobileNet V2	86.26 %
Inception V3	90.12
VGG16	**97.27 %**
VGG19	94.65 %
XCeption	88.96 %

suggested in this work [4, 14, 17, 18]. According to Table 3, VGG16 achieved the highest accuracy. Figure 3b is a comparison of all CNNs used in this work with different values of K. The value of K that achieved highest accuracy in each model is only a local maxima in the domain $K > 1$. However, to save time, space, and preserve feature space size, no further search has been carried out beyond $K > 10,000$. K-Means clustering has a random nature, meaning that every round of running the algorithms might result in different cluster centres and data distribution per cluster. To overcome this issue, a random state variable has been set as a constant so that all runs of K-Means have a controlled random initialisation.

3.3 *Feature Map Selection*

As mentioned in Sect. 2.2, regions of interest were fed into pre-trained CNNs to measure which feature maps resonate discriminating visual features the most. These regions of interest are defined as bordering areas that contain parts of subfigures in compound images. In ImageCLEF 2015 and 2016 medical datasets, approximately 6,000 compound figures were separated using horizontal and vertical lines. More than 10,000 of such regions (without the separating lines) were extracted and warped into square images. Then, these regions were fed into pre-trained CNNs. The mean value of each feature map at the final convolution layer was calculated and 128 feature maps with the highest activation mean were selected to formulate the final visual descriptors of training and test images.

3.4 *The Classifier*

Similar to selecting the optimised number of clusters for each CNN, fine-tuning SVM parameters enhances accuracy of the model. SVM training accuracy significantly depends on choice of kernel, the regularization parameter (C), and the kernel coefficient (Gamma) constants. All experiments have been carried out using the sigmoid kernel. C is set to 1. Gamma is scaled using the following formula:

$$\gamma = \frac{1}{\textit{Number of features} \times \textit{Variance of training vectors}} \tag{3}$$

In the procedure of training the SVM, 5-fold cross validation has been used.

3.5 *Ensemble Learning*

As mentioned in Sect. 2.3, final model for each CNN is a group of 5 models with the highest accuracy. These were chosen with respect to the number of clusters. The policy of the ensemble is maximum voting.

4 Discussion and Conclusion

In this paper an ensemble of predictive models for CFD was suggested. The base model is built on the concept of inductive knowledge transfer in learning by supervision and BoVW. The model combines pre-trained convolutional neural networks with the BoVW approach [5]. It uses CNNs as visual feature extractors, unsupervised learning in the form of K-Means clustering to reduce feature space to dominant visual descriptors, and supervised learning in the form of an SVM to classify images.

Unlike text, visual words do not bear any human understandable meaning. Feature maps from a pre-trained CNN that are used as visual words make the problem more challenging. In text corpora several pre-processing steps take place to filter stop words out. Clustering visual words is a similar step to reduce feature space size and help dominating features to show up. Bag of words is not able to detect polysemy in text. Therefore, BoVW cannot recognise spatial relationship among visual words. Despite the fact that pre-trained CNNs detect such a relationship, it is lost in clustering. Choice of K requires multiple rounds of training and depends on application and dataset. The essence of CFD, being binary classification discriminates input images to two very unequal input spaces. Compound figures in biomedical papers have very detailed visual features compared to non-compound figures.

The ensembled model in this paper could achieve 97.27% at the highest accuracy and outperformed all previously suggested models. It is concluded from results that more traditional convolutional neural networks such as VGG16 achieve better performance for specific transfer learning tasks as compared to their more recent counterparts such as XCeption. Also, a model like BoVW can provide a robust framework for transfer learning. Using a systematic approach to set number of visual words can even lead to higher accuracy.

Acknowledgements This research used the SPECTRE High Performance Computing Facility at the University of Leicester.

References

1. Azizpour, H., Razavian, A.S., Sullivan, J., Maki, A., Carlsson, S.: From generic to specific deep representations for visual recognition. In: 2015 IEEE Conference on Computer Vision and Pattern Recognition Workshops (CVPRW), pp. 36–45 (2015). https://doi.org/10.1109/CVPRW.2015.7301270
2. Cheplygina, V., Bruijne, M., Pluim, J.P.W.: Not-so-supervised: a survey of semi-supervised, multi-instance, and transfer learning in medical image analysis. Med. Image Anal. **54**, 280–296 (2019)
3. Chitroub, S.: Classifier combination and score level fusion: concepts and practical aspects. Int. J. Image Data Fusion **1**, 113–135 (2010)
4. Chollet, F.: Xception: Deep learning with depthwise separable convolutions. In: 2017 IEEE Conference on Computer Vision and Pattern Recognition (CVPR), pp. 1800–1807 (2017). https://doi.org/10.1109/CVPR.2017.195
5. Csurka, G., Dance, C.R., Fan, L., Willamowski, J., Bray, C.: Visual categorization with bags of keypoints. In: Workshop on Statistical Learning in Computer Vision ECCV, pp. 1–22 (2004)
6. Deng, J., Dong, W., Socher, R., Li, L., Kai Li, Li Fei-Fei: Imagenet: A large-scale hierarchical image database. In: 2009 IEEE Conference on Computer Vision and Pattern Recognition, pp. 248–255 (2009). https://doi.org/10.1109/CVPR.2009.5206848
7. Fox, E., Shaw, A.J.: Combination of multiple searches. In: Proceedings of The Second Text Retrieval Conferences (TREC-2), pp. 243–252 (1994)
8. Gonzalez, R., Richard, E.: Digital Image Processing. Prentice-Hall (2002)
9. Garcia Seco de Herrera, A., Schaer, R., Bromuri, S., Müller, H.: Overview of the medical tasks in image CLEF 2016. In: Proceedings of Image CLEF, pp. 219–232 (2016)
10. Hu, F., Xia, G., Yang, W., Zhang, L.: Mining deep semantic representations for scene classification of high-resolution remote sensing imagery. IEEE Trans. Big Data **6**(03), 522–536 (2020). https://doi.org/10.1109/TBDATA.2019.2916880
11. Lee, S.L., Zare, M.R.: Biomedical compound figure detection using deep learning and fusion techniques. IET Image Process. **12**, 1031–1037 (2018)
12. Li, P., Sorensen, S., Kolagunda, A., Jiang, X., Wang, X., Kambhamettu, C., Shatkay, H.: USEL CIS at image CLEF medical task 2016. In: Proceedings of Image CLEF (2016)
13. Pelka, O., Friedrich, C.M.: FHDO biomedical computer science group at medical classification task of image CLEF. In: Proceedings of Image CLEF (2015)
14. Sandler, M., Howard, A., Zhu, M., Zhmoginov, A., Chen, L.C.: Mobilenetv2: inverted residuals and linear bottlenecks. In: 2018 IEEE/CVF Conference on Computer Vision and Pattern Recognition, pp. 4510–4520 (2018)
15. Shapiro, L., Stockman, G.: Computer Vision. Prentice-Hall (2001)
16. Shin, H., Roth, H.R., Gao, M., Lu, L., Xu, Z., Nogues, I., Yao, J., Mollura, D., Summers, R.M.: Deep convolutional neural networks for computer-aided detection: CNN architectures, dataset characteristics and transfer learning. IEEE Trans. Med. Imaging **35**, 1285–1298 (2016)
17. Simonyan, K., Zisserman, A.: Very deep convolutional networks for large-scale image recognition. In: International Conference on Learning Representations (2015)
18. Szegedy, C., Liu, W., Jia, Y., Sermanet, P., Reed, S., Anguelov, D., Erhan, D., Vanhoucke, V., Rabinovich, A.: Going deeper with convolutions. In: 2015 IEEE Conference on Computer Vision and Pattern Recognition (CVPR), pp. 1–9 (2015). https://doi.org/10.1109/CVPR.2015.7298594
19. Tajbakhsh, N., Shin, J.Y., Gurudu, S.R., Hurst, R.T., Kendall, C.B., Gotway, M.B., Liang, J.: Convolutional neural networks for medical image analysis: full training or fine tuning? IEEE Trans. Med. Imaging **35**(5), 1299–1312 (2016). https://doi.org/10.1109/TMI.2016.2535302

20. Taschwer, M., Marques, O.: Automatic separation of compound figures in scientific articles. Multimed. Tools Appl. **77**, 519–548 (2018)
21. Wang, X., Jiang, X., Kolagunda, A., Shatkay, H., Kambhamettu, C.: Cis UDEL working notes on image CLEF 2015, compound figure detection task. In: Proceedings of Image CLEF (2015)
22. Xie, M., Jean, N., Burke, M., Lobell, D., Ermon, S.: Transfer learning from deep features for remote sensing and poverty mapping. In: Proceedings of the Thirtieth AAAI Conference on Artificial Intelligence, pp. 3929–3935. AAAI'16, AAAI Press (2016)
23. Yosinski, J., Clune, J., Bengio, Y., Lipson, H.: How transferable are features in deep neural networks? In: Proceedings of the 27th International Conference on Neural Information Processing Systems, vol. 2, pp. 3320–3328. NIPS'14, MIT Press, Cambridge, MA, USA (2014)
24. Zare, M.R., Müller, H.: Automatic detection of biomedical compound figure using bag of words. Int. J. Comput. Commun. Instrum. Eng. **4**, 6–10 (2017)
25. Zare, M.R., Mehtarizadeh, M.: An ensemble of deep semantic representation for medical x-ray image classification. In: 2021 55th Annual Conference on Information Sciences and Systems (CISS), pp. 1–6 (2021). https://doi.org/10.1109/CISS50987.2021.9400268

Love Wave in a Layered Magneto-Electro-Elastic Structure with Flexomagneticity and Micro-Inertia Effect

Olha Hrytsyna, Jan Sladek, and Vladimir Sladek

Abstract The higher-grade theory with flexomagneticity and micro-inertia effects is used to provide a theoretical framework for studying the propagation of Love wave along the free surface of semi-infinite piezoelectric substrate covered with a nano-thin guiding flexomagnetic layer. The phase velocity of Love wave is calculated for the magneto-electrically open boundary conditions. For tetragonal piezoelectric materials of point group 4 mm, the influence of piezoelectricity, flexomagneticity and micro-inertia characteristic length on phase velocity of Love wave is investigated. The effect of waveguide layer thickness on the dispersion curves is evaluated as well. It is found that the profile of dispersion curves depends on the material properties of the layer and substrate, the waveguide layer thickness, and the ratio between the values of flexomagnetic coefficients and the micro-inertia characteristic length. The obtained results can be useful in the design of nano-size sensors, actuators and acoustic devices where the high-frequency surface waves occur.

Keywords Love wave propagation · Strain gradient theory · Micro-inertia effect · Piezoelectricity · Flexomagneticity · Dispersion relation

1 Introduction

Within the framework of the classical theory, piezoelectricity and piezomagneticity describe the linear coupling effects arising in elastic solids under the action of external electromagnetic field and/or mechanical loading. Then, the electro/magneto-mechanical coupling between the electric/magnetic polarization and the uniform strain occurs only in noncentrosymmetric crystals. On the other hand, it is known that the nonuniform strain field (finite value of strain gradients) can induce an electric polarization in crystalline dielectrics (even in centrosymmetric ones). The electric polarization induced by the strain gradient is referred to as flexoelectricity [30, 34].

O. Hrytsyna (✉) · J. Sladek · V. Sladek
Institute of Construction and Architecture Slovak Academy of Sciences, Dúbravská cesta 9, 84503 Bratislava, Slovakia
e-mail: olha.hrytsyna@savba.sk

H. Dai (ed.), *Computational and Experimental Simulations in Engineering*, Mechanisms and Machine Science 119,
https://doi.org/10.1007/978-3-031-02097-1_18

The dependence between strain gradient and magnetic polarization is known as flexomagneticity [8, 9, 20]. Nowadays, due to the useful properties, the flexoelectric and flexomagnetic materials are widely used in sensors, actuators, filters, delay line resonators and other acoustic devices where the high-frequency surface waves may occur. In the recent decade, researches have paid special attention to the Love-type wave propagation in layered structures. The Love wave is a transverse surface wave having one component of mechanical displacement, which is parallel to the guiding layer surface and perpendicular to the direction of the wave propagation. Such kind of a surface wave in an isotropic layer deposited on isotropic substrate was originally studied by Love in 1911 [19]. The Love-wave problems in electro-magneto-elastic layered structures are widely investigated in recent studies, but the literature is mostly focused on the piezoelectric crystals [4, 7, 18, 35] and piezomagnetic or piezoelectromagnetic materials [1, 3, 5, 6, 10, 31]. The mentioned investigations were based on the postulates of the classical theory. However, to consider the effect of flexoelectricity/flexomagneticity on the surface wave propagation, the non-classical theories should be used.

To take into account the microscopic aspects of material structure and interatomic interactions, the generalized mathematical model with polarization gradient was used to investigate the Love wave propagation in centrosymmetric, isotropic, dielectric layer attached to an isotropic half-space [21]. Making use of non-classical theory with surface effects, the behavior of Love waves in an electrically-shorted piezoelectric nanofilm on an elastic substrate was studied by Zhang with co-workers [37]. Recently, the strain gradient theory of electroelastic media with flexoelectricity has been used to study the existence of Love wave in structure consisting of a flexoelectric layer rigidly linked to an elastic substrate [33]. In the above paper, in addition to the strain gradient, the effect of the high-order electric quadrupoles was considered as well. The results showed that the dispersion curves of Love waves are strongly dependent on the guiding layer thickness if its thickness reduces to nanometers. It was also derived that if the flexoelectricity is taken into account, the real part of the phase velocity can exceed the shear bulk wave velocity in the substrate and thus the 'cut-off wave numbers' can emerge [13, 33]. Using the governing equations of the strain gradient piezoelectricity, Singhal et al. [26, 27] analytically investigated the Love-type wave vibrations in a piezoelectric thin film overlying the pre-stressed elastic plate and studied the flexoelectricity effect in distinct piezoelectric materials (PZT-2, PZT-4, PZT-5H, $LiNbO_3$, $BaTiO_3$). These investigations proved that the flexoelectric effect is pronounced for sufficiently large wave numbers. On the other hand, series of studies [11, 12, 14, 15, 24, 25, 32] revealed that for high-frequency waves, it is very important to consider the micro-inertia effect. Using the strain gradient theory, a combined influence of the flexoelectric coefficients and micro-inertia characteristic length on electromechanical behavior of Love wave has been considered in [13]. It was found that flexoelectricity increases the phase velocity of Love wave while the micro-inertia effect decreases its value. The authors concluded that both the flexoelectricity and micro-inertia effect signifficantly influence the phase velocity of short-length waves and could not be omitted in layered structures with nano-scale dimensions.

Although the Love wave propagation with consideration of the flexoelectric effect has been investigated in several papers [13, 26, 27, 33], there are no studies using the mathematical models for Love waves in piezo-/flexo-magnetic structures. It should be noted that up to now, some results regarding the flexomagnetic effect in solids have been published (see for example, [8, 9, 20, 22, 36]). However, studies on the flexomagnetic effect are very seldom in literature. Since Love waves in magneto-electro-elastic materials have a practical importance, in this paper we study the influence of piezo-/flexo-magnetic effect combined with micro-inertia effect on Love wave propagation in a nano-sized wave-guiding piezomagnetic layer rigidly bonded to a piezoelectric substance. In order to separate the influence of the electric and magnetic effects on the Love waves, the piezo- and flexo-electric properties are omitted in the layer.

The paper structure is as follows. The linear equations of magneto-electro-elastic anisotropic media with flexomagneticity and micro-inertia effect are summarized in Sect. 2.1. Equations that describe the Love wave propagation (anti-plane motion) in piezoelectric and flexomagnetic ceramics are presented in Sects. 2.2 and 2.3, respectively. Section 2.4 contains the equations and general solution to the boundary problem in vacuum. Boundary conditions and dispersion relation are obtained in Sect. 2.5. Numerical results for the following material combination 'flexomagnetic ceramic $CoFe_2O_4$ and piezoelectric ceramic $BaTiO_3$' are presented in Sect. 3. The conclusions are drawn in final Sect. 4.

2 Formulation and Theoretical Treatment of the Problem

2.1 *Basic Relations*

Based on the results presented in works [5, 8, 17, 28], the free energy density function F of a magneto-electro-elastic continuum with piezo-/flexo-magnetic and piezoelectric effects can be generalized as:

$$\begin{aligned} F =& \frac{1}{2}c_{ijkl}\varepsilon_{ij}\varepsilon_{kl} - \frac{1}{2}\mu_{ij}H_iH_j - \frac{1}{2}a_{ij}E_iE_j + \frac{1}{2}g_{jklmni}\eta_{jkl}\eta_{mni} \\ & - d_{kji}\varepsilon_{ij}H_k - e_{kji}\varepsilon_{ij}E_k - q_{ij}E_iH_j - \xi_{ijkl}H_i\eta_{jkl}. \end{aligned}$$

Here, ε_{ij}, E_i, and H_i are the strain, electric field, and magnetic field components, respectively; η_{mni} is the component of strain-gradient tensor; c_{ijkl}, e_{ijk}, d_{ijk}, a_{ij}, μ_{ij} and q_{ij} represent the elastic, piezoelectric, piezomagnetic, electric permittivity, magnetic permeability and magneto-electric constants, respectively; g_{jklmni} is the higher order elastic coefficient representing the strain-gradient elasticity, and ξ_{ijkl} is the flexomagnetic coefficient.

For an anisotropic piezoelectric/piezomagnetic media with flexomagneticity, the linear coupled constitutive equations can be expressed as follows:

$$\sigma_{ij} = \frac{\partial F}{\partial \varepsilon_{ij}} = c_{ijkl}\varepsilon_{kl} - e_{kij}E_k - d_{kij}H_k, \tag{1}$$

$$\tau_{jkl} = \frac{\partial F}{\partial \eta_{jkl}} = -\xi_{ijkl}H_i + g_{jklmni}\eta_{mni}, \tag{2}$$

$$D_i = -\frac{\partial F}{\partial E_i} = e_{ijk}\varepsilon_{jk} + a_{ij}E_j + q_{ij}H_j, \tag{3}$$

$$B_i = -\frac{\partial F}{\partial H_i} = d_{ijk}\varepsilon_{jk} + q_{ij}E_j + \mu_{ij}H_j + \xi_{ijkl}\eta_{jkl}. \tag{4}$$

Here, symbols σ_{ij}, τ_{ijk}, D_i and B_i are used to denote the stress tensor, higher-order stress, electric displacement and magnetic induction tensors, respectively. Note that the last terms in Eqs. (2) and (4) are the contribution from the strain gradients.

The linear strain–displacement relations and expressions for the strain-gradient tensor are defined as:

$$\varepsilon_{ij} = \frac{1}{2}\left(u_{i,j} + u_{j,i}\right), \quad \eta_{ijk} = \varepsilon_{ij,k} = \frac{1}{2}\left(u_{i,jk} + u_{j,ik}\right), \tag{5}$$

where u_i is the component of the displacement vector, and comma stands for partial differentiation with respect to the indicated space coordinate.

Within the quasi-static approximation, the equations, which relate the electric field and magnetic field vectors to electric potential φ_e and magnetic potential ψ_m, are:

$$E_j = -\varphi_{e,j}, \quad H_j = -\psi_{m,j}. \tag{6}$$

When the micro-inertia effect is taken into account, the motion equation can be written as follows [2, 29]:

$$\sigma_{ij,j} - \tau_{ijk,jk} = \rho\left(1 - l_1^2\nabla^2\right)\ddot{u}_i, \tag{7}$$

where ρ is the mass density, l_1 is used to denote the micro-inertia characteristic length, ∇ is nabla operator and the dot over the vector component u_i refers to the time derivative. Note that the micro-inertia characteristic length l_1 is linked to the microstructure of the material [2].

For the electrostatics, in media without free electric charges, the electric and magnetic fields are governed by the Gauss-Coulomb and the Gauss–Faraday laws and are given by [16]:

$$D_{k,k} = 0, \quad B_{k,k} = 0. \tag{8}$$

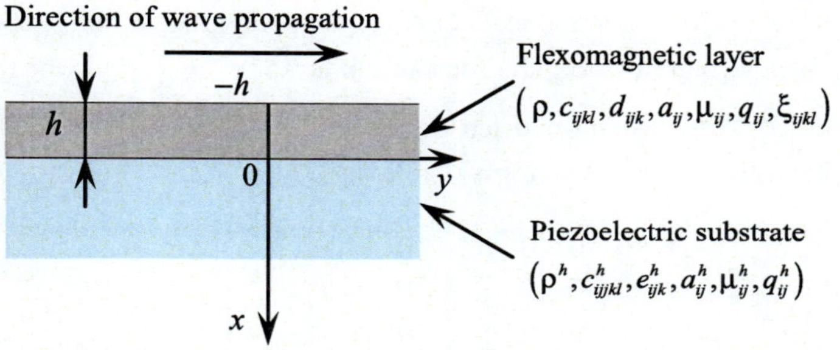

Fig. 1 Schema of the layered structure and choice of the coordinate system

Field equations (7), (8), constitutive and kinematic relations (1)–(6) are needed for a unique set of equations of linear strain gradient theory of piezo-flexomagnetic continua with micro-inertia effect.

Let us obtain the differential equations describing the Love wave propagation in a layered structure formed by a piezoelectric transversely isotropic semi-infinite substrate and a thin piezo/-flexo-magnetic guiding layer. The medium above the layer is air. A Cartesian coordinate system (*x, y, z*) is chosen in such a way that the *x*-axis is vertical to the substrate surface (see Fig. 1). Assume that the surface wave propagates in the *y*-direction and its amplitude decays with depth along the *x*-axis. We suppose that the upper surface of the piezo/-flexo-magnetic layer ($x = -h$) is traction free with open-circuit conditions for the electric and magnetic fields. Note that constitutive relations (1)–(4) take on a different form in the guiding layer and in the substrate because of the different material properties.

2.2 *Piezoelectric Substrate (Domain x > 0)*

The substrate is considered as a piezoelectric material. Because of huge dimensions of the substrate, the flexo-electric/magnetic and micro-inertia effects are supposed to be negligible. In this case, constitutive relations (1)–(4) can be written as:

$$\sigma^h_{ij} = c^h_{ijkl}\varepsilon^h_{kl} - e^h_{kij}E^h_k, \tag{9}$$

$$D^h_i = e^h_{ijk}\varepsilon^h_{jk} + a^h_{ij}E^h_j + q^h_{ij}H^h_j, \tag{10}$$

$$B^h_i = q^h_{ij}E^h_j + \mu^h_{ij}H^h_j. \tag{11}$$

Note that here and in what follows, all quantities related to the half-space are indicated by superscript '*h*'.

In the current work, the tetragonal crystal of the point group 4 mm is considered. Hence, by using Voigt notation, the fourth-rank tensor $\mathbf{c}^h = \left\{c_{ijkl}^h\right\}$, the third-rank tensor $\boldsymbol{e}^h = \left\{e_{kij}^h\right\}$, the second-rank tensors $\mathbf{a}^h = \left\{a_{ij}^h\right\}$, $\boldsymbol{\mu}^h = \left\{\mu_{ij}^h\right\}$ and $\mathbf{q}^h = \left\{q_{ij}^h\right\}$ can be represented in the matrix form as follows [23]:

$$\begin{bmatrix} c_{11}^h & c_{12}^h & c_{13}^h & 0 & 0 & 0 \\ c_{12}^h & c_{11}^h & c_{13}^h & 0 & 0 & 0 \\ c_{13}^h & c_{13}^h & c_{33}^h & 0 & 0 & 0 \\ 0 & 0 & 0 & c_{44}^h & 0 & 0 \\ 0 & 0 & 0 & 0 & c_{44}^h & 0 \\ 0 & 0 & 0 & 0 & 0 & c_{66}^h \end{bmatrix}, \begin{bmatrix} 0 & 0 & e_{31}^h \\ 0 & 0 & e_{31}^h \\ 0 & 0 & e_{33}^h \\ 0 & e_{15}^h & 0 \\ e_{15}^h & 0 & 0 \\ 0 & 0 & 0 \end{bmatrix}, \begin{bmatrix} b_{11}^h & 0 & 0 \\ 0 & b_{11}^h & 0 \\ 0 & 0 & b_{33}^h \end{bmatrix},$$

where the notation $b_{ij}^h \in \left\{a_{ij}^h, \mu_{ij}^h, q_{ij}^h\right\}$ is used.

The quantities that characterize the wave propagation within the substrate are:

$$\mathbf{u}^h = \left(0, 0, u_3^h(x, y, t)\right), \ \mathbf{E}^h = \left(E_1^h(x, y, t), E_2^h(x, y, t), 0\right)$$
$$\mathbf{B}^h = \left(B_1^h(x, y, t), B_2^h(x, y, t), 0\right),$$
$$\varphi_e^h = \varphi_e^h(x, y, t), \ \psi_m^h = \psi_m^h(x, y, t).$$

The non-vanishing strain and stress tensors components and electromagnetic field vectors can be given as:

$$\varepsilon_{13}^h = \frac{1}{2}\frac{\partial u_3^h}{\partial x}, \ \varepsilon_{32}^h = \frac{1}{2}\frac{\partial u_3^h}{\partial y}, \tag{12}$$

$$E_1^h = -\frac{\partial \varphi_e^h}{\partial x}, \ E_2^h = -\frac{\partial \varphi_e^h}{\partial y}, \ B_1^h = -\frac{\partial \psi_m^h}{\partial x}, \ B_2^h = -\frac{\partial \psi_m^h}{\partial y}, \tag{13}$$

$$\sigma_{31}^h = 2c_{44}^h \varepsilon_{31}^h - e_{15}^h E_1^h, \ \sigma_{32}^h = 2c_{44}^h \varepsilon_{32}^h - e_{15}^h E_2^h, \tag{14}$$

$$D_1^h = a_{11}^h E_1^h + q_{11}^h H_1^h + 2e_{15}^h \varepsilon_{13}^h, \ D_2^h = a_{11}^h E_2^h + q_{11}^h H_2^h + 2e_{15}^h \varepsilon_{32}^h, \tag{15}$$

$$B_1^h = q_{11}^h E_1^h + \mu_{11}^h H_1^h, \ B_2^h = q_{11}^h E_2^h + \mu_{11}^h H_2^h. \tag{16}$$

The general solution for waves spreading in y-direction of the infinite half-space can be obtained by the method of separation of variables as follows:
$u_3^h(x, y, t) = u^h(x)e^{ik(y-ct)}$, $\varphi_e^h(x, y, t) = \varphi^h(x)e^{ik(y-ct)}$, $\psi_m^h(x, y, t) = \psi^h(x)e^{ik(y-ct)}$.

Here, $u^h(x)$, $\varphi^h(x)$ and $\psi^h(x)$ are unknown functions which represent the amplitudes of the mechanical displacement, electric potential and magnetic potential in half-space; k denotes the wave number, c is the phase velocity, and i is the imaginary unit defined by formula $i = \sqrt{-1}$.

Hence, for deformable half-space without micro-inertia effect, the governing set of differential equations reduces to:

$$c_{44}^h \frac{d^2u^h}{dx^2} - (c_{44}^h - \rho^h c^2)k^2 u^h + e_{15}^h\left(\frac{d^2\varphi^h}{dx^2} - k^2\varphi^h\right) = 0, \tag{17}$$

$$\frac{d^2\varphi^h}{dx^2} - k^2\varphi^h = \frac{e_{15}^h}{\overline{a}_{11}^h}\left(\frac{d^2u^h}{dx^2} - k^2u^h\right), \tag{18}$$

$$\frac{d^2\psi^h}{dx^2} - k^2\psi^h = -\frac{q_{11}^h e_{15}^h}{\mu_{11}^h \overline{a}_{11}^h}\left(\frac{d^2u^h}{dx^2} - k^2u^h\right). \tag{19}$$

Here, ρ^h denotes the mass density of substrate, and $\overline{a}_{11}^h = a_{11}^h\left[1 - \left(q_{11}^h\right)^2 / a_{11}^h\mu_{11}^h\right]$.

Since the displacement, electric and magnetic potentials in the substrate should tend to zero far away from the interface (that is, $u_3^h \to 0$, $\varphi_e^h \to 0$, $\psi_m^h \to 0$ as $x \to +\infty$), a general solution to Eqs. (17)–(19) can be found as:

$$u_3^h(x, y, t) = C_1 e^{-\beta kx} e^{ik(y-ct)}, \tag{20}$$

$$\varphi_e^h(x, y, t) = \left(\frac{e_{15}^h}{\overline{a}_{11}^h} C_1 e^{-\beta kx} + C_2 e^{-kx}\right) e^{ik(y-ct)}, \tag{21}$$

$$\psi_m^h(x, y, t) = \left(-\frac{q_{11}^h e_{15}^h}{\mu_{11}^h \overline{a}_{11}^h} C_1 e^{-\beta kx} + C_3 e^{-kx}\right) e^{ik(y-ct)}. \tag{22}$$

Here, C_1, C_2, and C_3 are unknown constants, $\beta = \sqrt{1 - c^2/(c_{pe}^h)^2}$, and $c_{pe}^h = \sqrt{\overline{c}_{44}^h/\rho^h}$ is the velocity of the shear wave in piezoelectric substrate where $\overline{c}_{44}^h$ is given by $\overline{c}_{44}^h = c_{44}^h + \left(e_{15}^h\right)^2 / \overline{a}_{11}^h$.

2.3 Flexomagnetic Wave-Guide Layer (Domain $-h < x < 0$)

Studying the Love waves in a guiding layer, the displacement vector has an axial component u_3 only, that is $\mathbf{u} = (0, 0, u_3(x, y, t))$. The electric field vector, magnetic field vector, electric potential and magnetic potential are assumed to be as follows: $\mathbf{E} = (E_1(x, y, t), E_2(x, y, t), 0)$, $\mathbf{B} = (B_1(x, y, t), B_2(x, y, t), 0)$, $\varphi_e = \varphi_e(x, y, t)$, $\psi_m = \psi_m(x, y, t)$. The kinematic relations are:

$$\varepsilon_{31} = \frac{1}{2}\frac{\partial u_3}{\partial x},\ \varepsilon_{32} = \frac{1}{2}\frac{\partial u_3}{\partial y}, \tag{23}$$

$$\begin{aligned} \eta_{131} = \eta_{311} = \frac{1}{2}\frac{\partial^2 u_3}{\partial x^2},\ \eta_{232} = \eta_{322} = \frac{1}{2}\frac{\partial^2 u_3}{\partial y^2}, \\ \eta_{231} = \eta_{321} = \eta_{132} = \eta_{312} = \frac{1}{2}\frac{\partial^2 u_3}{\partial x \partial y}, \end{aligned} \tag{24}$$

$$E_1 = -\frac{\partial \varphi_e}{\partial x},\ E_2 = -\frac{\partial \varphi_e}{\partial y}, \tag{25}$$

$$H_1 = -\frac{\partial \psi_m}{\partial x},\ H_2 = -\frac{\partial \psi_m}{\partial y}. \tag{26}$$

For flexomagnetic ceramic, constitutive equations (1)–(4) can be rewritten as:

$$\sigma_{31} = 2c_{44}\varepsilon_{31} - d_{15}H_1,\ \sigma_{32} = 2c_{44}\varepsilon_{32} - d_{15}H_2, \tag{27}$$

$$\tau_{311} = \xi_{52}H_2,\ \tau_{321} = -\xi_{41}H_1,\ \tau_{321} = -\xi_{41}H_1,\ \tau_{322} = \xi_{41}H_2, \tag{28}$$

$$D_1 = a_{11}E_1 + q_{11}H_1,\ D_2 = a_{11}E_2 + q_{11}H_2, \tag{29}$$

$$B_1 = 2d_{15}\varepsilon_{31} + q_{11}E_1 + \mu_{11}H_1 + 2(\xi_{41} + \xi_{52})\eta_{321}, \tag{30}$$

$$B_2 = 2d_{15}\varepsilon_{32} + q_{11}E_2 + \mu_{11}H_2 - 2\xi_{52}\eta_{311} - 2\xi_{41}\eta_{322}. \tag{31}$$

The following notation is adopted here for the flexomagnetic coefficients $\xi_{2311} = \xi_{2131} = -\xi_{52},\ \xi_{1312} = \xi_{1132} = \xi_{52},\ \xi_{1231} = \xi_{1321} = \xi_{41},\ \xi_{2232} = \xi_{2322} = -\xi_{41}$ Following Yang et al. [33], the effect of strain gradient terms is neglected in the formulae (28), i.e. $g_{jklmni} = 0$.

Substitution of constitutive equations (27)–(31) and kinematic relations (23)–(26) into balance equations (7) and (8) yields:

$$\begin{aligned} &c_{44}\left(\frac{\partial^2 u_3}{\partial x^2} + \frac{\partial^2 u_3}{\partial y^2}\right) + d_{15}\left(\frac{\partial^2 \psi_m}{\partial x^2} + \frac{\partial^2 \psi_m}{\partial y^2}\right) + \xi_{41}\left(\frac{\partial^3 \psi_m}{\partial y^3} - \frac{\partial^3 \psi_m}{\partial x^2 \partial y}\right) \\ &= \rho\left[\frac{\partial u_3}{\partial t^2} - l_1^2\left(\frac{\partial^4 u_3}{\partial t^2 \partial x^2} + \frac{\partial^4 u_3}{\partial t^2 \partial y^2}\right)\right], \end{aligned} \tag{32}$$

$$\begin{aligned} \left(\frac{\partial^2 \varphi_e}{\partial x^2} + \frac{\partial^2 \varphi_e}{\partial y^2}\right) = -\frac{q_{11}}{a_{11}}\left(\frac{\partial^2 \psi_m}{\partial x^2} + \frac{\partial^2 \psi_m}{\partial y^2}\right) \\ -q_{11}\left(\frac{\partial^2 \varphi_e}{\partial x^2} + \frac{\partial^2 \varphi_e}{\partial y^2}\right) - \mu_{11}\left(\frac{\partial^2 \psi_m}{\partial x^2} + \frac{\partial^2 \psi_m}{\partial y^2}\right) \end{aligned} \tag{33}$$

$$+d_{15}\left(\frac{\partial^2 u_3}{\partial x^2}+\frac{\partial^2 u_3}{\partial y^2}\right)+\xi_{41}\left(\frac{\partial^3 u_3}{\partial x^2 \partial y}-\frac{\partial^3 u_3}{\partial y^3}\right)=0. \tag{34}$$

Governing set of equations (32)–(34) describes the propagation of the surface acoustic wave and its associated electromagnetic field in the piezo-/flexo-magnetic wave-guide layer. Comparing to the classical theory for piezo-magneticity, additional terms proportional to the flexomagnetic coefficient ξ_{41} and micro-inertia characteristic length l_1 appeared in governing equations (32) and (34).

The displacement component, electric and magnetic potentials are assumed as:

$$u_3(x, y, t)=u(x)e^{ik(y-ct)}, \tag{35}$$

$$\varphi_e(x, y, t)=\varphi(x)e^{ik(y-ct)}, \tag{36}$$

$$\psi_m(x, y, t)=\psi(x)e^{ik(y-ct)}. \tag{37}$$

where $u(x)$, $\varphi(x)$ and $\psi(x)$ are the amplitudes of the mechanical displacement, electric and magnetic potentials in a wave-guide layer.

Substitution of expressions (35)–(37) into governing equations (32)–(34) yields system of ordinary differential equations:

$$\begin{aligned}&\left(c_{44}-\rho l_1^2 k^2 c^2\right)\frac{d^2 u}{dx^2}+\left[\rho c^2\left(1+l_1^2 k^2\right)-c_{44}\right]k^2 u\\&+(d_{15}-ik\xi_{41})\frac{d^2\psi}{dx^2}-(d_{15}+ik\xi_{41})k^2\psi=0,\end{aligned} \tag{38}$$

$$\left(\frac{d^2\varphi}{dx^2}-k^2\varphi\right)=-\frac{q_{11}}{a_{11}}\left(\frac{d^2\psi}{dx^2}-k^2\psi\right), \tag{39}$$

$$\begin{aligned}&q_{11}\left(\frac{d^2\varphi}{dx^2}-k^2\varphi\right)+\mu_{11}\left(\frac{d^2\psi}{dx^2}-k^2\psi\right)-(d_{15}+ik\xi_{41})\frac{d^2 u}{dx^2}\\&+(d_{15}-ik\xi_{41})k^2 u=0.\end{aligned} \tag{40}$$

The general solution to equations (38)–(40) can be written as follows:

$$u(x)=S_1\left(N_1 e^{k\Lambda_1 x}+N_2 e^{-k\Lambda_1 x}\right)+S_2\left(N_3 e^{k\Lambda_2 x}+N_4 e^{-k\Lambda_2 x}\right), \tag{41}$$

$$\begin{aligned}\varphi(x)=&N_5 e^{kx}+N_6 e^{-kx}\\&-\frac{q_{11}}{a_{11}}\left(N_1 e^{k\Lambda_1 x}+N_2 e^{-k\Lambda_1 x}+N_3 e^{k\Lambda_2 x}+N_4 e^{-k\Lambda_2 x}\right),\end{aligned} \tag{42}$$

$$\psi(x) = N_1 e^{k\Lambda_1 x} + N_2 e^{-k\Lambda_1 x} + N_3 e^{k\Lambda_2 x} + N_4 e^{-k\Lambda_2 x}. \tag{43}$$

Here, N_j $(j = 1, ..., 6)$ are unknown constants to be determined and the following notations are used:

$$\Lambda_1^2 = \frac{2c_m^2 - c^2 - 2l_1^2c^2k^2 - 2k^2d_\xi^2 + \sqrt{D}}{2\left(c_m^2 - l_1^2k^2c^2 + k^2d_\xi^2\right)},$$

$$\Lambda_2^2 = \frac{2c_m^2 - c^2 - 2l_1^2c^2k^2 - 2k^2d_\xi^2 - \sqrt{D}}{2\left(c_m^2 - l_1^2k^2c^2 + k^2d_\xi^2\right)},$$

$$S_1 = -\frac{d_{15}\left(\Lambda_1^2 - 1\right) - i\xi_{41}k\left(\Lambda_1^2 + 1\right)}{\left(c_{44} - l_1^2\rho c^2k^2\right)\left(\Lambda_1^2 - 1\right) + \rho c^2},$$

$$S_2 = -\frac{d_{15}\left(\Lambda_2^2 - 1\right) - i\xi_{41}k\left(\Lambda_2^2 + 1\right)}{\left(c_{44} - l_1^2\rho k^2c^2\right)\left(\Lambda_2^2 - 1\right) + \rho c^2},$$

$$D = c^4 + 8d_\xi^2c^2k^2\left(1 + 2l_1^2k^2\right) - 16c_m^2d_\xi^2k^2,$$

$$d_\xi^2 = \frac{\xi_{41}^2}{\rho\overline{\mu}_{11}},\ c_m^2 = \frac{\overline{c}_{44}}{\rho},\ \overline{c}_{44} = c_{44} + \frac{d_{15}^2}{\overline{\mu}_{11}},\ \overline{\mu}_{11} = \mu_{11}\left(1 - \frac{q_{11}^2}{a_{11}\mu_{11}}\right).$$

2.4 *The Air (Domain $x < -h$)*

Since the layer is made of the piezomagnetic ceramic, we take the electromagnetic field in the air (domain $x < -h$) into account. We consider the air as a vacuum. Both the electric and magnetic potentials in the vacuum satisfy the Laplace equations, i.e., $\nabla^2\varphi_e^v = 0$ and $\nabla^2\psi_m^v = 0$. Here, ∇^2 is the two-dimensional Laplac operator, and superscript 'v' indicates the electric and magnetic potentials in the vacuum. The potentials tend to zero far away from the surface $x = -h$ along the negative x-direction, i.e., $\varphi_e^v \to 0$ and $\psi_e^v \to 0$ as $x \to -\infty$. Therefore, the electromagnetic field above the layer is given by the expressions:

$$\varphi_e^v(x, y, t) = C_4 e^{kx} e^{ik(y-ct)},\ \psi_m^v(x, y, t) = C_5 e^{kx} e^{ik(y-ct)}, \tag{44}$$

where C_4 and C_5 are the unknown constants. The electric displacement $D_i^v = a_0 E_i^v$ and magnetic induction $B_i^v = \mu_0 H_i^v$ in the air ($x < -h$) are as follows:

$$D_1^v = -a_0\frac{\partial\varphi_e^v}{\partial x} = -kC_4a_0e^{kx}e^{ik(y-ct)},\ D_2^v = -a_0\frac{\partial\varphi_e^v}{\partial y} = -ikC_4a_0e^{kx}e^{ik(y-ct)}, \tag{45}$$

$$B_1^v = -\mu_0 \frac{\partial \psi_m^v}{\partial x} = -kC_5\mu_0 e^{kx} e^{ik(y-ct)},\ B_2^v = -\mu_0 \frac{\partial \psi_m^v}{\partial y} = -ikC_5\mu_0 e^{kx} e^{ik(y-ct)}. \tag{46}$$

Here, a_0 and μ_0 are the electric permittivity and magnetic permeability of vacuum, respectively.

2.5 Boundary Conditions and Dispersion Equation

The eleven unknown constants $N_j\ (j = 1-6)$ and $C_l\ (l = 1-5)$ are determined by the boundary conditions at surfaces $x = -h$ and $x = 0$.

In case of the electrical and magnetic open-circuit conditions, we require the flux of electric displacements, magnetic inductions, as well as the electric and magnetic potentials to be continuous across the surface $x = -h$. Thus, for traction-free interfaces and continuity of displacements, we consider the complete boundary conditions as:

On the surface $x = -h$:

$$\left(\left(\sigma_{31} - \tau_{311,1} - \tau_{312,2}\right) - \tau_{321,2} + \rho l_1^2 \frac{\partial \ddot{u}_3}{\partial x}\right)\Big|_{x=-h} = 0, \tag{47}$$

$$\begin{aligned} \varphi_e|_{x=-h} &= \varphi_e^v\big|_{x=-h},\ D_1|_{x=-h} = D_1^v\big|_{x=-h}, \\ \psi_m|_{x=-h} &= \psi_m^v\big|_{x=-h},\ B_1|_{x=-h} = B_1^v\big|_{x=-h}. \end{aligned} \tag{48}$$

On the interface between the layer and half-space $x = 0$:

$$u_3|_{x=0} = u_3^h\big|_{x=0},\ \left(\left(\sigma_{31} - \tau_{311,1} - \tau_{312,2}\right) - \tau_{321,2} + \rho l_1^2 \frac{\partial \ddot{u}_3}{\partial x}\right)\Big|_{x=0} = \sigma_{31}^h\big|_{x=0}, \tag{49}$$

$$\varphi_e|_{x=0} = \varphi_e^h\big|_{x=0},\ D_1|_{x=0} = D_1^h\big|_{x=0},\ \psi_m|_{x=0} = \psi_m^h\big|_{x=0},\ B_1|_{x=0} = B_1^h\big|_{x=0}. \tag{50}$$

Here, $\tau_{31} = \left(\sigma_{31} - \tau_{311,1} - \tau_{312,2}\right) - \tau_{321,2} + \rho l_1^2 \frac{\partial \ddot{u}_3}{\partial x}$ is the z-component of generalized tractions on the surface $x = \text{const}$, $y, z \in (-\infty, \infty)$.

Thus, the propagation problem of the Love wave in the layered structure turns into the solution of (20)–(22), (35)–(37), (41)–(43) and (44)–(46) under boundary conditions (47)–(50).

Substitution of the general solutions into boundary conditions (47)–(50) produces eleven homogeneous algebraic linear equations to find unknown constants $N_j\ (j = 1-6)$ and $C_l\ (l = 1-5)$. Eliminating $C_l\ (l = 1-5)$, N_5 and N_6 from the obtained set of equations we get four equations with respect to $N_j\ (j = 1-4)$ which can be written in a matrix form as follows: $\mathbf{MN} = 0$, where $\mathbf{N}^T = \begin{bmatrix} N_1 & N_2 & N_3 & N_4 \end{bmatrix}$, and $\mathbf{M}$ is a 4×4 coefficient matrix. The elements of matrix $\mathbf{M}$ are given by formulae:

$m_{11} = b_1 e^{-kh\Lambda_1}$, $m_{12} = -b_1 e^{kh\Lambda_1}$, $m_{13} = b_2 e^{-kh\Lambda_2}$, $m_{14} = -b_2 e^{kh\Lambda_2}$,

$m_{21} = (n_1 + \mu_0) e^{-kh\Lambda_1}$, $m_{22} = -(n_1 - \mu_0) e^{kh\Lambda_1}$,

$m_{23} = (n_2 + \mu_0) e^{-kh\Lambda_2}$, $m_{24} = -(n_2 - \mu_0) e^{kh\Lambda_2}$,

$m_{31} = \left[b_1 + \left(c_{pe}^h\right)^2 \beta \rho^h S_1 \right] - \frac{(n-1)}{(n+1)} \frac{a_{11}}{a_{11}^h} \left(b_1 - b \rho^h S_1 \right)$,

$m_{32} = -\left[b_1 - \left(c_{pe}^h\right)^2 \beta \rho^h S_1 \right] + \frac{(n-1)}{(n+1)} \frac{a_{11}}{a_{11}^h} \left(b_1 + b \rho^h S_1 \right)$,

$m_{33} = \left[b_2 + \left(c_{pe}^h\right)^2 \beta \rho^h S_2 \right] - \frac{(n-1)}{(n+1)} \frac{a_{11}}{a_{11}^h} \left(b_2 - b \rho^h S_2 \right)$,

$m_{34} = -\left[b_2 - \left(c_{pe}^h\right)^2 \beta \rho^h S_2 \right] + \frac{(n-1)}{(n+1)} \frac{a_{11}}{a_{11}^h} \left(b_2 + b \rho^h S_2 \right)$,

$m_{41} = n_1 - \mu_{11}^h$, $m_{42} = -\left(n_1 + \mu_{11}^h\right)$, $m_{43} = n_2 - \mu_{11}^h$, $m_{44} = -\left(n_2 + \mu_{11}^h\right)$.

Note that in the above formulae, for simplicity the terms with magneto-electric constants q_{11}^h and q_{11} are neglected and the following notations are adopted:

$b_1 = \left[\left(c_{44} - \rho l_1^2 k^2 c^2 \right) S_1 + d_{15} - ik\xi_{41} \right] \Lambda_1$,

$b_2 = \left[\left(c_{44} - \rho l_1^2 k^2 c^2 \right) S_2 + d_{15} - ik\xi_{41} \right] \Lambda_2$,

$n_1 = [d_{15} S_1 - \mu_{11} + ik(\xi_{41} + \xi_{52}) S_1] \Lambda_1$,

$n_2 = [d_{15} S_2 - \mu_{11} + ik(\xi_{41} + \xi_{52}) S_2] \Lambda_2$,

$b = \left(\frac{e_{15}^h e_{15}^h}{\rho^h a_{11}^h} - \left(c_{pe}^h\right)^2 \beta \right)$, $n = \frac{(a_{11} - a_0)}{(a_{11} + a_0)} e^{-2kh}$.

From the above equations one can observe that the flexomagneticity-related terms are dependent on the wave number k, while the micro-inertia-related terms are dependent on k^2.

Non-trivial solution to the boundary-value problem can be obtained if the determinant of matrix $\mathbf{M}$ is equal to zero. This condition leads to the transcendental dispersion equation $\det[\mathbf{M}(c, k)] = 0$, which determines the dependence of the Love-wave phase velocity c on the wave numbers k, i.e., $c = c(k)$. Since for the guiding layer, the influence of flexo-/piezo-magnetic and micro-inertia effects is taken into account and due to the consideration of piezoelectric properties of the substrate, the dispersion relation becomes very complicated and numerical methods should be used to solve it.

3 Numerical Results

Note that the Love wave exhibits a multimode character. Since the first mode is characterized by the largest amplitude, in current work, the attention has been focused on the electro-magneto-mechanical properties of this mode only. Following [33], we assume the wave number to be positive real quantity while the phase velocity of Love wave is considered as a complex one, $c = c_1 + ic_2$. The imaginary part of velocity c_2 characterizes the wave attenuation. The negative value of c_2 means that the modified wave amplitude (wave amplitude $\times e^{-ic_2 t}$) drops, whereas the positive value of c_2 implies that the modified wave amplitude grows with increasing time.

In this section, the numerical results regarding the dispersion relation of Love wave are provided for a layered structure with the following material properties 'flexomagnetic ceramic $CoFe_2O_4$ and piezoelectric ceramic $BaTiO_3$'. The material coefficients for barium titanate and cobalt ferrite are chosen as follows [5, 6, 28]:

$\rho^h = 5.8 \times 10^3$ kg/m^3, $c_{44}^h = 4.3 \times 10^{10}$ N/m^2, $e_{15}^h = 11.6$ C/m
$a_{11}^h = 1.12 \times 10^{-8}$ F/m, $\mu_{11}^h = 0.5 \times 10^{-5}$ Ns^2/C^2
$\rho = 5.3 \times 10^3$ kg/m^3, $c_{44} = 4.53 \times 10^{10}$ N/m^2
$d_{15} = 550$ N/Am, $a_{11} = 0.8 \times 10^{-10}$ F/m, $\mu_{11} = 5.9 \times 10^{-4}$ Ns^2/C^2.

For air, the electric permittivity and magnetic permeability coefficients are $a_0 = 8.85 \times 10^{-12}$ F/m and $\mu_0 = 4\pi \times 10^{-7}$ H/m. In numerical calculations, it is assumed that flexomagnetic coefficients ξ_{41} and ξ_{52} are equal to each other, i.e., $\xi_{41} = \xi_{52} \equiv \xi$, and their order increases from 10^{-6} to 10^{-5} [8, 28]. Usually, the dynamic characteristic length l_1 is set to be proportional to the material lattice parameter [32]. In this study, we assume l_1 to range from 0.4 Å to 6 Å. In calculations, it is also assumed that the thickness of the guiding layer is equal to 40 nm.

Figure 2 illustrate the influence of micro-inertia characteristic length on Love wave phase velocity when the flexomagneticity is neglected, i.e., $\xi_{41} = \xi_{52} = 0$. To study the influence of micro-inertia characteristic length on the dispersion curve, we assume that the above parameter ranges from 0.4 Å to 1.2 Å (Fig. 2a) or 1 Å to 3 Å (Fig. 2b). In this case, the imaginary part of phase velocity is equal to zero. In Fig. 2, the classical electro-magneto-elasticity solution ($l_1 = 0$) is also shown for comparison (see the solid line). It can be seen that for sufficiently large wave numbers, the micro-inertia characteristic length parameter has an effect on the phase velocity of the wave. Increasing the micro-inertia length parameter from 0.4 Å to 3 Å, the wave velocity decreases. The effect becomes more pronounced for larger values of micro-inertia characteristic length (Fig. 2b). The numerical investigations also showed that when the dynamic characteristic length l_1 does not exceed

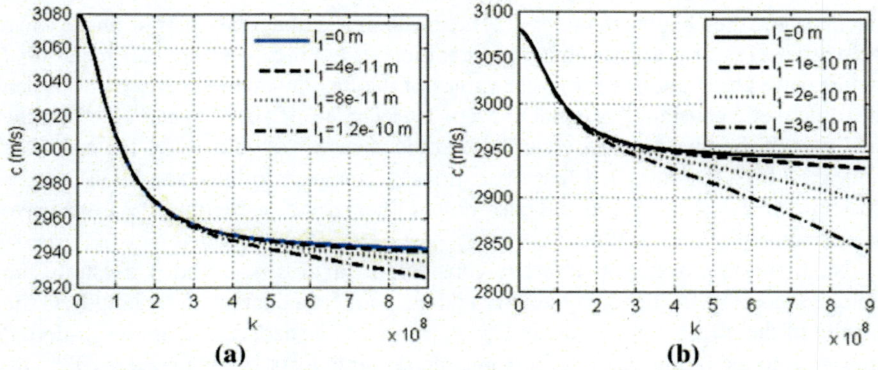

Fig. 2 The phase velocity of Love wave versus wave number k for the guiding layer thickness 40 nm and various micro-inertia characteristic lengths. The influence of flexomagneticity is neglected (i.e., $\xi_{41} = \xi_{52} = 0$)

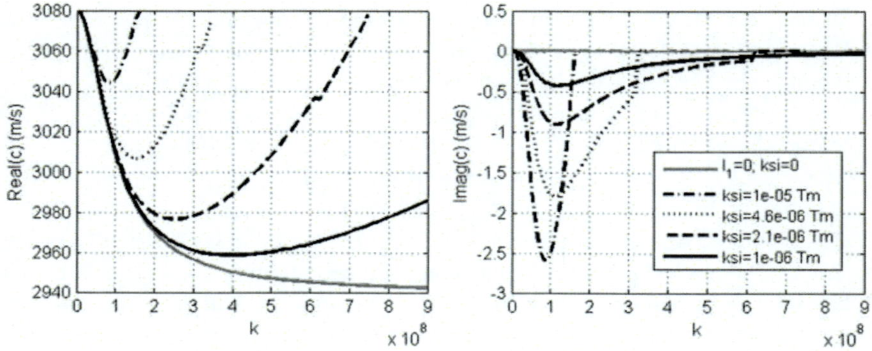

Fig. 3 Real and imaginary parts of phase velocity of Love wave versus wave number k for the guiding layer thickness 40 nm and various values of the flexomagnetic coefficients. The influence of micro-inertia terms is neglected (i.e., $l_1 = 0$)

$0.1 \div 0.2$ Å, the dispersion curve coincides with the results obtained from the classical theory. In that case, the influence of micro-inertia effect can be neglected.

The effect of flexomagnetic properties of the guiding layer on the wave velocity is illustrated in Fig. 3, where the influence of micro-inertia effect is not considered, i.e., it is assumed that $l_1 = 0$. Grey solid lines present the corresponding classical solution for piezomagnetic layer on a piezoelectric substrate. Within the classical theory, the phase velocity of Love wave is a monotonously decreasing function with the wave number. The presence of flexomagneticity leads to a complex function of phase velocity with negative imaginary part. In this case, from Fig. 3 it is observed that the real part of phase velocity first decreases for lower values of the wave number, reaches the minimum and then begins to rise. For a sufficiently large wave number, the phase velocity is higher than the one predicted by the classical theory. The imaginary part of the wave velocity, imag(c), displays the same trends, that is, it first declines and then begins to increase. The minimum of imaginary part of phase wave velocity decreases if the flexomagnetic coefficients ξ_{41} and ξ_{52} grow. This means that better wave attenuation is for larger values of the flexomagnetic coefficients. When flexomagnetic constant ξ is equal to 2.1×10^{-6}Tm, 4.6×10^{-6}Tm and 10^{-5}Tm, the real part of the phase velocity can exceed the shear wave velocity in the substrate and, therefore, the so-called 'cut-off regions' can appear in the considered layered structure (see dashed, dotted and dash-dotted lines). In these regions, the Love wave is not capable of propagating.

Next, the micro-inertial effect is considered. Figures 4, 5 and 6 illustrate the coupled effect of the flexomagneticity and micro-inertia characteristic length on the profile of the dispersion curves. In Fig. 4, the solid, dashed, dotted and dash-dotted lines are plotted for the values of flexomagnetic coefficients 10^{-6} Tm, 2.1×10^{-6} Tm, 4.6×10^{-6} Tm and 10^{-5} Tm, respectively. In the calculations, the micro-inertia characteristic length is assumed to be 4 Å. A solid grey line presents the result obtained from the classical theory of elastic electromagnetic media without flexomagneticity

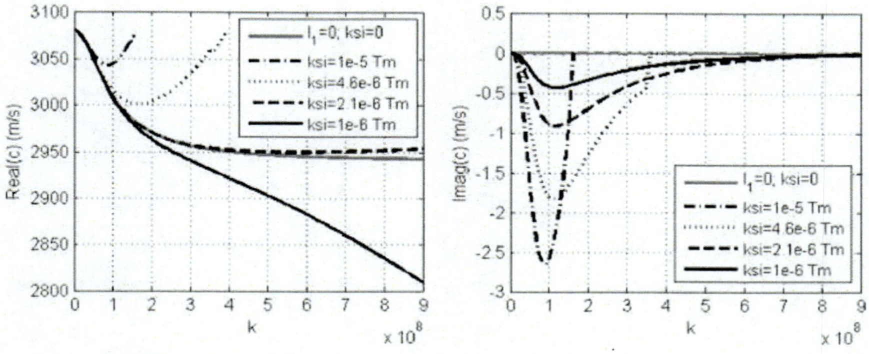

Fig. 4 Real and imaginary parts of phase velocity of Love wave versus wave number k for the guiding layer thickness 40 nm, $l_1 = 0.4$ nm and various values of the flexomagnetic coefficients

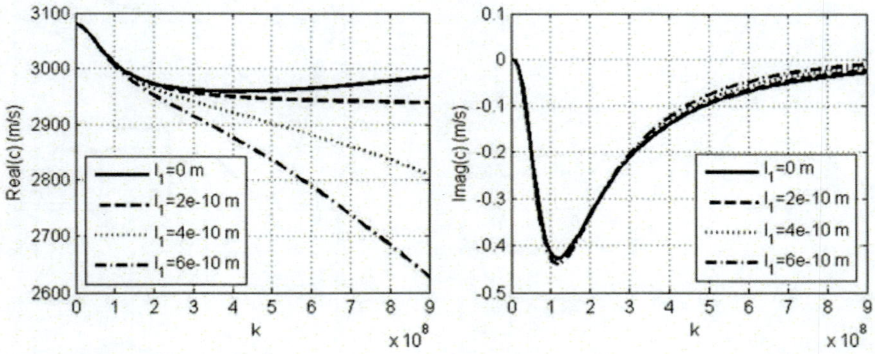

Fig. 5 Real and imaginary parts of phase velocity of Love wave versus wave number k for the guiding layer thickness 40 nm, $\xi_{41} = \xi_{52} = 10^{-6}$Tm and various values of the micro-inertia characteristic length

and micro-inertia effect. We found that the profile of the dispersion curves changes significantly if flexomagnetic and micro-inertia effects are considered. As can be seen from Fig. 4, due to combining influence of flexomagneticity and micro-inertia effect, for large wave numbers, the real part of phase velocity increases with the increase of the flexomagnetic coefficient ξ if $\xi = 2.1 \times 10^{-6}$ Tm, 4.6×10^{-6} Tm and 10^{-5} Tm. However, if $\xi = 10^{-6}$ Tm, the influence of micro-inertia terms becomes dominant and the real part of the phase velocity is smaller than the ones predicted by the classical theory. We also found that if the micro-inertia effect is taken into account, the cut-off regions do not occur when the $\xi = 2.1 \times 10^{-6}$ Tm. This means that the profile of dispersion curve significantly depends on the ratio of the flexomagnetic coefficients and micro-inertia characteristic length.

Figure 5 gives profiles of dispersion curves for flexomagnetic coefficient 10^{-6} Tm and values of the micro-inertia characteristic length 2 Å, 4 Å and 6 Å. From Fig. 5

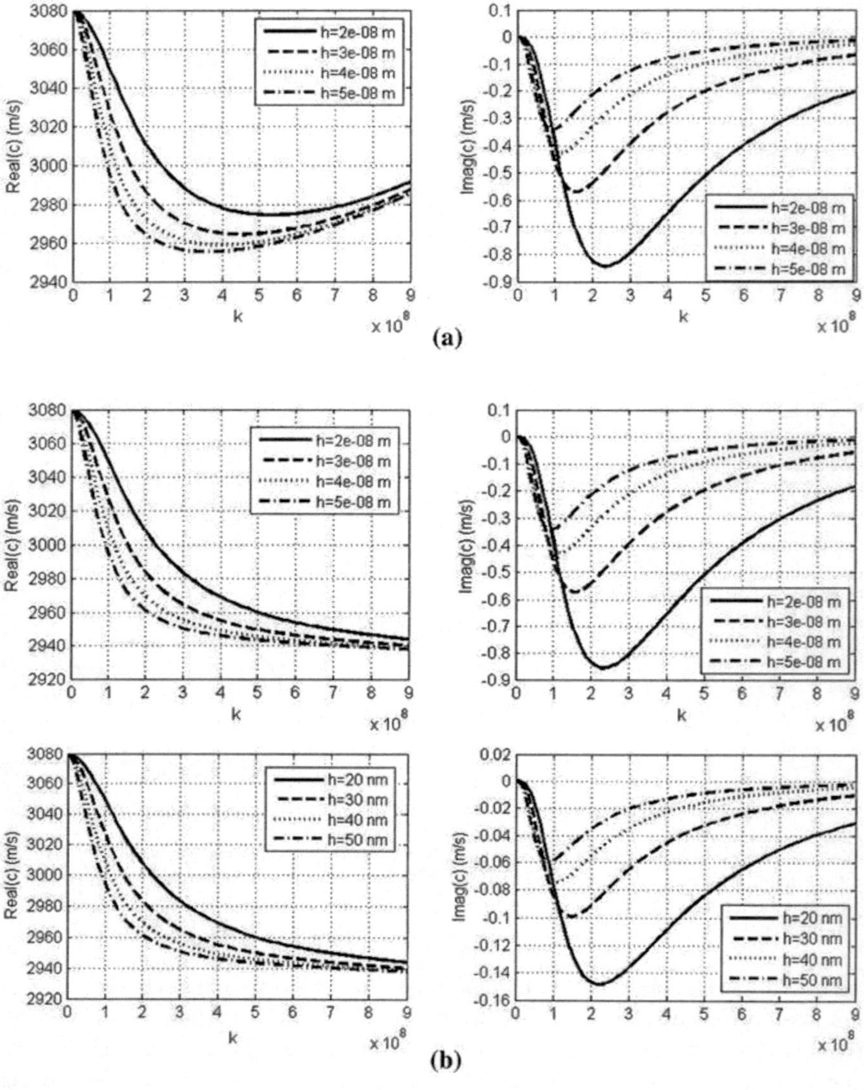

Fig. 6 The effect of the guiding layer thickness on the real and imaginary parts of phase velocity for flexomagnetic coefficients $\xi_{41} = \xi_{52} = 10^{-6}$ Tm, $l_1 = 0$ (figure **a**) and $l_1 = 0.2$ nm (figure **b**)

it is observed that micro-inertia terms have a significant influence on the real part of phase velocity c. This effect is stronger for high wave number (short wavelengths). When the wave number k is small, the imaginary parts of phase velocity calculated from various values of the micro-inertia characteristic length are close to each other.

Decreasing distinctions in dispersion curves for imag(c) can be observed only at large wave numbers.

Figure 6 shows the dispersion curves for the values of the guiding layer thickness 20 nm, 30 nm, 40 nm and 50 nm. The result obtained from the generalized theory of flexomagnetic media without micro-inertia effect is presented in Fig. 6a. The dispersion curves in Fig. 6b are plotted with the assumption that characteristic length is equal to 2 Å. As can be seen from the curves, the effect of the layer thickness is more pronounced for narrower layers. The minimum of real(c) decreases whereas the minimum of imag(c) increases if the guiding layer thickness increases. Reducing the layer thickness, the wave attenuation reaches its maximum at higher wave numbers *k*. From Fig. 6a, b one can also observe that the influence of micro-inertia terms becomes important for sufficiently high wave numbers.

4 Conclusions

The classical theories are not capable of appropriately describe the magneto-electro-mechanical behavior of Love waves in a nano-scale flexo-piezomagnetic layer overlying the piezoelectric half-space. The influence of flexomagneticity and micro-inertia effect should be considered in this case. In this study, the behavior of magneto-electro-elastic surface Love waves in a structure consisting of piezoelectric substrate of crystal class 4 mm and flexo-piezomagnetic elastic layer is studied. Mathematical model of a substrate takes the piezoelectric properties of the material into account while the relations for a nano-thin layer accommodate the influence of flexomagneticity and micro-inertia effect. A solution to dispersion relation is found for magneto-electrically open boundary conditions. The dependence of phase wave velocity on the wave number is numerically studied in detail for piezomagnetic ceramics $CoFe_2O_4$ and piezoelectric ceramics $BaTiO_3$ for various values of flexomagnetic coefficients and micro-inertia characteristic length.

The study proved that both flexomagnetic and micro-inertia effects play an important role in layered structures at the micro-/nano-scales and can significantly affect the profiles of dispersion curves. Numerical investigations showed that growing flexomagnetic coefficient increases the Love wave phase velocity, while its value decreases with increasing the micro-inertia length parameter. The influence of flexomagnetic and micro-inertia effects is remarkable for sufficiently high wave numbers. This influence is more pronounced for a smaller thickness of wave-guide layer as well as for larger values of micro-inertia characteristic lengths and flexomagnetic coefficients. Contrary to the prediction of the classical theories, if the flexomagnetic properties of a layer are taken into account, the cut-off regions can occur in the considered layered structure in case of large values of flexomagnetic coefficients and a small micro-inertia characteristic length. The profile of dispersion curves and the presence or absence of cut-off regions in these curves, depend on the guiding layer thickness, and on the ratio between the values of flexomagnetic coefficients and the micro-inertia characteristic length.

The obtained results can be helpful in mathematical modelling and engineering applications of new small-scale acoustic devices made of smart piezoelectric and piezomagnetic materials.

Funding This research was supported by the Slovak Research and Development Agency (grant number APVV-18-0004) and Ministry of Education, Science, Research and Sport of the Slovak Republic (grant number VEGA-2/0061/20).

References

1. Alshits, V.I., Darinskii, A.N.: On the existence of surface waves in half-infinite anisotropic elastic media with piezoelectric and piezomagnetic properties. Wave Motion **16**, 265–283 (1992)
2. Askes, H., Aifantis, E.C.: Gradient elasticity in statics and dynamics: An overview of formulations, length scale identification procedures. Int. J. Solids Struct. **48**, 1962–1990 (2011)
3. Chen, J.Y., Pan, E., Chen, H.L.: Wave propagation in magneto-electro-elastic multilayered plates. Int. J. Solids Struct. **44**, 1073–1085 (2007)
4. Danoyan, Z.N., Piliposian, G.T.: Surface electro-elastic Love waves in a layered structure with a piezoelectric substrate and a dielectric layer. Int. J. Solids Struct. **44**, 5829–5847 (2007)
5. Du, J.K., Jin, X., Wang, J.: Love wave propagation in layered magnetoelectro-elastic structures. Sci. China, Ser. G **51**(6), 617–631 (2008)
6. Du, J., Jin, X., Wang, J.: Love wave propagation in layered magneto-electro-elastic structures with initial stress. Acta Mech. **192**, 169–189 (2007)
7. Du, J., Xian, K., Wang, J., Yong, Y.-K.: Love wave propagation in piezoelectric layered structure with dissipation. Ultrasonics **49**, 281–286 (2009)
8. Eliseev, E.A., Glinchuk, M.D., Khist, V., Skorokhod, V.V., Blinc, R., Morozovska, A.N.: Linear magnetoelectric coupling and ferroelectricity induced by the flexomagnetic effect in ferroics. Phys. Rev. B **84**, 174112 (2011)
9. Eliseev, E.A., Morozovska, A.N., Glinchuk, M.D., Blinc, R.: Spontaneous flexoelectric/flexomagnetic effect in nanoferroics. Phys. Rev. B **79**, 165433 (2009)
10. Ezzin, H., Amor, M.B., Ghozlen, M.H.B.: Love waves propagation in a transversely isotropic piezoelectric layer on a piezomagnetic half-space. Ultrasonics **69**, 83–89 (2016)
11. Georgiadis, H.G., Vardoulakis, I., Lykotrafitis, G.: Torsional surface waves in a gradient-elastic half-space. Wave Mot. **31**, 333–348 (2000)
12. Georgiadis, H.G., Velgaki, E.G.: High-frequency Rayleigh waves in materials with microstructure and couple-stress effects. Int. J. Solids Struct. **40**, 2501–2520 (2003)
13. Hrytsyna, O., Sladek, J., Sladek, V.: The effect of micro-inertia and flexoelectricity on Love wave propagation in layered piezoelectric structures. Nanomaterials **11**(9), 2270 (2021)
14. Hu, T.T., Yang, W.J., Liang, X., Shen, S.P.: Wave propagation in flexoelectric microstructured solids. J. Elast. **130**, 197–210 (2018)
15. Jiao, F.Y., Wei, P.J., Li, Y.Q.: Wave propagation through a flexoelectric piezoelectric slab sandwiched by two piezoelectric half-spaces. Ultrasonics **82**, 217–232 (2018)
16. Landau, L.D., Lifshitz, E.M.: Electrodynamics of Continuum Media, 2nd edn. Butterworth-Heinemann, Oxford (1984)
17. Li, G.-E., Kuo, H.-Y.: Effects of strain gradient and electromagnetic field gradient on potential and field distributions of multiferroic fibrous composites. Acta Mech. **232**, 1353–1378 (2021)
18. Liu, J., He, S.: Properties of Love waves in layered piezoelectric structures. Int. J. Solids Struct. **47**, 169–174 (2010)
19. Love, A.E.H.: Some Problems of Geodynamics. Cambridge University Press, London (1911)

20. Lukashev, P., Sabirianov, R.F.: Flexomagnetic effect in frustrated triangular magnetic structures. Phys. Rev. B **82**, 094417 (2010)
21. Majorkowska-Knap, K., Lenz, J.: Piezoelectric Love waves in non-classical elastic dielectrics. Int. J. Eng. Sci. **27**, 879–893 (1989)
22. Malikan, M., Eremeyev, V.A.: On nonlinear bending study of a piezo-flexomagnetic nanobeam based on an analytical-numerical solution. Nanomaterials **10**(9), 1762 (2020)
23. Nowacki, W.: Efekty elektromagnetyczne w stałych ciałach odkształcalnych. Warszawa, Państwowe Wydawnictwo Naukowe (1983). In Polish
24. Ottosen, N. S., Ristinmaa, M., Ljung, C.: Rayleigh waves obtained by the indeterminate couple-stress theory. Eur. J. Mech. A/Solids **19**, 929–947 (2000)
25. Shodja, H.M., Goodarzi, A., Delfani, M.R., Haftbaradaran, H.: Scattering of an antiplane shear wave by an embedded cylindrical micro-/nano-fiber within couple stress theory with micro inertia. Int. J. Solids Struct. **58**, 73–90 (2015)
26. Singhal, A., Sahu, S.A., Nirwal, S., Chaudhar, S.: Anatomy of flexoelectricity in micro plates with dielectrically highly/weakly and mechanically complaint interface. Mater. Res. Express **6**, 105714(17) (2019)
27. Singhal, A., Sedighi, H.M., Ebrahimi, F., Kuznetsova, I.: Comparative study of the flexoelectricity effect with a highly/weakly interface in distinct piezoelectric materials (PZT-2, PZT-4, PZT-5H, $LiNbO_3$, $BaTiO_3$). Waves Random Complex Med. (2019). https://doi.org/10.1080/17455030.2019.1699676
28. Sladek, J., Sladek, V., Xu, M., Deng, Q.: A cantilever beam analysis with flexomagnetic effect. Meccanica **56**, 2281–2292 (2021)
29. Sladek, J., Sladek, V., Repka, M., Deng, Q.: Flexoelectric effect in dielectrics under a dynamic load. Compos. Struct. **260**, 113528 (2021)
30. Tagantsev, A.: Theory of flexoelectric effect in crystals. JETP Lett. **88**(6), 2108–2122 (1985)
31. Yang, J.S.: Love waves in piezoelectromagnetic materials. Acta Mech. **168**, 111–117 (2004)
32. Yang, W., Liang, X., Deng, Q., Shen, S.: Rayleigh wave propagation in a homogeneous centrosymmetric flexoelectric half-space. Ultrasonics **103**, 106105 (2020)
33. Yang, W., Liang, X., Shen, S.: Love waves in layered flexoelectric structures. Phil. Mag. **97**, 3186–3209 (2017)
34. Yudin, P., Tagantsev, A.: Fundamentals of flexoelectricity in solids. Nanotechnology **24**(43), 432001 (2013)
35. Zakharenko, A.: Love-type waves in layered systems consisting of two cubic piezoelectric crystals. J. Sound Vib. **285**, 877–886 (2005)
36. Zhang, N., Zheng, S., Chen, D.: Size-dependent static bending of flexomagnetic nanobeams. J. Appl. Phys. **126**, 223901 (2019)
37. Zhang, S., Gu, B., Zhang, H., Feng, X.-Q., Pan, R., Alamusi, Hu, N.: Propagation of Love waves with surface effects in an electrically-shorted piezoelectric nanofilm on a half-space elastic substrate. Ultrasonics **66**, 65–71 (2016)

Research on Risk Evaluation of New Infrastructure PPP Projects Based on PSO-FAHP

Ren Yingwei, Ma Rong, and Wang Boxun

Abstract In view of the complex characteristics of PPP project risk factors under the background of new infrastructure construction, this paper proposes a risk evaluation model based on improved particle swarm optimization-Fuzzy Analytic Hierarchy Process (PSO-FAHP); the fuzzy complementary judgment matrix of Analytic Hierarchy Process Consistency has always been a difficult problem in academia. This paper introduces particle swarm algorithm to transform the core content of FAHP-consistency judgment into objective function and constraint conditions, so that the weight result is more accurately revised on the basis of the maximum retention of expert judgment. In the end, this method was introduced into the new infrastructure PPP project example and found that although the newly added risk indicator—project management is inferior to construction risk and operation risk, its criticality cannot be ignored.

Keywords New infrastructure construction · PSO-FAHP · PPP · Risk

From the initial proposal of the concept of "new infrastructure construction" ("new infrastructure") at the Central Economic Work Conference to the current series of intensive deployments by the central government after the epidemic, the connotation and scope of "new infrastructure" have been continuously enriched and improved. The National Development and Reform Commission approved approximately 860 billion yuan in infrastructure investment in the areas of inter-city high-speed railways and urban rail transit. Different from traditional infrastructure construction, new infrastructure construction centers on intelligent scientific information technology. Because its service method is a network, the method of embedding individuals or teams into projects is more frequent than traditional infrastructure projects, resulting in more complex risk factors and obvious interactions. It is precisely because of this characteristic that most projects can only be completed by market entities with

R. Yingwei (✉) · M. Rong · W. Boxun
College of Civil Engineering and Architecture, Shandong University of Science and Technology, Qingdao 266590, Shandong, China
e-mail: 1107505405@qq.com

H. Dai (ed.), *Computational and Experimental Simulations in Engineering*, Mechanisms and Machine Science 119,
https://doi.org/10.1007/978-3-031-02097-1_19

scientific and technological capabilities. Therefore, this article aims to bring some new options for the risk weight assignment of new infrastructure PPP projects.

From the perspective of the international situation, China has become the world's largest PPP market, and my country has also been in the forefront of the world in the study of PPP models [1]. In recent years, there are many ways to evaluate the risks of PPP projects in China. Ma Weihai used the matter-element model to analyze the PPP water conservancy project. In the step of calculating the weight, the FAHP method combining qualitative and quantitative was used instead of the original subjective expert to directly score, and finally the level of each risk was obtained, which is clear at a glance [2]. Zhao Hui adopted the DEMATEL and information entropy weighting model based on the minimum deviation combination for the PPP sewage treatment project. This model not only draws on expert experience, but also weakens the deviation caused by subjectivity [3]. Wang Jianbo introduces the combined weighting of OWA operator and ER algorithm into the urban rail risk assessment under the PPP mode [4]. But the only shortcoming of this combination is that it is based on simple multiplicative synthesis. The multiplication effect obtained by this combined method is very unreasonable [5]. Wang Songjiang uses the combination of set pair theory and AHP to evaluate the risk of expressway PPP projects. This combination method not only uses the five-element connection number to analyze the development status of each risk, but also uses the characteristics of partial derivatives to make the development trend of each risk. Detailed forecast [6]. After Lai Zhixuan used AHP to determine the weights, he introduced the gray clustering method to evaluate the risks of water conservancy project irrigation area projects under the PPP mode [7]. Wang Shuai used rough sets and system dynamics to identify the risks of sponge city PPP projects, and used linear weighting to combine the objective entropy method and subjective G1 method to evaluate. The subjective and objective weight coefficient of the combined method was directly adopted as 0.5, which was not carried out. Weight coefficient allocation optimization [8]. Huang Guilin used covariance instead of the 1–9 scale method to bring into the level analysis, combined with the entropy method to weight the PPP shed reform project based on the game theory combination, and finally determined the risk level through fuzzy comprehensive evaluation [9].

1 The Characteristics of New Infrastructure PPP Projects and the Construction of Risk Evaluation Index System

1.1 Features of New Infrastructure PPP Projects

Due to the essential characteristics of new infrastructure innovation, engineering project management is no longer the waterfall management model in which Party B arrives at a solution based on Party A's needs. Instead, it is constantly experimenting, innovative, and rapidly iterating agile projects. Management mode. The change of this model will inevitably bring various new risks to the project.

New infrastructure projects have high technology content, high intelligence and digital content, and free is the basic attribute of digital virtual infrastructure. Therefore, compared with traditional PPP projects for profit, the profit risk in the operation of new infrastructure projects Subject to change.

During the operation of the new infrastructure PPP project, a large investment is required. Since the construction of the new infrastructure mainly uses various types of power and electronic equipment with faster technological upgrading and limited service life, the energy demand in the later stage of operation is relatively large. Therefore, the equipment maintenance and operation management risks of the new infrastructure are different from those of traditional PPP projects.

The biggest feature of the new infrastructure PPP project is that it has no industry boundaries, because it is a network and uses virtual numbers as a means of communication and service, which leads to a comprehensive management method, rather than being managed by any department alone.

1.2 Project Risk System Construction

This article combines objective and subjective, qualitative and quantitative methods, and uses literature reading and questionnaire surveys to identify the risk list. For the new infrastructure PPP project, due to its huge engineering volume, long construction period, advanced machinery and equipment, high scientific and technological content, difficult and complex operating environment, high construction risks and difficulties in project management organization and coordination, the new infrastructure urban rail PPP Project risk factors are complex and diverse, and the author of this article added project management risks through field investigations. The identification of risk factors is shown in Table 1.

2 Evaluation Model Based on PSO-FAHP

2.1 FAHP Principle and Basic Ideas

First, the experienced experts in the new infrastructure PPP project are asked to score the FAHP's initial fuzzy complementary judgment matrix according to the index scale, and the result is a positive and negative matrix as follows $A = (a_{ij})_{n\times n}$; a_{ij} is the attribute value,w_i is the weight vector that composes the attribute value.

Table 1 New type of infrastructure PPP project risk identification table

Criterion level one	Criterion level two	Criterion level three
System risk	Political risk U_1	Approval delay U_{11}
		Policy stability risk U_{12}
		Completion of the law and risk of change U_{13}
	Economic risk U_2	Inflation risk U_{21}
		Interest rate risk U_{22}
		Financing risk U_{23}
	FORCE MAJEURE力 U_3	Plague disease U_{31}
		Natural disaster U_{32}
	Construction risk U_4	Safety and environmental risks U_{41}
		Engineering quality U_{42}
		Cost overrun U_{43}
		Delay in construction period U_{44}
	Operation and maintenance risk U_5	Risk of insufficient fees or returns U_{51}
		Equipment maintenance U_{52}
		Operation management U_{53}
	Credit risk U_6	Financing party defaulted U_{61}
		Government breach of contract U_{62}
	Project management risk U_7	Management ability U_{71}
		Communication and negotiation U_{72}
		Technical ability U_{73}

$$A = \begin{bmatrix} a_{11} \; a_{12} \cdots a_{1n} \\ a_{21} \; a_{22} \cdots a_{2n} \\ \cdots \cdots \cdots \\ a_{n1} \; a_{n2} \cdots a_{nn} \end{bmatrix} = A = \begin{bmatrix} \frac{w_1}{w_1} & \frac{w_1}{w_2} & \cdots & \frac{w_1}{w_n} \\ \frac{w_2}{w_1} & \frac{w_2}{w_2} & \cdots & \frac{w_2}{w_n} \\ \cdots & \cdots & \cdots & \cdots \\ \frac{w_n}{w_1} & \frac{w_n}{w_2} & \cdots & \frac{w_n}{w_n} \end{bmatrix}$$

The premise of using fuzzy complementary judgment matrix is that the matrix must meet the requirements of complete consistency. Since complete consistency is difficult to achieve, the traditional FAHP method requires a consistency of less than 0.1 to be acceptable. When the calculation does not meet the conditions of the fuzzy boundary, it needs to return Adjust the judgment matrix until the condition is met. In this way, the method not only requires a lot of work, but also the result is not necessarily accurate. In order to simplify complex problems and improve the accuracy of the results, the fuzzy boundary conditions can be transformed and optimized by combining the intelligent algorithm to find the optimal solution. The specific ideas are as follows.

When the positive and negative matrix meets the complete consistency, it can be expressed as the following formula:

$$a_{ij} \cdot a_{jk} = a_{ik}$$

$$\mathrm{Is}a_{jk} = \frac{a_{ij}}{a_{ik}} = \frac{w_i/w_j}{w_i/w_k} = w_j/w_k$$

$$\sum_{k=1}^{n}(a_{jk} \cdot \omega_k) = \sum_{k=1}^{n}(\omega_j/\omega_k) \cdot \omega_k = n \cdot \omega_j$$

$$\sum_{j=1}^{n} |\sum_{k=1}^{n}(a_{jk}\omega_k - n\omega_j)| = 0$$

However, in reality, due to various factors, the matrix cannot fully satisfy the consistency, and can only be infinitely close to the consistency, that is, the above formula tends to zero infinitely, which can be transformed into a problem of seeking optimal value: $CIF = min \sum_{j=1}^{n} | \sum_{k=1}^{n}(x_{jk} \cdot \omega_k - n \cdot \omega_j)|$.

In practice, the differences in the subjective consciousness of experts should also be considered, and the gap between the revised judgment matrix and the initial judgment matrix should be reduced as much as possible. which is $min \sum_{i=1}^{n} \sum_{j=1}^{n} (x_{jk} - a_{jk})^2$.

Therefore, this paper optimizes two goals, but the particle swarm algorithm cannot directly optimize multiple fitness functions, so the method of dynamic weights is reduced to single-objective optimization. Finally, the fitness function of the particle swarm is determined to be

$$minY = \sum_{k=1}^{n}\sum_{j=1}^{n} \lambda_1(x_{jk} - a_{jk})^2 + \lambda_2(x_{jk} - \omega_j/\omega_k)].$$

2.2 Principles and Basic Ideas of PSO

PSO is a kind of random search intelligent algorithm inspired by bird flocks that represent candidate solutions to forage and evolve the solution. It is mostly used to solve the optimization of complex nonlinear problems with only position and corresponding speed, without mass and volume. This article is applied to FAHP Consistent optimization solution [10], so that the risk assessment results of new infrastructure PPP projects are more accurate. After the above analysis, the objective function and constraint conditions of the PSO algorithm are:

$$minY = \sum_{k=1}^{n}\sum_{j=1}^{n}\lambda_1(x_{jk} - a_{jk})^2 + \lambda_2(x_{jk} - \omega_j/\omega_k)]$$

$$\begin{aligned}
&\text{s.t. } \lambda_1 + \lambda_2 = 1, \lambda_1, \lambda_2 \geq 0;\\
&w_k > 0, \sum_{i=1}^{n} w_k = 1\\
&x_{jk} \in [(1-\theta)a_{jk}, (1+\theta)a_{jk}], k, j = 1 \sim 7\\
&x_{jk} = \frac{1}{x_{kj}}
\end{aligned}$$

The key to applying the PSO algorithm lies in the selection of the global optimum, that is, the position and speed of the candidate solution. The speed and position formulas used in this paper to solve the optimal solution of the objective function are as follows:

$$\begin{cases} V_i^{k+1} = wV_i^k + c_1r_1(P_i^k - X_i^k) + c_2r_2\left(P_g^k - X_i^k\right) \\ X_i^{k+1} = X_i^k + V_i^{k+1} \end{cases}$$

where $X_i = (x_{i1}, x_{i2}, \cdots, x_{iN})$ Is the flight position of the bird swarm (particle swarm), which is the decision vector of the candidate solution; $V_i = (v_{i1}, v_{i2}, \cdots, v_{iN})$ Is the speed of flight, is the direction and length change of the decision vector corresponding to the candidate solution; k Is the number of iterations; P_i^k Are particles i $(i = 1, 2, \ldots, M)$ The best position passed in the k iteration; P_g^k It is the best position that the particle swarm passes by at time k.; c_1 Is the step size for the particle to move to its optimal position; c_2 Is the step size of the particle moving to the global optimal position; r_1 and r_2 Is a constant between $[0, 1]$. In particle swarm algorithm, In order to control the components of V_i^k and X_i^k in a reasonable area, Specify $V_{\max}, X_{\max}, V_i \in [-V_{\max}, V_{\max}], X_i \in [-X_{\max}, X_{\max}]$, Generally, the maximum value of the speed vector is limited, When $V_i \geq V_{\max}$, take $V_i = V_{\max}$; when $V_i \leq -V_{\max}$, take $V_i = -V_{\max}$.

3 Empirical Analysis of New Infrastructure PPP Projects

The urban rail transit line 1 of a certain city is one of the new infrastructure PPP projects approved by the National Development and Reform Commission in the field of urban rail transit. It is a multi-industry interlaced subway based on big data and wireless communication, integrated monitoring, automatic fare collection

Table 2 Risk expert judgment matrix of each method of the first criterion layer

	Political risk	Economic risk	Force majeure	Construction completion risk	Operation and maintenance risk	Credit risk	Project management risk
Political risk	1	5.5/6.5	5.5/5	5.5/8	5.5/7.5	5.5/6	5.5/7
Economic risk	6.5/5.5	1	6.5/5	6.5/8	6.5/7.5	6.5/6	6.5/7
Force majeure	5/5.5	5/6.5	1	5/8	5/7.5	5/6	5/7
Construction completion risk	8/5.5	8/6.5	8/5	1		8/7.5	8/7
Operation and maintenance risk	7.5/5.5	7.5/6.5	7.5/5	7.5/8		1	7.5/7
Credit risk	6/5.5	6/6.5	6/5	6/8	6/7.5	1	6/7
Project management risk	7/5.5	7/6.5	7/5	7/8	7/7.5	7/6	1

and other intelligent functions. Its complex modern design makes the construction and management of this project more difficult than ever before. Take the first-level indicators as an example to establish the initial fuzzy judgment matrix. Get Table 2.

After obtaining the initial fuzzy complementary judgment matrix scored by the expert, it is substituted into the particle swarm intelligent optimization model, and each parameter is set as follows: $\lambda_1 = 0.4,\ \lambda_2 = 0.6,\ \theta = 0.3;\ c_1 = c_2 = 2,\ w = 1,\ m = 80$, The number of iterations is 5000. Randomly set initial position and initial speed.

Using MATLAB 6.5 software to program and solve, the adjusted expert judgment matrix can be obtained as follows (Fig. 1).

The optimal position in the particle swarm optimization process is as shown in the figure below. The optimal position is reached when the number of iterations reaches about 3400, and the minimum value is 1.1324e-15, which is infinitely close to zero. It shows that the subjective error of the expert is consistent with the judgment matrix Sex has reached a minimum 1.1324e-15 (Fig. 2).

In the end, the weight value of the second-level judgment criterion is [0.1180310.1423650.1050640.180590.1682710.1302330.155444], From the analysis of this result, it is concluded that: construction risk > operation and maintenance risk > project management risk > economic risk > credit risk > political risk > force majeure risk.

The complementary judgment matrix of pentagonal fuzzy numbers is (example of political risk) (Table 3).

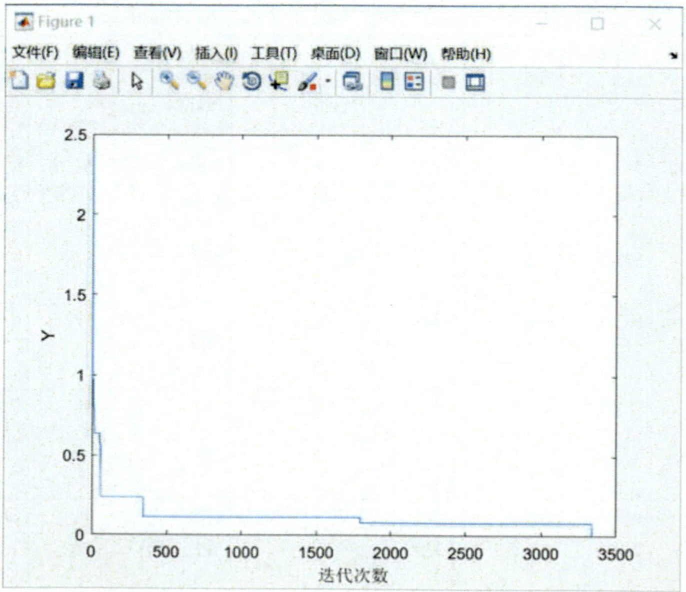

Fig. 1 The judgment matrix of the first criterion level after PSO optimization

```
Xbest =

    1.0000    0.8462    1.1000    0.6875    0.7333    0.9167    0.7857
    1.1818    1.0000    1.3000    0.8125    0.8667    1.0833    0.9286
    0.9091    0.7692    1.0000    0.6250    0.6667    0.8333    0.7143
    1.4545    1.2308    1.6000    1.0000    1.0667    1.3333    1.1429
    1.3636    1.1538    1.5000    0.9375    1.0000    1.2500    1.0714
    1.0909    0.9231    1.2000    0.7500    0.8000    1.0000    0.8571
    1.2727    1.0769    1.4000    0.8750    0.9333    1.1667    1.0000

时间已过 3.460315 秒。
```

Fig. 2 The optimal value of Y and the point corresponding to the optimal value

Table 3 The third criterion level pentagonal fuzzy number complementary judgment matrix

Criterion level three	Impossible	Lower	Medium	Higher	Extremely high
Approval delay	0	0.2	0.5	0.3	0
Policy stability risk	0.2	0.3	0.2	0.2	0.1
Completion of the law and risk of change	0.2	0.3	0.2	0.2	0.1

The final score value of the three-level fuzzy evaluation matrix is as follows; (0.0813239810.3213160340.3348805120.1951889420.067288531) The maximum degree of membership is moderately lower, which is acceptable.

4 Concluding Remarks

Based on the above analysis, this paper constructs a new infrastructure PPP risk assessment model based on PSO-FAHP. The revised expert judgment matrix not only meets the consistency requirements better than traditional methods, but also obtains a more reasonable risk of new infrastructure PPP projects. The relative ranking proportions of the indicators, from the weight value of the second-level judgment criterion, the project management risk weight in the evaluation of new infrastructure PPP projects is ranked in the middle. For the new infrastructure PPP project, the technical difference between the project and the project Concepts are very different. Past successful experience can hardly be a guide for future success. The interlacing of industry boundaries, the high integration of technology and management, and the contract management and quality management of projects may be fundamentally different depending on the project. Therefore, future new infrastructure PPP projects are increasingly inseparable from compound management talents.

References

1. A holistic review of public-private partnership literature published between 2008 and 2018
2. Weihai, M., Fanhui, Z.: Research on the risk of water conservancy PPP projects based on F-AHP and matter-element model. Math. Pract. Knowl. **49**(19), 80–90 (2019)
3. Hui, Z., Zehui, B., Shengbin, M.: Study on the selection of operation mode of sewage treatment PPP project based on combined weighting and GRA-TOPSIS. Water Resour. Hydropower Technol. **51**(09), 143–153 (2020)
4. Jianbo, W., Longbiao, P., Na, L., Shuai, Z.: Urban rail transit PPP project risk assessment based on OWA-ER. J. Civ. Eng. Manag. **34**(05), 46–51 (2017)
5. Ying, S., Xinzhong, B.: A combination weighting evaluation method based on maximizing variance and its application. Chin. Manag. Sci. **19**(06), 141–148 (2011)
6. Songjiang, W., Zhongkui, C.: The application of multiple connection number method in the risk assessment of highway PPP projects. J. Kunming Univ. Sci. Technol. (Nat. Sci. Ed.) **45**(02), 130–142 (2020)
7. Zhixuan, L., Zhao, L., Tan, Z.: Research on PPP project risk assessment of water conservancy project irrigation district based on grey clustering method. J. Eng. Manag. **32**(03), 75–80 (2018)
8. Shuai, W., Shengyue, H.: Sponge city PPP project risk dynamic evaluation based on system dynamics. J. Eng. Manag. **33**(03), 63–68 (2019)
9. Guilin, H., Xiulu, W.: Risk assessment of PPP shed reform project based on combined weighting method. J. Civ. Eng. Manag. **36**(04), 40–46 (2019)
10. Zhuangkuo, L., Youtian, X.: Research on the improvement and application of fuzzy analytic hierarchy process based on particle swarm optimization. Oper. Res. Manag. **22**(04), 139–143+219 (2013)

The Dynamic Coupling of Heterogeneous Robotic Systems for Spacecraft Motion Emulation

Ryan Ketzner, Hunter Quebedeaux, and Tarek A. Elgohary

Abstract Increasing access to space has driven demand for low cost, portable, and highly specialized robotic platforms to accurately simulate multi-dimensional space missions. Presented is an effective, low cost, 9 degrees of freedom heterogeneous robotic system that can emulate orbital motion using mobile manipulator motion planning; the dynamical models for a mobile manipulator are derived, and studied for an "orbit-like" trajectory.

Keywords Mobile manipulator · Dynamics · Test bed · Orbit emulation

1 Introduction

The design, implementation and testing of control algorithms for spacecraft motion is a challenging task and has been the subject of research and development by aerospace engineers for decades. An effective platform that simulates spacecraft motion, rendezvous, servicing, and probing is increasingly demanded by the multidimensional space missions of today. Many researching robotic platforms have begun the steps to undertake this challenge.

For example, in order to emulate the contact tasks carried out by the special purpose dexterous manipulator, a robotic arm on the ISS, The Canadian Space Agency developed a task verification facility utilizing a 6-DOF hydraulic robot [1]. The European Proximity Operations Simulator developed by the German Aerospace Center expands on the idea of using robotic platforms to emulate motion by using two robotic manipulators, allowing real time simulation of docking and rendezvous [2]. As the need for more complex, multi-body space maneuvers increases, so does the need for using multiple robotic agents, which have been approach by Cortes and Egerstedt; they proved the viability of using multi-robotic systems in decentralizing control and coordination strategies through a gradient descent algorithm [3]. An efficient

R. Ketzner · H. Quebedeaux (✉) · T. A. Elgohary
Mechanical and Aerospace Engineering,12760 Pegasus Drive, Orlando, FL, USA
e-mail: hunterq@knights.ucf.edu

H. Dai (ed.), *Computational and Experimental Simulations in Engineering*, Mechanisms and Machine Science 119,
https://doi.org/10.1007/978-3-031-02097-1_20

and reliable detumbling mission for a swarm of CubeSats to capture and control Near-Earth Asteroids was also addressed [4]. Furthermore, less sophisticated reinforcement learning swarm robots trained by a multi-agent deep deterministic policy gradient algorithm was studied to traverse and explore the surface of Mars [5]. A common theme of these platforms are the use of dynamics-driven motion.

These international research advancements address progressing controls and multi-staged robots. High cost, singular purpose systems are often the result. There is a need for a flexible and highly portable control testing simulation platform that satisfies precise demands of varying deep-space missions for both academia and industry. Seleit et al. begin to address this need as they develop orbital kinematic control models for a robotic platform called ROME [6]. Hardware experiments using control algorithms derived from the kinematic relations of the test bed's subsystems presented good preliminary results for a spacecraft motion test bed, however the kinematic models proved too inaccurate for motion recreation. These subsystems are a ground vehicle robotic manipulator which when coupled define a 9 DOF system; a 6 DOF robotic manipulator is fixed on top of a 3 DOF ground vehicle accompanied with on board computer and sensors [6]. Presented are the dynamical models for the same heterogeneous robotic platform. Firstly, the following dynamical models and controllers are derived with this hardware in mind, then the interial coupling model is derived based on the premise of a mobile manipulator system. Finally, the closed loop dynamics simulation for a mock two body problem is demonstrated.

2 Ground Vehicle Dynamical Model

The dynamical equations for the ground vehicle motion are derived from the work of Liu et al. [7] using the following fundamental assumptions.

1. In the traction force direction, the wheels have no slippage.
2. The friction on the motor shaft and gears are considered to be viscous friction forces only.
3. Friction forces that are not aligned with the traction force are neglected.
4. The electrical time constant of the motor is neglected.

Two frames of references are used to formulate the equations of motion: the ground vehicle body frame $\{GV\}$, which is fixed to the center of mass of the vehicle, and the inertial frame $\{I\}$, as shown in Fig. 1.

Since the vehicle can translate and rotate simultaneously, linear and rotational accelerations are considered. The equations of motion are derived from the geometry and Newton's second law $\sum F = ma$.

In the X direction, the force balance evaluates to,

$$m\dot{u} - m\Omega v = f_2 \cos\left(\frac{\pi}{6}\right) - f_3 \cos\left(\frac{\pi}{6}\right) \tag{1}$$

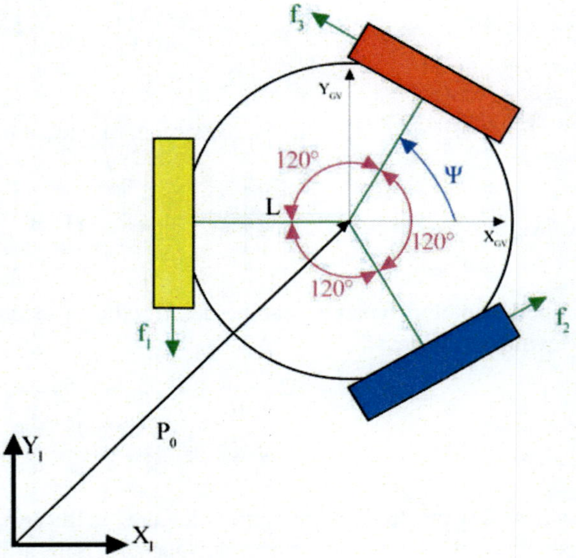

Fig. 1 Force analysis on the wheels

where $\dot{u}$ is the linear acceleration and the rotational acceleration term is $\Omega v = \Omega^2 L$, where $v = \Omega L$. Similarly, in the Y direction,

$$m\dot{v} + m\Omega u = -f_1 + f_2 \sin\left(\frac{\pi}{6}\right) + f_3 \sin\left(\frac{\pi}{6}\right) \tag{2}$$

Similarly, the equation of motion in the Ψ direction can be written as,

$$I_z\dot{\Omega} = f_1 L + f_2 L + f_3 L \tag{3}$$

The three equations of motion can be rewritten as follows:

$$\begin{aligned}
\dot{u} &= \Omega v + \frac{f_1}{m} + \frac{f_2}{m}\cos\left(\frac{\pi}{6}\right) - \frac{f_3}{m}\cos\left(\frac{\pi}{6}\right) \\
\dot{v} &= -\Omega u + \frac{f_1}{m} + \frac{f_2}{m}\sin\left(\frac{\pi}{6}\right) + \frac{f_3}{m}\sin\left(\frac{\pi}{6}\right) \\
\dot{\Omega} &= \frac{f_1}{I_z} + \frac{f_2}{I_z} + \frac{f_3}{I_z}
\end{aligned} \tag{4}$$

Let

$$\mathbf{H} = \begin{bmatrix} 1/m & 0 & 0 \\ 0 & 1/m & 0 \\ 0 & 0 & 1/I_z \end{bmatrix} \tag{5}$$

and

$$\mathbf{B} = \begin{bmatrix} 0 & \cos\left(\frac{\pi}{6}\right) & -\cos\left(\frac{\pi}{6}\right) \\ -1 & \sin\left(\frac{\pi}{6}\right) & \sin\left(\frac{\pi}{6}\right) \\ L & L & L \end{bmatrix} \tag{6}$$

From Eqs. (4), (5), and (6), the system of equations can be written as,

$$\begin{bmatrix} \dot{u} \\ \dot{v} \\ \dot{r} \end{bmatrix} = \begin{bmatrix} rv \\ -ru \\ 0 \end{bmatrix} + \mathbf{H} \cdot \mathbf{B} \begin{bmatrix} f_1 \\ f_2 \\ f_3 \end{bmatrix} \tag{7}$$

The DC motor dynamics can be described as a second order system using the following equations:

$$\begin{aligned} L_a \frac{di_a}{dt} + R_a i_a + k_3 \omega_m &= E \\ J_0 \dot{\omega}_m + b_0 \omega_m + \frac{Rf}{n} &= K_2 i_a \end{aligned} \tag{8}$$

where E is the applied armature voltage, i_a is the armature current, ω_m is the motor shaft speed, L_a is the armature inductance, R_a is the armature resistance, k_3 is the back emf constant, k_2 is the motor torque constant, J_0 is the combined inertia of the motor, gear train and wheel referred to the motor shaft, b_0 is the viscous-friction coefficient of the motor, gear and wheel combination, **R** is the wheel radius, f is the wheel traction force, and n is the motor to wheel gear ratio. The motor electric circuit dynamics can be neglected because the time constant of the motor is very small compared to the mechanical time constant. Consequently, $L_a \frac{di_a}{dt} = 0$ and $i_a = \frac{1}{R_a}(E - k_3 \omega_m)$. Now, the dynamics of the three motors can be written as

$$J_0 \begin{bmatrix} \dot{\omega}_{m1} \\ \dot{\omega}_{m2} \\ \dot{\omega}_{m3} \end{bmatrix} + b_0 \begin{bmatrix} \omega_{m1} \\ \omega_{m2} \\ \omega_{m3} \end{bmatrix} + \frac{R}{n} \begin{bmatrix} f_1 \\ f_2 \\ f_3 \end{bmatrix} = \frac{k_2}{R_a} \begin{bmatrix} E_1 \\ E_2 \\ E_3 \end{bmatrix} - \frac{K_2 K_3}{R_a} \begin{bmatrix} \omega_{m1} \\ \omega_{m2} \\ \omega_{m3} \end{bmatrix} \tag{9}$$

By combining Eqs. 7 and 9 in the body frame, we get

$$\begin{bmatrix} \dot{u} \\ \dot{v} \\ \dot{r} \end{bmatrix} = G^{-1} \begin{bmatrix} rv \\ -ru \\ 0 \end{bmatrix} - G^{-1} H B B^T \left(\frac{K_2 \cdot k_3}{R_a} + b_0 \right) \times \frac{n^2}{R^2} \begin{bmatrix} r \\ v \\ r \end{bmatrix} + G^{-1} H B \frac{K_2 n}{R \cdot R_a} \begin{bmatrix} E_1 \\ E_2 \\ E_3 \end{bmatrix} \tag{10}$$

where, $G = (1 + HBB^T \frac{n^2 J_0}{R^2})$. The dynamic model describes the motion of the robot. The kinematics can be obtained by a coordinate transformation of the body rates in to the inertial frame. The friction constant b_0 is obtained experimentally. The presented nonlinear coupled dynamics model becomes linear only if the robot does not rotate while in translation or rotates around a fixed point with translation. A linear controller can be applied to the nonlinear dynamics by treating the nonlinear motion as a disturbance to the linear model, as in the works by Kalmar, Samani [8–10].

3 Robotic Manipulator Dynamical Model

Trajectory tracking problems for robotic manipulators using any dynamical model relies on the use of kinematic propagation in order to determine the attitude of the tracked end effector. The kinematic model has been excluded for brevity, but has been defined in the past [6].

3.1 Dynamical Model

Using the well known Newton-Euler Formulation for the joint space dynamics of a 6 DOF robotic arm, we can define the equations of motion for the inverse and forward kinematics. In the past kinematics work, closed form solutions have been preferred; however, the Recursive Newton-Euler algorithm for solving the equations of motion are much more beneficial in this application where run time is a constraint of the hardware system [11].

3.1.1 Inverse Dynamics

The inverse dynamics is a problem that demands for any joint angle, joint angle velocity, and joint angle acceleration, the torque applied at that joint can be found. Where q indicates the joint angles for the manipulator, the torques applied at each joint can be described as in Eq. (11).

$$\tau = M(q)\ddot{q} + C(q, \dot{q}) + F(\dot{q}) + G(q) + J(q)^T W \tag{11}$$

where $q, \dot{q}, \ddot{q}$ are the respective derivatives of the generalized coordinates, M is the mass matrix, C is the Coriolis and Centripetal vector, F is the friction vector, G is the gravity loading vector, J is the Jacobian of the system, and W is the wrench load applied to the end effector.

Assuming the effects of joint function are neglected and the wrench applied to the end effector is not applied, we can simplify Eq. (11) further.

$$\tau = M(q)\ddot{q} + C(q, \dot{q}) + G(q) \tag{12}$$

Considering each link of an open chain manipulator is considered a rigid body, and the center of mass of each link has a coordinate frame the coincides with each link as well as a frame $n + 1$ at the end effector and a frame 0 fixed in the world. Further, ν_i is defined as the twist of link i in its respective coordinate frame. To preform the Recursive Newton-Euler algorithm, the forward and backward iterations of the algorithm are preformed. The forward iterations which calculate the configuration, twist and acceleration of each link starting at the base of the manipulator. The backward iterations calculate the required joint forces and torques starting from joint n working backward to joint 1.

Defining $M_{i,i-1}$ as the transform defining the frame $i-1$ relative to frame i when joint i is at its zero position or $\theta_i = 0$. A_i is defined as the screw axis of join i in frame i. F_{n+1} is defined as the wrench applied by the end effector, otherwise known as W. Finally, gravity is modelled by defining $\dot{\nu_0}$ as the vector the acceleration vector of the base. In the case of a non-moving base $\dot{\nu_0} = \begin{bmatrix} 0 & 0 & -g \end{bmatrix}$ where g is the acceleration due to gravity.

The forward iterations pseudo-code is written as follows: Given $\theta, \dot{\theta}, \ddot{\theta}$ and for $i = 1$ to n, the configuration of frame $i-1$ relative to i is given by

$$T_{i,i-1} = e^{-[A_i]\theta_i} M_{i,i-1} \tag{13}$$

Next the twist of link i is the sum of the twist of link $i-1$ expressed in frame i using the matrix adjoint of $T_{i,i-1}$ defined as

$$\nu_i = [Ad_{T_{i,i-1}}]\nu_{i-1} + A_i\dot{\theta}_i \tag{14}$$

Finally, the acceleration of link i is defined as

$$\dot{\nu_i} = [Ad_{T_{i,i-1}}]\dot{\nu_{i-1}} + [ad_{\nu_i}]A_i\dot{\theta}_i + A_i\ddot{\theta}_i] \tag{15}$$

The backward iterations pseudo-code is written as follows: For $i = n$ to 1, the wrench required by the the link i as

$$F_i = [Ad_{T_{i,i+1}}]^T F_{i+i} + G_i\dot{\nu_i} - [ad_{\nu_i}]^T G_i \nu_i \tag{16}$$

where G is defined as the spatial inertia matrix. We can then finally compute the torque at each joint as

$$\tau = F_i^T A_i. \tag{17}$$

3.1.2 Forward Dynamics

Finding the corresponding forward dynamics is a very simple problem that relies on the inverse dynamics, Eq. (12). First the inverse dynamics solved for when $\ddot{q} = 0$ in order to solve for the C and G vectors. Then, again using the inverse dynamics we solve for the mass matrix M by setting the all joint velocities and gravity equal to zero, and all joint accelerations to zero except for one which is set to unity for $i = 1$ to n.

$$M_i(q) = \tau_i \tag{18}$$

Then knowing the M, C, and G matrices, we can rearrange Eq. 12 for $\ddot{q}$ results in the following.

$$\ddot{q} = M(q)^{-1}(\tau - C(q, \dot{q}) - G(q)) \tag{19}$$

Then Eq. (19) can be numerically integrated using to find the configuration angles and velocities for each joint in a robotic manipulator.

3.2 Dynamic Control

For a 6 axis revolute manipulator, an inverse dynamic control algorithm derived by Siciliano [12], can be used to employ an exact linearization of system dynamics obtained by means of a nonlinear state feedback using the Recursive Newton-Euler algorithm mentioned in Eq. (12). Control of the robotic manipulator can be developed either in the arm joint space, or the Cartesian task space. Equation (20) demonstrates the joint space control law. This simple PD controller can also be manipulated for the resultant task space control law, Eq. (21), which has a very similar PD controller. The control structure can be seen in the block diagram in Fig. 2.

$$\ddot{q}_r = \ddot{q}_d + K_D(\dot{q}_d - \dot{q}_r) + K_P(q_d - q_r) \tag{20}$$

$$\ddot{x}_r = \ddot{x}_d + K_D(\dot{x}_d - \dot{x}_r) + K_P(x_d - x_r) \tag{21}$$

With the goal of ROME to be accurate Cartesian space trajectory tracking, the use of the task space control law will be used. However, interior control of the joints is used in the control of a mobile manipulator, and the ability to convert between

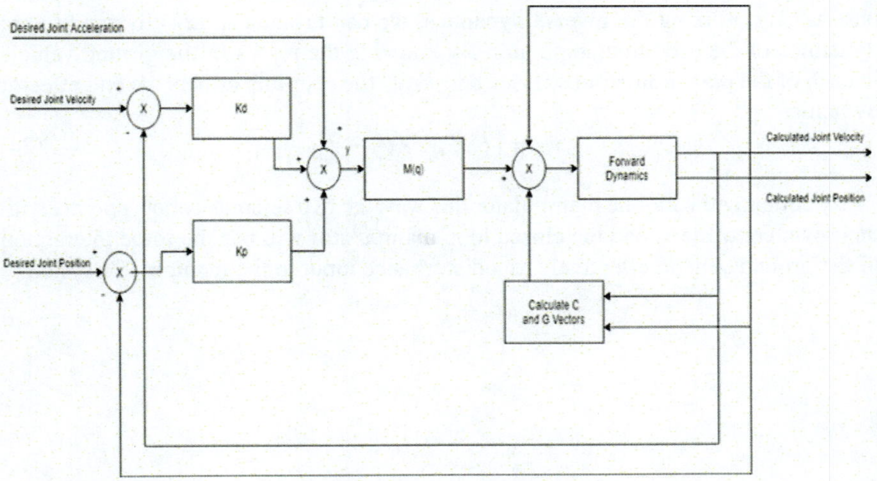

Fig. 2 Inverse dynamics control block diagram

the two accelerations is desirable for driving motors based on torques. Equation (22) demonstrates the conversion from joint space to Cartesian space, while Eq. 23 demonstrates the conversion from task space to joint space. Then using the equations derived for inverse dynamics, the joint accelerations can be converted to torques.

$$\ddot{x} = J(q)\ddot{q} + \dot{J}(q, \dot{q})\dot{q} \tag{22}$$

$$\ddot{q} = J(q)^{-1}(\ddot{x} - \dot{J}(q, \dot{q})\dot{q}). \tag{23}$$

4 Mobile Manipulator Dynamical Model

The coupling action of the ground vehicle and the robotic manipulator is a simple problem that can be solved in a variety of ways. When two robots are considered as two systems coupled together to produce a singular motion instead of a single robot, a heterogeneous robotic system is considered.

The coupling of two robots can be either simple, where the inertia transfer is not considered, or inertial coupling, where the force interaction of the two robots are considered. Using the simple coupling, Seleit et al. demonstrated this simple coupling using kinematics. Using this method, large tracking errors are seen. In contrast, inertial coupling techniques consider the transfer of inertia between two robots. With the desired motion of the robotic system being a stack of robotic systems, the motion of the bottom robot will propagate through the top mounted robot.

The coupling between a ground vehicle and a top mounted robotic arm, otherwise known as a mobile manipulator, can be modeled using the same definition used for modeling the gravity effects of the robotic manipulator on the robotic manipulator base $\dot{v_0}$. By utilizing the inverse dynamics, we can utilize Eq. (24) to input to the dynamics of the robotic manipulator. By knowing the forces of the ground vehicle in each of the planar directions, F_{GVx} and F_{GVy} the coupling affects the end effector dynamics.

$$\dot{v_0} = \begin{bmatrix} F_{GVx} & F_{GVy} & -g \end{bmatrix} \tag{24}$$

By coupling the mobile manipulator this way, the two separate robots operate with their own controllers, and the closed loop manipulator will use the force interaction of the ground vehicle effectively as a disturbance input to the manipulator system.

5 Simulations

5.1 *Ground Vehicle*

The dynamics of the ground vehicle, described in Eq. (10), was solved using ODE45. For this demonstration, we used a circular reference trajectory with a constant acceleration in θ. The circular trajectory is described in Eq. (25).

$$x(t) = \cos(0.25t) \qquad y(t) = \sin(0.25t) \qquad \theta(t) = 0.5t^2 \tag{25}$$

To track this trajectory, we implemented the dynamics-based trajectory linearization controller described in [7] using MATLAB. An initial position error of $\left[.1\ .1\ .087\right]$ was specified. Figure 3 shows a top-down view of the ground vehicle trajectory. Figure 4a shows the error in position for the same simulation, and Fig. 4b shows the error in velocity.

The dynamics-based trajectory linearization controller demonstrates accurate tracking in position and velocity even for simultaneous motion in x, y, and θ. This is an advantage over controllers based only on kinematics, which are unable to account for the coupling introduced by angular motion about the central axis.

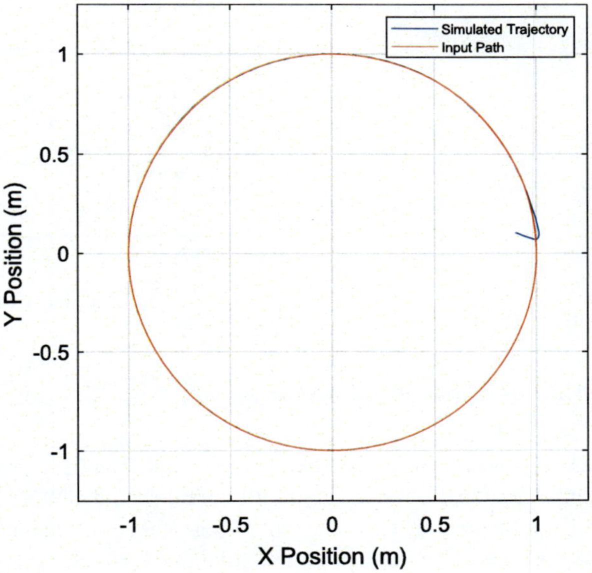

Fig. 3 Circular trajectory, XY View

Fig. 4 Tracking error for circular trajectory

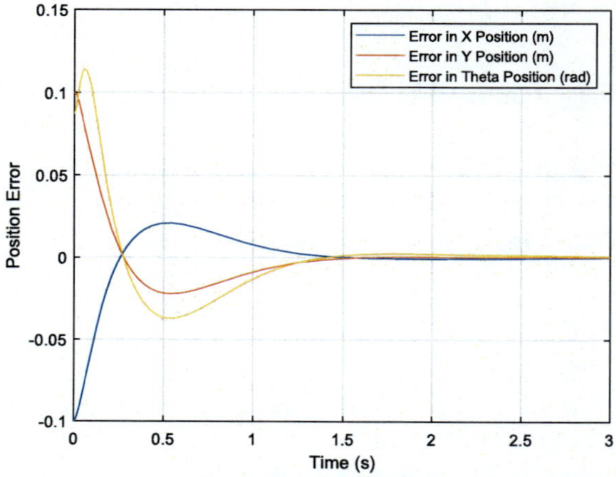

(a) Circular Trajectory, Position Error

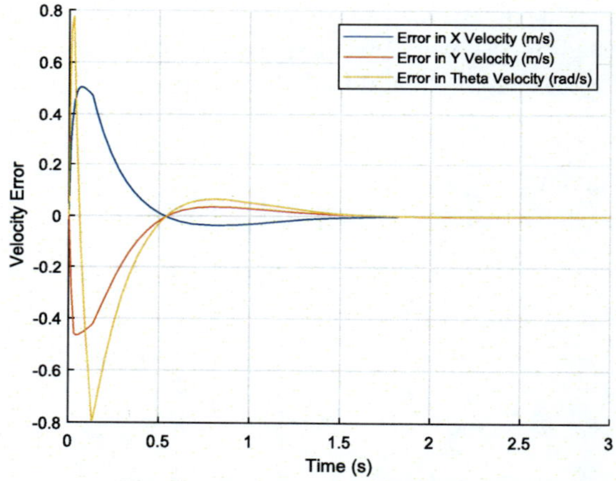

(b) Circular Trajectory, Velocity Error

5.2 Robotic Manipulator

Using the control algorithm derived in equation (21), an experiment was designed to emulate similar motion capabilities of the previous ROME kinematic works. Isolating all but the vertical Cartesian coordinate, the robotic arm follows a 200 mm sinusoidal wave over a 30 s time period relative to the robotic arm base. Specifically, equation (26) were supplied to the control scheme, with the initial conditions in equation (27)

Fig. 5 Closed loop position and velocity tracking of the vertical axis

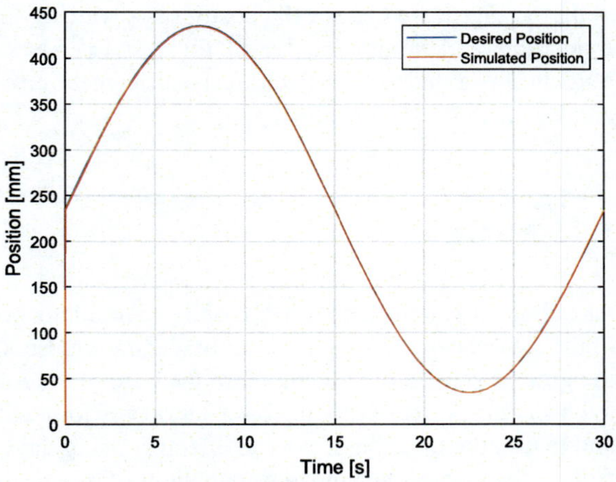

(a) Task Space Position

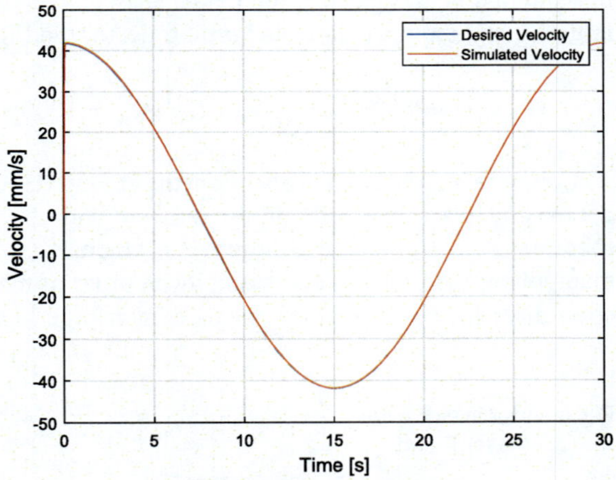

(b) Task Space Velocity

$$\begin{aligned} x &= \left[-173.45\ 0\ 200\ \sin(\tfrac{2\pi t}{30}) + 234.5\ 0\ 0\ 0\right] \\ \dot{x} &= \left[0\ 0\ 200\tfrac{2\pi}{30}\ \cos(\tfrac{2\pi t}{30})\ 0\ 0\ 0\right] \\ \ddot{x} &= \left[0\ 0\ -200\tfrac{2\pi}{30}^2\ \sin(\tfrac{2\pi t}{30})\ 0\ 0\ 0\right] \end{aligned} \tag{26}$$

$$\begin{aligned} x_{IC} &= \left[-173.45\ 0\ 234.5\ 0\ 0\ 0\right] \\ \dot{x}_{IC} &= \left[0\ 0\ 0\ 0\ 0\ 0\right] \\ \ddot{x}_{IC} &= \left[0\ 0\ 0\ 0\ 0\ 0\right] \end{aligned} \tag{27}$$

In the Fig. 5, we can see the position and velocity tracking between the reference trajectories and the simulated result for x_3 and $\dot{x}_3$. Over this time period, the controller quickly converged to the trajectories, with an mean error of less than 0.25% in position and less than 0.01% in velocity.

5.3 ROME

As a coupled system, the ROME mobile manipulator combines the dynamical models of the two systems. Using the coupling equation, Eq. (24), the dynamic effects from the ground vehicle propagating into the base of the robotic manipulator can be used to observe the effect a moving base has on any joint angle of the robotic manipulator, and in turn the location of the end effector. Using the ground vehicle force equations from equation (28), and the same equations of motion used by the robotic manipulator from equations (26), the coupled system is simulated over the course of 30 s. From this simulation, the acceleration model error can be seen in Fig. 6, the position error in Fig. 7, and the final simulated orbit trajectory in Fig. 8.

$$GVx(t) = 1000\ \cos(\frac{t}{30}) \qquad GVy(t) = 1000\ \sin(\frac{t}{30}) \qquad \theta(t) = 0 \tag{28}$$

Using the inertial coupling between the ground vehicle and the robotic manipulator demonstrates a viable solution for working with the dynamics of a heterogeneous robotic system. Compared to the previous kinematic work, positional error for mobile manipulator systems has decreased by an order of magnitude [6]. Additionally, the controllers for both robotic systems demonstrate higher robustness with dynamic

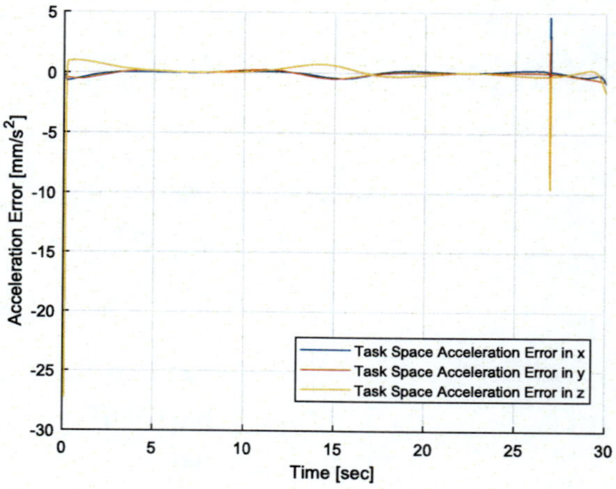

Fig. 6 Error in task space acceleration for ROME

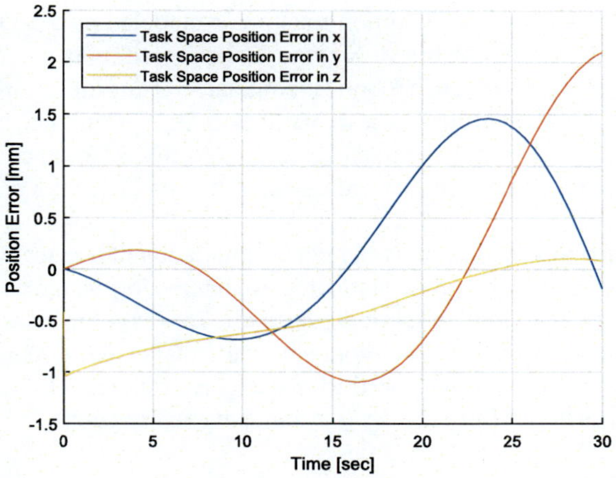

Fig. 7 Error in task space position for ROME

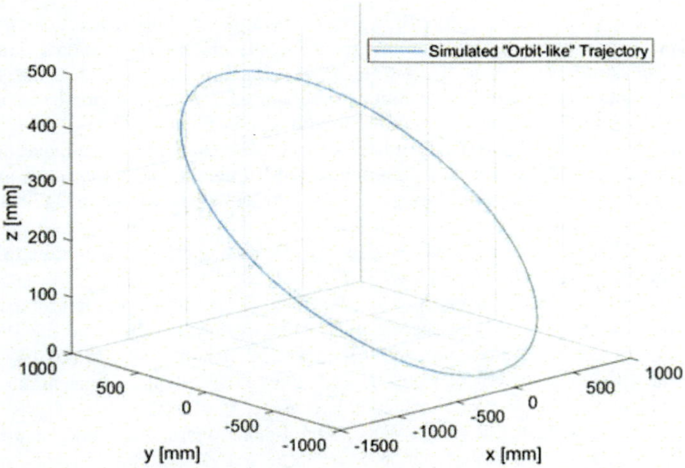

Fig. 8 Simulated "Orbit-like" Trajectory

control compared to kinematic control. The positional error less severe, now that the force interaction between the robotic systems are modeled.

6 Conclusion

Continuing the previous kinematics work from [6], the dynamical models for a heterogeneous, 9-DOF robotic system are being developed. Using a feedback linearization model, accurate sinusoidal motion tracking is modeled on a 3-DOF holonomic ground vehicle, while the Recursive Newton-Euler algorithm is used to model sinusoidal motion vertical motion of a robotic manipulator. Using these combined models, mobile manipulators can utilize these dynamical models to simulate and test various space missions, sensor performance and control algorithms like docking maneuvers, servicing missions, or accelerations of celestial bodies under the constraint of the two-body problem.

With the dynamics models derived here, an emulation of a rendezvous mission is planned with two ROME dynamic platforms. An interesting design problem revolves around the ability to emulate a servicing mission. This experiment will be designed such that these two mobile manipulators are emulating a chief-deputy flying formation. The chief manipulator will need to match speeds and reach the target deputy, manipulate any material from the deputy, and leave the target.

References

1. Piedboeuf, J.-C., De Carufel, J., Aghili, F., Dupuis, E.: Task verification facility for the canadian special purpose dextrous manipulator. In: Proceedings 1999 IEEE International Conference on Robotics and Automation (Cat. No. 99CH36288C), vol. 2, pp. 1077–1083. IEEE (1999)
2. Mietner, C.: European proximity operations simulator 2.0 (epos): a robotic-based rendezvous and docking simulator. J. Large-Scale Res. Facilities JLSRF **3**, 04 (2017)
3. Egerstedt, N., Hu, X.: Coordinated trajectory following for mobile manipulation. In: Proceedings 2000 ICRA. Millennium Conference. IEEE International Conference on Robotics and Automation. Symposia Proceedings (Cat. No. 00CH37065), vol. 4, pp. 3479–3484. IEEE (2000)
4. Jafari Nadoushan, M., Ghobadi, M., Shafaei, M.: Designing reliable detumbling mission for asteroid mining. Acta Astronautica **174**, 270–280 (2020)
5. Huang, Y., Wu, S., Mu, Z., Long, X., Chu, S., Zhao, G.: A multi-agent reinforcement learning method for swarm robots in space collaborative exploration. In: 2020 6th International Conference on Control, Automation and Robotics (ICCAR), pp. 139–144 (2020)
6. Seleit, A.E., Ketzner, R., Quebedeaux, H., Elgohary, T.A.: Rapid orbital motion emulator (rome): Kinematics. In: AIAA Scitech 2020 Forum, p. 1597 (2020)
7. Liu, Y., Zhu, J.J., Williams, R.L., II., Wu, J.: Omni-directional mobile robot controller based on trajectory linearization. Robot. Auton. Syst. **56**(5), 461–479 (2008)
8. Kalmár-Nagy, T., D'Andrea, R., Ganguly, P.: Near-optimal dynamic trajectory generation and control of an omnidirectional vehicle. Robot. Auton. Syst. **46**(1), 47–64 (2004)
9. Kalmár-Nagy, T., Ganguly, P., D'Andrea, R.: Real-time trajectory generation for omnidirectional vehicles. In: American Control Conference, 2002. Proceedings of the 2002, vol. 1, pp. 286–291. IEEE (2002)
10. Samani, H.A., Abdollahi, A., Ostadi, H., Rad, S.Z.: Design and development of a comprehensive omni directional soccer player robot. Int. J. Adv. Rob. Syst. **1**(3), 20 (2004)
11. Lynch, K.M., Park, F.C.: Modern Robotics. Cambridge University Press (2017)
12. Siciliano, B., Sciavicco, L., Villani, L., Oriolo, G.: Robotics: Modelling, Planning and Control. Springer Science & Business Media (2010)

A Development of Marketing Business Game-Overview of Agent-Based Models

Satoru Kawakami, Megumi Aibara, and Masakazu Furuichi

Abstract Research on computer business games forcusing on the theme of marketing is an issue that has been tackled by many researchers for about 65 years (Andlinger in Harvard Bus Rev 36(1):115–125, 1958). On the other hand, marketing has undergone major changes in the last 65 years (Kotler et al. in MARKETING 5.0, Jhon Wiley & Sons, Inc., New Jersey, 2021). We have to say that business games have not been able to keep up with this transformation. We will solve the problems that must be overcome in order to incorporate new technologies into business game research and follow them. Agent-Based Modeling (ABM) simulates the market space by individually modeling consumers, which are the most important targets in marketing. Through this simulation, market demand is expressed as changes in individual consumer demand, and a business game for learning marketing strategy decision making and strategy formulation is prototyped and tested. The experiment result confirmed that the consumers individually modeled by ABM expressed the market demand close to reality in the prototype marketing business game,. In addition, it was confirmed that players who performed simulated management can make management decisions repeatedly and freely formulate strategies.

Keywords Business game · Agent-based modeling · Marketing · Statistical modeling · Strategic decision-making

S. Kawakami (✉) · M. Furuichi
Graduate School of Industrial Technology, Nihon University, Tokyo, Japan
e-mail: cisa18001@g.nihon-u.ac.jp

M. Aibara
College of Science and Technology, Nihon University, Tokyo, Japan

H. Dai (ed.), *Computational and Experimental Simulations in Engineering*, Mechanisms and Machine Science 119,
https://doi.org/10.1007/978-3-031-02097-1_21

1 Introduction

1.1 Background

In 1999, Marketing Principles simulation was announced [3]. It was a marketing game that improved Segmentation and Targeting Skill while considering the Product Life cycle in the digital camera market 2009, "E-Commerce Market Agent-Based Simulation" reported the first business game by ABM at ABSEL [4] 2010, "Marketing Simulation Game of Cascade Demand Phenomenon" formulated and expressed the relationship between consumers [5] 2014, "Internet Marketing Simulation" was announced [6].

2013, Palia and Ryck investigated market segmentation and brand relocation by plotting consumer preferences for products/services as a basic axis in a two-dimensional plane positioning map [7]. We thought that by overlaying this product positioning map and plotting the daily-changing preferences of individual consumers, it would be possible to express a market that showed the intensity of demand from the plot intensity.

Consumers in marketing are the main actors who show their preference for 4P. Morioka and Imanishi, who are practitioners of management, call Product and Price out of 4P "The preference of consumers determined by brand equity, product performance and price", and "essence" that determines the market structure [8].

Statistical modeling models marketing as follows.

When considering the purchasing behavior of individual consumers in discrete time, the judgment of "buying" or "not buying" is made every hour to reach the purchase, and this probability follows the binomial distribution.

When the consumer observation period is observed in N-divided discrete time, when there is a product that the consumer meets probability of p, and the probability of purchases r times Pr is

$$P_r = \frac{N!}{r!(N-r)!} \cdot p^r \cdot (1-p)^{N-r} \tag{1}$$

When this consumer is observed for continuous time, the probability of purchasing a product r times is calculated when the average number of purchases ($N \times p$) is μ as $N \to \infty$.

$$Poisson(r|\mu) = \frac{\mu^r}{r!} \cdot e^{-\mu} \tag{2}$$

When a consumer finds a product that matches (or is close to) its preference, consumer purchases it.

In this study, we prototyped a marketing business game (MBG) by superimposing consumer preferences according to statistical modeling on a positioning map.

1.2 Hypothesis

In order to guide strategic decision-making, we prototype a business game that incorporates display utilizing ABM and statistical modeling, and formulate the following three hypotheses.

Hypothesis1 (H1) ABM allocates a number of consumer agents according to the age-specific population pyramid. The Consumer Agent is plotted on the Consumer Preference Space Map. Market demand can be represented by plotting the consumer agent on the consumer preference space map. Furthermore, a market system can be constructed by defining the relationship between the product agent and the company agent.

Hypothesis2 (H2) Dynamic market demand can be expressed by moving consumers on a map by simulating changes in attributes according to age and consumer behavior as a distribution based on statistical modeling.

Hypothesis3 (H3) By managing a product sales company that responds to market demand and making it a game that achieves profit targets, etc., it guides players to repeat management decisions and formulate strategies.

2 Method

For the derivation of the core structure of the market, We use the Agent-Based Model (ABM) to simulate market components with a simple algorithm, and by displaying the market demand formed by individual consumer preferences with a positioning map.

The advantage of simulating the artificial market of business games with ABM is the economic phenomenon that emerges in the artificial market formed by multiple agents with different characteristics through complex interactions. In ABM, changes in individual consumer demand can be expressed as market demand. By constructing a business game that develops the market by STP Steps in response to this economic phenomenon, it can be expected to develop a business game that learns marketing strategies.

We have developed Marketing Business Game (MBG) on artisoc4 [9], a multi-agent simulation construction execution environment that allows ABM programming.

2.1 Market System

The market system built by ABM consists of consumers, products, and companies that develop products (Fig. 1), and the companies have a player-owned company and four agent-played competitors. Consumers form demand according to their preferences, and companies supply products according to demand. Consumers who

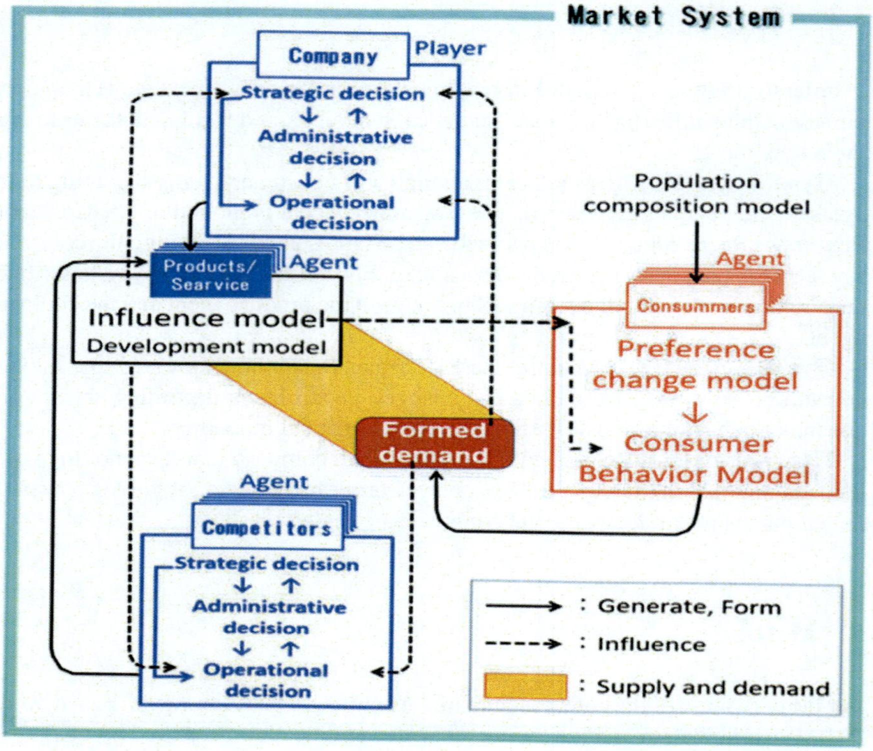

Fig. 1 Market system

purchase a product are greatly influenced by the community of similar products and change their preferences. The heterogenous consumer preferences significantly change market demand, and changes in demand affect a company's product development. The targeting of a company's product development is also influenced by the sales situation of other companies' products. Such a market system is implemented in a Marketing Business Game according to the phenomenon of the actual market.

2.2 Consumers Model

Marketing focuses on products/services being purchased according to individual consumer preferences [10]. Consumer preferences vary depending on product purchasing experience and age-appropriate interests. We will model this consumer and try to build a market that creates demand for goods and services.

By expressing individual consumers with these heterogeneity of preferences as agents in ABM, we can simulate a market of 40 million people comprising individual consumers with heterogenous preferences.

We model consumers in business games so that they can be distributed by age group according to the population pyramid by default.

The age of consumers is from 0 to 100 years old, the birth rate is set, the mortality rate is set according to age, and finally the population increase/decrease and the transition of the population pyramid are simulated as a model in which all consumers die at 100 years old.

As shown in Fig. 1, consumers are modeled by a preference change model and a consumer behavior model.

Consumer preferences change with age in the preference change model. Consumers after purchasing a product significantly change their preferences based on their purchasing experience in the consumer behavior model [11]. This forms an Negative Binomial Distribution (NBD) model of statistical modeling techniques in which the purchase frequency of consumers is almost the same as that of real consumers [8].

Consumer preferences are strongly influenced by the empathy of participating in the community after purchasing a product. Due to this effect, consumer preferences create a flow of movement in the same direction and begin solidarity movement. Therefore, selection concentration occurs on products around a specific preference.

When selection concentration occurs in products around a specific preference, the purchase probability increases among all consumers, and the long-term average purchase frequency μ is gamma-distributed.

The probability that a product with the number of purchases λ per unit time will be purchased α times is

$$Gamma(\mu|\alpha, \lambda) = \frac{\lambda^{\alpha}}{\Gamma(\alpha)}\mu^{\alpha-1}e^{-\lambda\mu}. \tag{3}$$

The probability that this product will be purchased α times by Poisson-distributed consumers is the frequency with which the product is purchased, which is almost the same as the reality in the NBD model as shown in the following equation [8].

$$Poisson(r|\mu)\widehat{\mu}Gamma(\mu|\alpha, \lambda) = \frac{\Gamma(\alpha+r)}{r!\cdot\Gamma(\alpha)}\left(\frac{\lambda}{\lambda+1}\right)^{\alpha}\left(\frac{1}{\lambda+1}\right)^{r} \tag{4}$$

where r is the number of product purchases, and μ is the average number of purchases.

2.3 Products/Services Model

For products that consumers purchase according to Preference, in addition to product characteristics that satisfy Preference, price, fascination (entertaining) time, and product life are the elements of the product model.

Each element of the product model changes depending on the development investment. In fact, the quality of a product is proportional to the development cost. Here, in consideration of game characteristics, we simulate a products market that cannot be easily selected. For the development cost of the 4th class, we prepared the fascination time of the 5th class, the life of the 3rd class, and the price of the 4th class, and divided the ease of gathering of consumer preferences into the 3rd class and considered 720 cases. We found 11 cases in which we were uncertain about the selection, and set the characteristics of the products to be sold in the segmented market.

When product elements are divided into merits that contribute to profits and merits that contribute to market share, the factors that affect profits are development costs, prices, and consumers, and the factors that affect market share are fascination time, product life, and consumers. Acquiring consumers, which fall into both merits, is the most important marketing goal.

Profit (G) is expressed by the following formula (5) from the cost (C) including the development cost, the selling price (V), and the number of units sold (N) [12].

$$G = \sum_i \left\{ \sum_t (V_i \cdot N_{it}) - C_i \right\} \tag{5}$$

where i is the subscript of the product, and t is the subscript of the unit time.

The market share is expressed by the following equation by the Dirichlet NBD model when the number of purchases of product j for the total purchases number R per unit time is rj [8].

$$P(R, r_1, r_2, \ldots, r_j, \ldots, r_g) \\ = \{Multinomial(r|p, R)\hat{p}Dirichlet(p|\beta)\}\hat{R}\{Poisson(R|\mu)\hat{\mu}Gamma(\mu|\alpha, \lambda)\} \tag{6}$$

where α is the product purchased times, λ is the number of purchases per unit time, μ is the average purchases number, β is the product selected times, and p is the product selected probability.

The fascination time lowers the β of other products (relatively raises the β of that product), and the product life increases α. This will increase the market share of company products.

2.4 Enterprise Management Process

Company management in the MBG is a player-centered process, and it adjust resource allocation to control development progress, maintain business management, and develop sales and advertisements during marketing. The STP steps of the target marketing [10] was progressed in the marketing. The STP steps is the flow that involves segmentation of the market by market research (market segmentation: S), defining the segmented market to be targeted based on information of the competitors' products/services market expanding status and others (market targeting: T), develop a product/service that matches the segment market (market positioning: P), and releasing the product. Resource allocation adjustments require increasing/decreasing expenses to accelerate/slow progress during the development period and employment/dismissal when a certain adjustment range is exceeded. Employment/dismissal increases costs and decreases profits. With STP steps and cost adjustments, launching products/services with good positioning and timing will increase consumer purchasing and increase profits. Products/Services aimed at segment markets with higher consumer purchasing rates were assigned higher development costs, and the played manager was required to contrive targeting.

2.5 Competitors

Shinoda, Ryouke, Terano, Nakamori (2006) investigated the use of agent players as opponents with the ability to choose simple decision-making rules and strategies in business games. The results indicated that the agent player could be a viable competitor to the human player [13].

In the MBG, four agent companies with market strategies of a leader, challenger, follower, and nicher [14] compete in the same market with the subject. We expect them to serve as a model for cost adjustments at the beginning of the game and as formidable competitors in the middle of the game. Unlike the agents in Shinoda et al., they do not have the function of selecting strategies, but the company with the most effective strategy in the market situation became a formidable competitor to gain market share as a result of the four agent companies competing with different strategies.

3 Result

3.1 Market System

A market system was constructed by ABM, and the MBG shown in Fig. 2 was prototyped. As a result of running MBG, consumers have formed market demand due to slight differences between individuals, and competitors have launched products targeting that demand. Consumers who purchased the product experienced solidarity movements that changed their preferences according to the consumer behavior model. Competitors have launched more products in segments where their preferences are concentrated in response to changes in consumer preferences. In the company management process by the player, market information collection, resource allocation adjustment, and the product characteristics selection to be released were smoothly performed, and the company's products could be released in a competitive manner with the competitors. By the above operation verification, it was verified

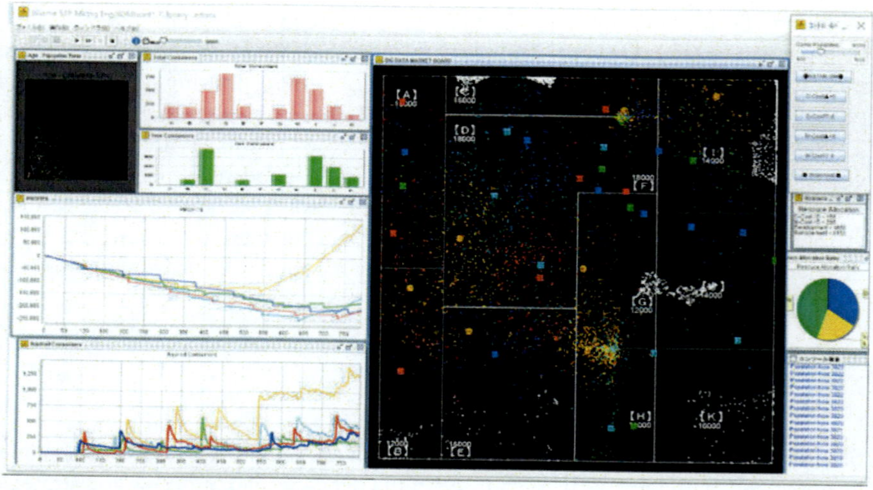

Fig. 2 Overview of marketing business game

that the relationship between Agents in the market system sufficient to simulate the market environment was established, and H1 was confirmed to be Positive.

3.2 Consumers Model

In order for the prototype MBG to observe the formulation of future management strategies, it is necessary to be able to simulate the market up to several decades before the results will appear. The population composition model mounted on the MBG can reduce the increase and decrease in the consumer population by accurately setting the birth rate and mortality rate, and with this setting (2017 Japan's demographic values [15]), it is a business equivalent to more than 100 years. It turned out that the game can be played.

As shown in Fig. 3, the market at the initial stage of MBG startup was set by the consumer preference change model, and a slight bias was observed in the consumer preference when the product was not released, simulating the appearance of an undeveloped market. It simulated the appearance of an undeveloped market. After the product is launched, an active market in which consumers dynamically move between products according to the consumer behavior model is simulated (Fig. 4), and the consumer Preference after purchase forms a community and sympathizes with the preference. Was born and was seen moving in solidarity. As a result, there was a concentration of preferences for some products. From the above operation verification, it was verified that the consumer preference expressed by Agent was able to form market demand according to statistical modeling, and H2 was confirmed to be Positive.

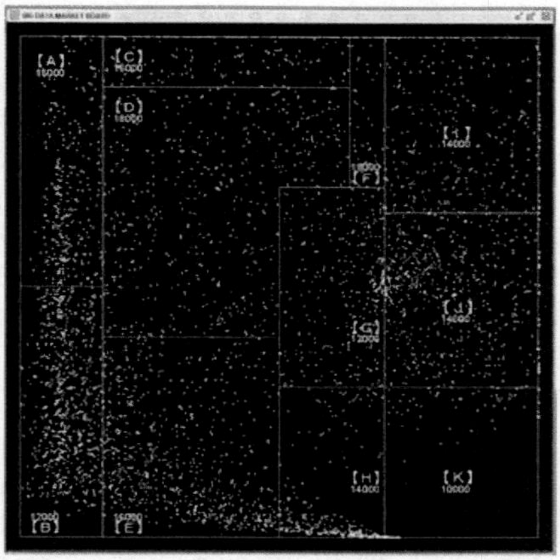

Fig. 3 The initial stage of MBG startup

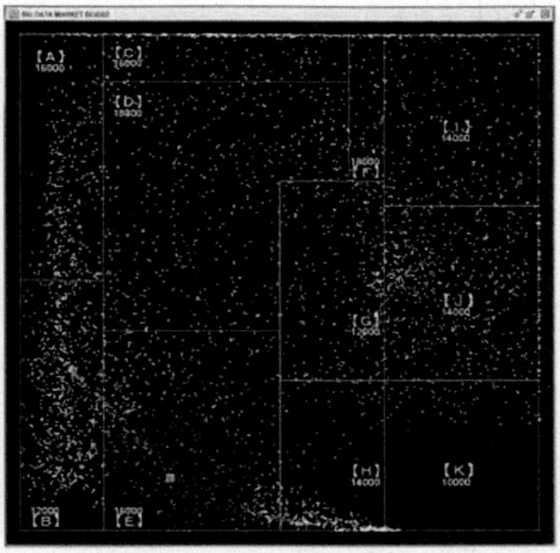

Fig. 4 Consumers dynamically move

3.3 *Products/Services Model*

MBG has simplified the selection of developed products to enhance the game. Even so, there are many measures that can be taken to increase profits or increase market share, and even among competitors who have set strategies and played Agents, products to be released due to market changes have occurred in different situations every time.

It was confirmed that the effect of launching series products is also supported because the direction of movement of consumer preferences can be changed and the effect of strategy can be changed depending on how the product characteristics are arranged.

3.4 Enterprise Management Process

As shown in Fig. 5, the player can control the development speed and the characteristics of the released products by changing the company's resource allocation in MBG. The decision of this control element is based on operational decision making, but changing the resource allocation for that purpose is based on administrative decision making. Furthermore, strategic decision making is required to secure continuous profits and increase market share (Fig. 6). In the operation verification, since the debt at the beginning of the business cannot be recovered with one product, it is necessary to plan and release products with characteristics that increase profits one after another, and strategic decision making for that purpose has become indispensable (Fig. 7). From the above verification results, it was confirmed that MBG repeated management decision-making while grasping changes in market demand, and in the background it became a game that required strategic decision-making, and H3 became Positive.

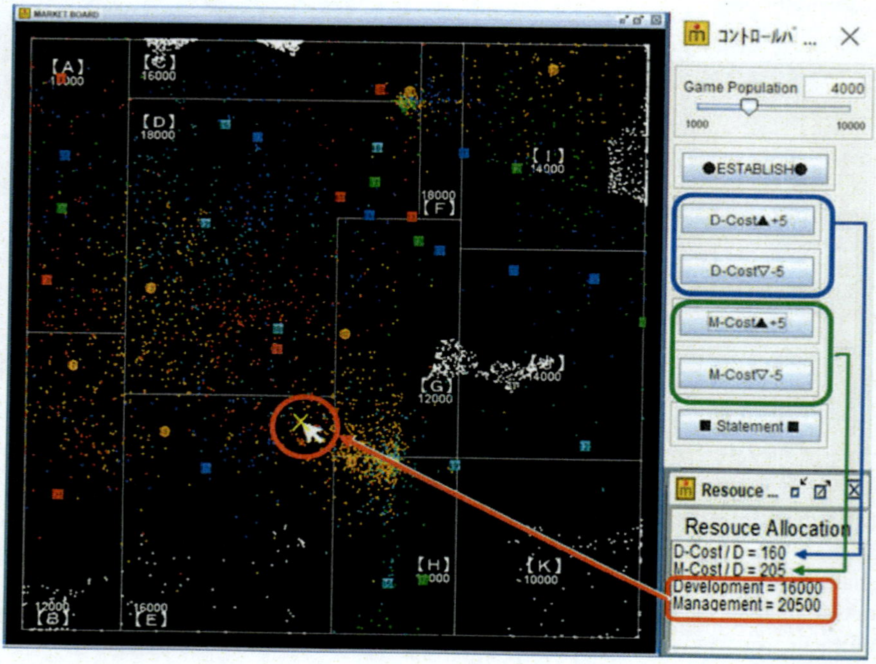

Fig. 5 Player's control

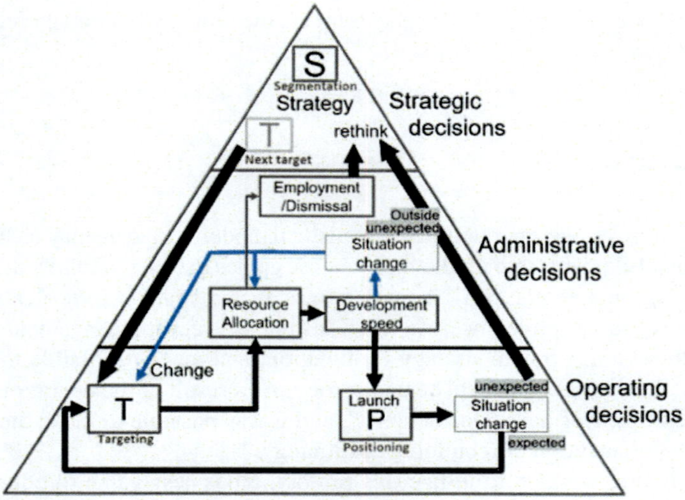

Fig. 6 Management decision-making hierarchy

Fig. 7 Initial debt recovered

3.5 Competitors

Moderate product development speed and product consumer attraction range vary from market to market. The development speed and product launch interval that were initially granted to suit the market conditions served as a model for competitors to avoid oversupply and overcompetition of products first, and to eliminate debt and return to profitability in advance. In addition, after the middle of the game, it was possible to implement a strategy orientation that makes the player think about a plan by launching a product prior to the player's target. As a result, it became a game that strengthened H3 and made the player repeat the thinking of strategy formulation during the game.

4 Conclusion

In this research, we incorporated ABM, statistical modeling and display techniques as new technical elements into marketing business games in order to follow the changes in marketing, and prototyped business games that display market demand from individual consumer preferences. In MBG, as players repeat management decision-making, they came up with strategy formulation, and made it possible to simulate decades so that the results could be understood. As a result of the operation verification, these objectives have been achieved, and it was possible to make the business game with the intention of rebuilding the strategy.

In the future, we plan to utilize this prototype business game to education for students. We are also interested in to extend this business game by utilizing real market data. Furthermore, research on the participation of multiple MBG players and remote play via the Internet will be another issues for the future.

References

1. Andlinger, G.R.: Business games—play one! Harvard Bus. Rev. **36**(1), 115–125 (1958)
2. Kotler, P., Kartajaya, H., Setiawan, I.: MARKETING 5.0, Jhon Wiley & Sons, Inc., New Jersey (2021)
3. Schaffer, R.W.: The Marketing Game: A Marketing Principles Simulation, Developments in Business Simulation & Experiential Exercises Volume 26, ABSEL, p. 223 (1999)
4. Umeda, T., Ichikawa, M., Koyama, Y., Deguchi, H.: Evaluation of Collaborative Filtering by Agent-Based Simulation Considering Market Environment, Developments in Business Simulation & Experiential Exercises Volume 36, ABSEL, pp. 214–222 (2009)
5. Cannon, H.M., Cannon, J.N., Andrews, C.R.: Modeling Cascading Demand: Accounting for the Effects of Captive Consumer Relationships, Developments in Business Simulation & Experiential Exercises Volume 37, ABSEL, pp. 33–41 (2010)
6. Draper, S.: Stukent Real Deal Simulation: An Internet Marketing Simulation, Developments in Business Simulation & Experiential Exercises Volume 41, ABSEL, p. 397 (2014)
7. Palia,A.P., Ryck, J.D.: Repositioning Brands with the Web-Based Product Positioning Map Graphics Package, Developments in Business Simulation & Experiential Exercises Volume 40, ABSEL, pp. 207–228 (2013)
8. Morioka, T., Imanishi, S.: Probability Strategy for Marketing, 22nd edn. ISBN978-4-04-104142-0. KADOKAWA, Japan (2016)
9. Kozo Keikaku Engineering Inc.: artisoc4. https://mas.kke.co.jp/artisoc4/ (2020)
10. Kotler, P., Armstrong, G.: Marketing: An Introduction, 4th edn. Pearson Education Inc. (1997)
11. Schmitt, B.H.: Experiential Marketing: How to Get Customers to Sense. Think, Act, Relate, The Free Press, Feel (2000)
12. Toyota, Y.: Data Driven Marketing by R, Ohmsha, Japan (2017)
13. Shinoda, Y., Ryouke, M., Terano, T., Nakamori, Y.: Design of a Software Agent for Business Gaming Simulation. JAIST Press. http://hdl.handle.net/10119/3868 (2005)

14. Kotler, P., Keller, K.L.: Marketing Management, 12th edn., Pearson Education Inc. (2006)
15. Statistics Bureau of Japan. https://www.stat.go.jp/english/data/jinsui/2.html. Last accessed 2021/11/1

Application of Particle Group Optimization Algorithm Based on Environmental Policy in Environmental Management

Shuwei Lu

Abstract In the rapid development of social economy, the problems of resource and environmental management have been highlighted. In order to effectively shorten the development difference between urban and rural areas in China, the environmental management should be deeply discussed according to the current environmental policy using optimization algorithm. Therefore, on the basis of understanding the current development status of resources, environment and management, this paper makes an in-depth discussion on the application of particle swarm optimization algorithm and its improved PSO algorithm in environmental planning and management, and finally further optimizes the environmental management mode.

Keywords Environmental policy · The particle swarm · Optimization algorithm · Environmental management · Solid waste disposal · PSO algorithm

1 Introduction

In the development of environmental planning and management, many decision-making problems can be summarized as constrained optimization problems (COP), such as allocation of air pollution emission allowances, solid waste treatment, rational use of water resources, etc. These problems not only need to achieve the expected set of environmental goals during the calculation and analysis, but also will be affected by resources, society, economy and other aspects of the limit. Therefore, in order to maximize the realization of environmental goals under the constraint conditions, it is necessary to strengthen environmental planning and resource management, and pay attention to the existing environmental management policies to put forward effective solutions. At the same time, according to the previous research experience, the analysis and calculation of extreme value problems with constraints can also be called as constrained optimization problems. From a practical point of view, in the continuous improvement of China's social economy, the gap between urban and

S. Lu (✉)
University of International Business and Economics, Beijing, China
e-mail: lswuibe@163.com

H. Dai (ed.), *Computational and Experimental Simulations in Engineering*, Mechanisms and Machine Science 119,
https://doi.org/10.1007/978-3-031-02097-1_22

rural areas and ecological balance and other issues have been effectively solved. And under the background of new era, in order to better explore the impact of related issues, improve ecological efficiency of resource application practice, guarantee the stability of the ecological environment development, to fully grasp the current status of environmental management, on the one hand, practice management system is not perfect, from the point of China's existing environmental management system, the corresponding management scheme does not have perfect and scientific, There is a certain difference with the actual development, the relevant preferential policies can not be implemented orderly in the practical environment, which directly limits the development of resources, environment and economic management; Environmental management, on the other hand, is in a state of flux [1]. This unstable state is reflected in the application and management of various resources, and directly limits the development speed of regional urbanization construction, which can not only improve the level of social and economic development, but also cause more resource problems. Therefore, researchers propose to use optimization algorithm to conduct in-depth discussion on environmental management. This paper mainly starts with the planning of solid waste treatment plant, and studies the feasibility and effectiveness of the improved example group optimization algorithm in dealing with environmental management constrained optimization problems [2, 3]. By putting a constraint condition respectively in the basic particle swarm optimization algorithm is used for processing, the main function and fitness calculation function and add a penalty term in the objective optimization function, constrained optimization into unconstrained optimization problems, in order to deal effectively with the corresponding constraint optimization problems, lay a foundation for the environmental protection in the practice of city construction [4–6].

2 Method

2.1 Particle Swarm Optimization Algorithm

Since the early 1990s, researchers have proposed a bionic intelligent algorithm to simulate the swarm behavior of natural organisms in practical exploration, including ACO, PARTICLE swarm optimization (PSO) and other theories, which are called swarm intelligence algorithm system when integrated together [7]. This content can obtain intelligence in the simulation and imitation of the cooperation, competition and other complex behaviors between biological individuals, so as to put forward effective solutions to some specific problems. The earliest application of particle swarm optimization algorithm is mainly to imitate the simple social system, and from this analysis and discussion of overly complex social behavior. Later, it was found in practical exploration that this kind of algorithm could be used to solve complex optimization problems, and because the practical operation was very simple and convenient to realize, once popularized, it got the attention of scholars around the world, and began to be widely used in many fields such as natural science and engineering management.

2.2 Improved Algorithm

Assuming that the solid waste output of city A and City B reaches 700 t/wk and 1200 t/wk, incineration plant is planned to be built in treatment plant 1, which is 15 km away from the local city and 10 km away from city B. And the disposal site 0 is to build A garbage dumping dock, which is 5 km away from city A and 15 km away from city B. At the same time, the distance between the sanitary landfill site and city A is 30 km, and the distance between the sanitary landfill site and city B is 25 km. The cost of solid waste transportation is 0.5 yuan/t km. Fixed and variable costs for specific treatment plants are shown in Table 1. At this stage, we should first understand how to plan and study the facilities for solid waste treatment, so as to ensure that the annual treatment cost of city A and City B can be minimized.

Combination of the following as shown in Fig. 1 solid waste disposal system analysis, this article in view of the environmental management of urban planning, the purpose is mainly divided into two aspects, on the one hand, refers to solid waste shipping and handling cost to meet minimum overall spending, on the other hand, to guarantee solid waste produced by the two cities, can get effective treatment.

Table 1 Expenses and treatment methods of simulated solid waste treatment plants

Disposal site code	The disposal way	Fixed fee (RMB/a)	Fixed fee (RMB/a)	Variable cost (YUAN/t)	Processing capacity (T/WK)
I	Burning	200,000	3850	12	1000
O	For the sea	60,000	1150	16	500
L	Sanitary landfill	100,000	1920	6	1300

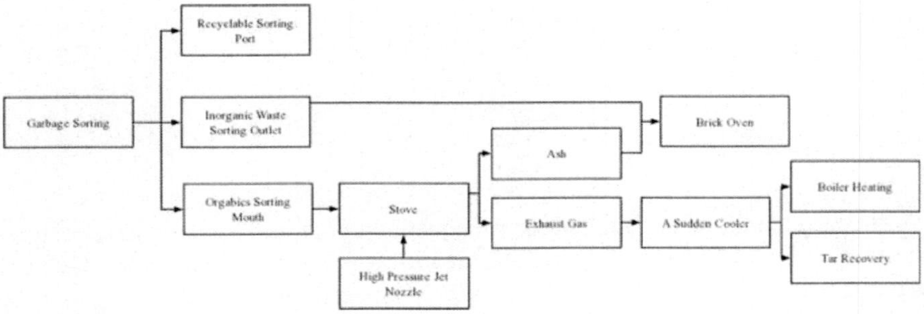

Fig. 1 Solid waste disposal system

In practical research, it is necessary to first study whether to choose a certain treatment method, which requires that a class of discrete variables are (y1, Y2, y3). This discrete variable should be analyzed by selecting integers. In other words, the ith treatment mode should be selected if yi = 1 is met, and the ith treatment mode should not be selected if yi = 0 is met. Meanwhile, we define xij = (I = 1, 2; J = 1, 2, 3) represents the amount of garbage incinerated into the sea and sanitary landfill in all cities, where xij represents the amount of solid waste treated from city I to disposal site j, in the specific unit of t/wk. Nine decision variables are obtained by calculation and analysis, three of which represent integer variables. And we want to do it in 100t, not in the tens place. Hence, the following constraints can be obtained:

$$y_i = \begin{cases} 0 \; x_{1j} + x_{2j} = 0 \\ 1 \; x_{1j} + x_{2j} > 0 \end{cases}$$

$$x_{ij} = n \times 100, \, (n = 0, 1, 2)$$

$$x_{1j} \leq 700, x_{2j} \leq 1200, \, (j = 1, 2, 3)$$

Starting from the second research objective of planning analysis, solid waste generated by the two cities should be completely treated, and the following balance equation can be obtained:

$$\sum_{j=1}^{3} x_{1i} = 700$$

$$\sum_{j=1}^{3} x_{2i} = 1200$$

All treatment plants have restriction conditions, and the corresponding constraint condition formula is:

$$\sum_{i=1}^{2} x_{i1} \leq 1000$$

$$\sum_{i=1}^{2} x_{i2} \leq 500$$

$$\sum_{i=1}^{2} x_{i3} \leq 1300$$

The next step is to study the objective equation to minimize the total cost expenditure of solid waste transportation and treatment capacity. At this time, the solid waste treatment costs can be divided into three aspects:

First, solid cost. This kind of expenditure belongs to the solid maintenance and protection cost of the treatment plant. The specific formula is as follows:

$$3850_{y1} + 1150_{y2} + 1920_{y3}$$

Second, transportation costs. This type of expenditure needs to utilize the transportation cost per unit distance of solid cost per unit quantity × the volume of pulled traffic. The specific formula is as follows:

$$7.5_{x11} + 2.5_{x12} + 15.0_{x13} + 5.0_{x21} + 7.5_{x22} + 12.5_{x23}$$

Third, variable costs, which are based on the cost of waste disposal per unit weight × the number of waste disposal per week, as shown below:

$$12.0_{x11} + 16.0_{x12} + 6.0_{x13} + 12.0_{x21} + 16.0_{x22} + 6.0_{x23}$$

Combining these three costs is the total cost of transportation and treatment of two MSW:

$$Z = 3850_{y1} + 1150_{y2} + 1920_{y3} + 19.5_{x11} + 18.5_{x12} + 21.0_{x13} + 17.0_{x21} + 23.5_{x22} + 18.5_{x23}$$

3 Result Analysis

Based on the above analysis, it can be seen that the key to solve the problem is to find the best xij = (I = 1, 2; J = 1, 2, 3), and the decision variable yj (j = 1, 2, 3) needs to be specified according to xij. Therefore, the improved particle swarm optimization algorithm is used for processing, and the particles need to be formed according to xiJ. At this time, the mansa number of the particles is 16. In the program design and operation, it is mainly divided into two aspects: on the one hand, it refers to the main program of particle swarm optimization algorithm main m; on the other hand, it refers to the calculation function of fitness value fitcal m. The former is mainly used for particle emergence and petrochemicals processing, and the corresponding algorithm is iteratively analyzed. The calculation function fitcal m of the latter should be combined with the optimization model and penalty term obtained above to obtain the specific value of fitness, which is also the cost of various solutions. The number of iterations selected for the design program in this paper is 100, the inertia weight W drops from 0.9 to 0.4 with the increase of the number of iterations, and the number of particles is 100 with c1 = C2 = 2. After five iterations, you get the results shown in Table 2.

Combined with the above results, analysis, program, every time can obtain the global optimal solution, the minimum total cost expense 41,070 yuan/tw, and expected results are consistent, this example of demonstrating improved after swarm

Table 2 Results of iteration

x_{ij}	x_{ij}	x_{ij}	x_{ij}	x_{ij}	x_{ij}	Total cost Z (Yuan/tw)	The number of iterations I
0	500	200	1000	0	200	41,070	1
100	500	100	900	0	300	41,070	19
0	500	200	1000	0	200	41,070	11
100	500	100	900	0	300	41,070	1
0	500	200	1000	0	200	41,070	17

optimization algorithm procedures, can be used to deal with the current environmental policy of environmental management problems, can not only guarantee to effectively deal with solid waste, you can also control unnecessary costs. At the same time, the program is looking for the optimal solution, and only 20 iterations are carried out, which proves that the algorithm program is simple to operate and the efficiency is improved.

Combination shown in Fig. 2, the following algorithm application in finding the global optimal solution of the particle swarm optimal solution during the process and the change of the global optimal solution fitness between analysis shows that process visual rendering particle swarm algorithm iteration is infeasible solution in the first 21 generation, and then the particle swarm will not appear in the infeasible solution, population convergence.

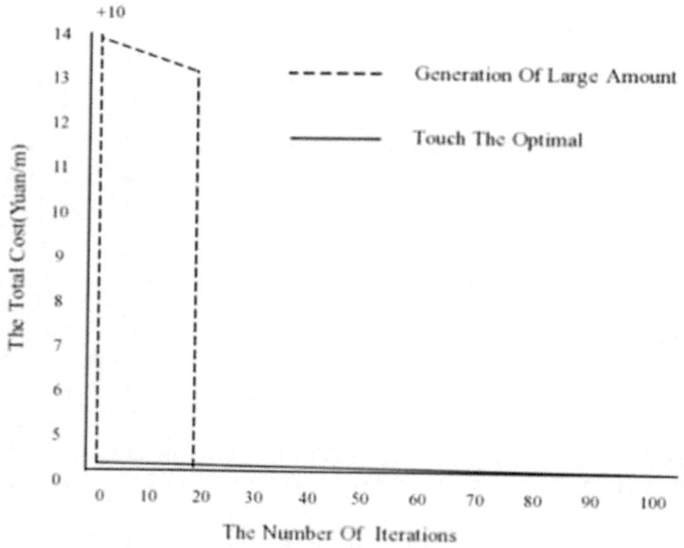

Fig. 2 The changing relationship between the fitness of the pSO optimal solution and the global optimal solution

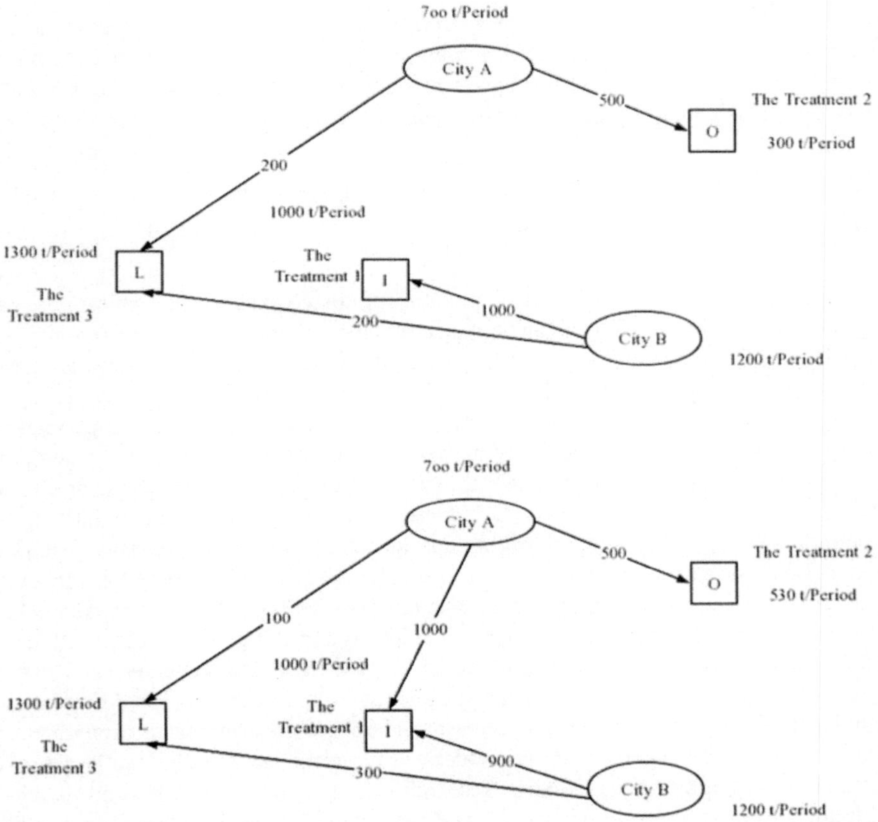

Fig. 3 Two planning methods based on particle swarm optimization algorithm

Combined with the analysis of Fig. 3, it can be seen that as the two planning methods obtained by the algorithm program in this study, they can not only guarantee the effective treatment of municipal solid waste in practical application, but also control the expenditure in a certain sense. By simply improving the particle swarm optimization algorithm, adding a penalty term to the fitness calculation function, the infeasible solution can be eliminated effectively. Although the initial particles are all feasible solutions, a certain proportion of infeasible solutions will remain during the practice update. The final experimental results show that the improved PARTICLE swarm optimization algorithm plays a positive role in solving the environmental management problems under the current environmental policy, and can improve the practical operation efficiency on the basis of ensuring the optimization and effective processing of resources. Nowadays, resources and environment management as the focal point of the urbanization development in our country, strengthen the management of resources in urban and rural construction and development, and can effectively shorten the development gap between rural and urban areas, the full implementation of environmental protection and resources conservation and sustainable development strategy, can further speed up the economic development of our country not only, also can optimize the environment of social economy in an all-round way.

4 Conclusion

To sum up, based on the analysis of China's accumulated experience in environmental governance in recent years, it can be seen that the most important thing to strengthen environmental management according to existing environmental policies is to get the best solution under multiple constraints. From a practical point of view, the implementation of environmental management based on environmental policies is a systematic project. It is not only necessary to optimize the relevant systems of environmental governance, but also to carry out innovative research according to the current situation of practice and development. Based on the understanding of particle swarm optimization algorithm and its advantages, on the basis of planning for solid waste treatment to delve into this practice issues, and by using the improved example swarm optimization algorithm is put forward the scheme of the planning, to master the self-help group of optimal solution and the global optimal solution, the relationship between the fitness of this for the current related research in the field of environmental management has a positive role. In construction development under the background of new era, therefore, the local government in strengthening environmental management in the moderate and at the same time, should actively organize to participate in scientific research scholars combined with optimization algorithm to explore relevant issues, and put forward accord with the sustainable development of the research plan, study more excellent application at home and abroad for reference, the algorithm of accumulated experience, vigorously develop related professionals, Emphasis should be placed on strengthening the technical concepts needed for research and exploration, so as to obtain more effective treatment scheme according to the selected optimization algorithm on the basis of integrating the accumulated experience data of environmental management [8–10].

References

1. Liu, J., Zhang, N., Guo, W.: Parallel particle swarm optimization algorithm based on optimal particle sharing and its application in classification. Machinery **036**(002), 32–34, 37 (2009)
2. Wang, M.: Application of GQPSO algorithm in dynamic environment optimization problem. Softw. Guide **017**(008), 35–39 (2018)
3. Liu, L., Wang, D.: Composite particle swarm optimization and its application in dynamic environment. J. Syst. Eng. **26**(002), 269–274 (2011)

4. Huang, M., Chen, Z., Guo, Z.: Application of particle swarm optimization algorithm based on JADE platform in environmental economic scheduling. Guangdong Electr. Power **000**(004), 51–56 (2015)
5. Wang, D., Li, F., Chen, D.: Optimization of land use allocation based on Pareto optimality and multi-objective particle swarm optimization. Resour. Environ. Yangtze Basin V **28**(09), 3–13 (2019)
6. Song, X., Xiang, T., Xiong, H., et al.: Low carbon generation expansion planning based on carbon emission right allocation. Autom. Electr. Power Syst. **036**(019), 47–52 (2012)
7. Xu, D., Ren, Y., Wang, R., et al.: Ecological security of water resources in Heilongjiang Province based on PSO-PPE model. China Environ. Monitor. **035**(004), 109–114 (2019)
8. Gao, Y., Li, J.: Price prediction of international carbon finance market based on EMD-PSO-SVM error correction model. China Popul. Resour. Environ. (6), 163–170 (2014)
9. Hu, S.: Research on the interactive relationship between China's economic growth and industrial pollution. Res. Fin. Econ. Issues (06), 19–25 (2015)
10. Yu, J., Ji, X., Xia, A.: Power system protection and control (01), 30–33 (2009)

A Computation Process for the Higher Order State Transition Tensors of the Gravity and Drag Perturbed Two-Body Problem Using Adaptive Analytic Continuation Technique

Tahsinul Haque Tasif and Tarek A. Elgohary

Abstract In this work, the Taylor series based integration approach, Analytic Continuation, has been implemented to compute Higher Order State Transition Tensors for the $J_2 - J_6$ gravity and drag perturbed Two-body problem. Analytic Continuation is an integration method applied to solve different fundamental problems of astrodynamics. In this method, two scalar quantities f and g_p are defined and differentiated to arbitrary order using the Leibniz product rule to obtain higher order time derivatives of the variables which are implemented in the Taylor series expansion of the solution. Previously, this method has been proved to be highly precise and computationally efficient in propagating Two-body trajectories with full spherical harmonics gravity and atmospheric drag perturbation. More recently, this method has been implemented in propagating gravity and drag perturbed State Transition Matrix (STM) with machine precision level of accuracy. An expansion of the procedure to compute J_2 gravity perturbed Higher Order State Transition Tensors has also been presented in the subsequent work. In this paper, the method is further expanded to incorporate up to J_6 gravity and drag perturbation in the computation of Higher Order State Transition Tensors. Four types of orbits are considered for numerical simulations: LEO, MEO, GTO, and HEO. First, RMS error of the unperturbed STMs comparing to the closed form solution of Battin are presented. Then, the error in the symplectic nature of the gravity perturbed STM and Higher Order State Transition Tensors are checked, showing double precision accuracy of the STMs and tensors. Finally, initial error of the states of the $J_2 - J_6$ gravity and drag-perturbed orbits are propagated using the computed perturbed Higher Order State Transition Tensors and compared against the results obtained using the perturbed STM, showing at least 3 to 4 digits of accuracy improvement while using Higher Order Tensors.

T. Haque Tasif (✉) · T. A. Elgohary
Mechanical and Aerospace Engineering, University of Central Florida, Orlando 32816, USA
e-mail: thtasif@Knights.ucf.edu

T. A. Elgohary
e-mail: elgohary@ucf.edu

H. Dai (ed.), *Computational and Experimental Simulations in Engineering*, Mechanisms and Machine Science 119,
https://doi.org/10.1007/978-3-031-02097-1_23

1 Introduction

Space Situational Awareness(SSA) is the process of tracking the near Earth celestial objects, satellites, and their debris and predicting their future states at a given time frame. In the past few years, Kessler syndrome has become a very familiar word in the field of SSA and point of concern of the future space missions [1]. Kessler syndrome was named after Donald J. Kessler and refers to the scenario where in the Low Earth Orbit (LEO) the space debris collide with each other and create a rapid increase of space debris around the Earth creating more possibility for further collisions. The uncertainty quantification of the near Earth objects is an active field of research in SSA to overcome challenges like conjunction analysis, tracking and probability of collision [2, 3]. Two major processes for uncertainty propagation in astrodynamics are: Linear models [4, 5], and nonlinear Monte Carlo simulation, [6, 7]. The linear models are computationally efficient but their accuracy level is limited. On the other side, the Monte Carlo simulation is highly precise but computationally very expensive, as it may need thousands of orbit propagations to get a reasonable solution for the orbital uncertainty problem [8, 9]. An alternate approach to uncertainty propagation using the State Transition Tensors based approach is the numerical solution of Fokker–Planck–Kolmogorov Equation (FPKE). Adaptive Gaussian Mixture Model (GMM) has been added with the solution of FPKE by Giza et al. to increase the accuracy of the orbit uncertainty propagation [10]. However, a major drawback of this approach is that, it assumes the final Probability Density Function is represented by the set of mixed Gaussians [2].

Several researchers addressed the procedure for deriving the higher order state transition tensors in several different ways and used them to propagate initial state errors, with the application of orbit error propagation as well as solving Lambert's problem and orbit transfers. Junkins et al. derived up to fifth order State Transition Tensors using higher order partial derivatives of the Lagrange Coefficients F and G [11]. However, in that work, gravity and drag perturbations were not considered [12]. Lantoine and Russell implemented higher order state transition tensors in the optimization of low-thrust problems. They used first and second order state transition tensors to develop a hybrid differential dynamic programming algorithm to increase the efficiency and accuracy of the solution [13]. Bani Younes et al. used OCEA (Object-oriented Coordinate Embedding Algorithm) to derive higher order State Transition Tensors and implemented the results in uncertainty propagation of initial state errors, [14]. OCEA is a software package which uses computational differentiation method to compute the higher order partials [15]. The approach was then expanded to solve the Lambert's problem, which resulted in showing increased convergence rate with the addition of the higher order tensors to up to 95% of an orbit [16]. The results of the symplectic nature of the higher order state transition tensors with higher order spherical harmonics gravity model were presented in an engineering note [17].

The semi-analytic integration approach, Analytic Continuation, is based on Taylor series and Leibniz product rule. Previously, it had been proven to be highly precise in generating the unperturbed as well as $J_2 - J_6$ gravity and atmospheric drag perturbed two-body problem trajectories [18–20]. In the subsequent works, this method was compared against *RKN1210* for the Spherical Harmonics (up to 70 x 70) gravity perturbed Two-body propagation showing superiority in performance for both accuracy and computation time [21]. Recently, gravity and drag perturbed State Transition Matrix (STM) has been computed using the Analytic Continuation method showing machine precision level of accuracy [22, 23]. In the subsequent development, the method has been expanded to derive the Higher Order State Transition Tensors for the unperturbed and J_2 gravity perturbed Two-body problem [24]. The gravity and drag perturbed STM, developed using Analytic Continuation method has also been implemented to improve the results of a space based perturbed orbit estimation method [25].

In this paper, the method has been further expanded to incorporate $J_2 - J_6$ gravity and drag perturbation in the derivation process of the Higher Order State Transition Tensors of the Two-body problem. In Sect. 2, the procedure of deriving Adaptive Analytic Continuation technique for unperturbed Two-body problem is shown. In Sect. 3 and 4, the procedure for deriving State Transition Matrix and Higher Order State Transition Tensor using Analytic Continuation method are presented respectively. In Sects. 5 and 6, the derivation process of $J_2 - J_6$ gravity and drag perturbation are described. Section 7 presents the numerical results of LEO, MEO, GTO and HEO. Finally, the discussion and concluding remarks on the paper are presented in Sects. 8 and 9.

2 Adaptive Analytic Continuation Technique

In the unperturbed Two-body problem, the acceleration vector is defined as,

$$\mathbf{r}^{(2)}(t) = -\mu \frac{\mathbf{r}(t)}{(\mathbf{r}(t).\mathbf{r}(t))^{3/2}} \tag{1}$$

where, $\mathbf{r}(t)$ is the position vector at time t and μ is the standard gravitational parameter.

Analytic Continuation technique has been developed based on two scalar variables f and g_p and these variables are defined as,

$$f(t) = \mathbf{r}(t).\mathbf{r}(t) \quad and \quad g_p(t) = f^{-p/2} \tag{2}$$

where, p is an integer. Using the newly defined variables, f and g_p, the accelerating vector of the unperturbed Two-body problem can be simplified as,

$$\mathbf{r}^{(2)}(t) = -\mu \boldsymbol{r}(t) g_3(t) \tag{3}$$

The advantage of expressing the acceleration vector as of Eq. (3) is that, the Leibniz product rule can now be readily implemented to compute arbitrarily higher order time derivatives of the variables via the following recursive formulas:

$$f^{(n)}(t) = \sum_{m=0}^{n} \binom{n}{m} \mathbf{r}^{(m)}(t).\mathbf{r}^{(n-m)}(t) \tag{4}$$

$$g_p^{(n+1)}(t) = -\frac{1}{f(t)} \left\{ \frac{p}{2} f^{(1)}(t) g_p^{(n)}(t) + \sum_{m=1}^{n} \binom{n}{m} \left(\frac{p}{2} f^{(m+1)}(t) g_p^{(n-m)}(t) + f^{(m)}(t) g_p^{(n-m+1)}(t) \right) \right\} \tag{5}$$

$$\mathbf{r}^{(n+2)}(t) = -\mu \sum_{m=0}^{n} \binom{n}{m} \mathbf{r}^{(m)}(t) g_3^{(n-m)}(t) \tag{6}$$

To obtain the position and velocity at the next time step, the higher order time derivatives of the position vector are used in the following Taylor series expansions:

$$\mathbf{r}(t + dT) = \sum_{m=0}^{n} \mathbf{r}^{(m)}(t) \frac{dT^m}{m!} \tag{7}$$

$$\mathbf{r}^{(1)}(t + dT) = \sum_{m=1}^{n} \mathbf{r}^{(m)}(t) \frac{dT^{m-1}}{(m-1)!} \tag{8}$$

Computational efficiency is achieved by introducing adaptive time step and expansion order to the method based on the modification of the research work done by Barrio et al. [26]. In this work the expansion order of the first step is guessed as,

$$N = round(-\delta_{fac} log(\delta) + k_{inc}) \tag{9}$$

where, $\delta = min(\delta_a, |\mathbf{r}|\delta_r)$ in the first step and δ_{fac}, h_{fac}, k_{inc}, δ_a and δ_r are tuning parameters. The time step to be used in the Taylor series expansion is determined by using the following set of equations:

$$dT_a = \delta \frac{(N-1)!}{|\mathbf{r}^{(N)}|_\infty^{\frac{1}{N-1}}} \tag{10}$$

$$dT_b = \delta \frac{(N-2)!}{|\mathbf{r}^{(N-1)}|_\infty^{\frac{1}{N-2}}} \tag{11}$$

$$dT = h_{fac} \times min(dT_a, dT_b) \tag{12}$$

In the subsequent time steps, δ used in Eqs. (9), (10), and (11) are updated using,

$$\delta = max(\delta_a, \delta_a \times fac) \tag{13}$$

where, $fac = ||\mathbf{r}(t_0)| - |\mathbf{r}(t)|| \times \frac{DU}{10000}$ and DU is the distance unit in canonical unit system.

3 State Transition Matrix Derivation Using Analytic Continuation

The two-body problem STM is defined as, [4, 27]

$$\phi^1 = \begin{bmatrix} \phi^1_{11}(t+dT,t) & \phi^1_{12}(t+dT,t) \\ \phi^1_{21}(t+dT,t) & \phi^1_{22}(t+dT,t) \end{bmatrix} = \begin{bmatrix} \frac{\partial \mathbf{r}(t+dT)}{\partial \mathbf{r}(t)} & \frac{\partial \mathbf{r}(t+dT)}{\partial \mathbf{r}^{(1)}(t)} \\ \frac{\partial \mathbf{r}^{(1)}(t+dT)}{\partial \mathbf{r}(t)} & \frac{\partial \mathbf{r}^{(1)}(t+dT)}{\partial \mathbf{r}^{(1)}(t)} \end{bmatrix} \tag{14}$$

where, ϕ^1_{11} and ϕ^1_{12} are the sensitivity of the next position to the current position and velocity respectively and ϕ^1_{21} and ϕ^1_{22} are the sensitivity of the next velocity to the current position and velocity respectively.

The elements of the STM are expanded using the Taylor series as follows,

$$\phi^1_{11}(t+dT,t) = \frac{\partial \mathbf{r}(t+dT)}{\partial \mathbf{r}(t)} = \sum_{m=0}^{n} \frac{\partial \mathbf{r}^{(m)}(t)}{\partial \mathbf{r}(t)} \frac{dT^m}{m!} \tag{15}$$

$$\phi^1_{12}(t+dT,t) = \frac{\partial \mathbf{r}(t+dT)}{\partial \mathbf{r}^{(1)}(t)} = \sum_{m=0}^{n} \frac{\partial \mathbf{r}^{(m)}(t)}{\partial \mathbf{r}^{(1)}(t)} \frac{dT^m}{m!} \tag{16}$$

$$\phi^1_{21}(t+dT,t) = \frac{\partial \mathbf{r}^{(1)}(t+dT)}{\partial \mathbf{r}(t)} = \sum_{m=1}^{n} \frac{\partial \mathbf{r}^{(m)}(t)}{\partial \mathbf{r}(t)} \frac{dT^{m-1}}{(m-1)!} \tag{17}$$

$$\phi^1_{22}(t+dT,t) = \frac{\partial \mathbf{r}^{(1)}(t+dT)}{\partial \mathbf{r}^{(1)}(t)} = \sum_{m=1}^{n} \frac{\partial \mathbf{r}^{(m)}(t)}{\partial \mathbf{r}^{(1)}(t)} \frac{dT^{m-1}}{(m-1)!} \tag{18}$$

where, dT is the time difference between two consecutive steps.

The partial derivatives of $\mathbf{r}(t)$, $\mathbf{r}^{(1)}(t)$, f and g_3 with respect to the initial position and velocity are as follows:

$$\begin{aligned} &\frac{\partial \mathbf{r}(t)}{\partial \mathbf{r}(t)} = \frac{\partial \mathbf{r}^{(1)}(t)}{\partial \mathbf{r}^{(1)}(t)} = \mathbf{I}_{3\times3} \quad \frac{\partial \mathbf{r}(t)}{\partial \mathbf{r}^{(1)}(t)} = \frac{\partial \mathbf{r}^{(1)}(t)}{\partial \mathbf{r}(t)} = \mathbf{0}_{3\times3} \quad \frac{\partial f(t)}{\partial \mathbf{r}(t)} = 2\mathbf{r}(t) \\ &\frac{\partial f(t)}{\partial \mathbf{r}^{(1)}(t)} = \frac{\partial g_p(t)}{\partial \mathbf{r}^{(1)}(t)} = \mathbf{0}_{1\times3} \quad \frac{\partial g_p(t)}{\partial \mathbf{r}(t)} = -p\frac{g_p(t)}{f(t)}\mathbf{r}(t) \end{aligned} \tag{19}$$

The partial derivatives of the higher order time derivatives of the variables with respect to the current position and velocity are calculated recursively using Eq. (20)–(22),

$$\frac{\partial f^{(n)}(t)}{\partial \chi} = \sum_{m=0}^{n} \binom{n}{m} \left\{ \psi_{\chi}^{(m)} \mathbf{r}^{(n-m)}(t) + \mathbf{r}^{(m)}(t) \psi_{\chi}^{(n-m)} \right\} \tag{20}$$

$$\begin{aligned} \frac{\partial g_p^{(n+1)}(t)}{\partial \chi} = -\frac{1}{f(t)} \Bigg\{ & g_p^{(n+1)}(t) F_{\chi} + \frac{p}{2} \left(F_{\chi}^{(1)} g_p^{(n)}(t) + f^{(1)}(t) G_{\chi}^{(n)} \right) \\ & + \sum_{m=1}^{n} \binom{n}{m} \frac{p}{2} \left(F_{\chi}^{(m+1)} g_p^{(n-m)}(t) + f^{(m+1)}(t) G_{\chi}^{(n-m)} \right) \\ & + \sum_{m=1}^{n} \binom{n}{m} \left(F_{\chi}^{(m)} g_p^{(n-m+1)}(t) + f^{(m)}(t) G_{\chi}^{(n-m+1)} \right) \Bigg\} \end{aligned} \tag{21}$$

$$\frac{\partial \mathbf{r}^{(n+2)}(t)}{\partial \chi} = -\mu \sum_{m=0}^{n} \binom{n}{m} \left\{ \psi_{\chi}^{(m)} g_p^{(n-m)}(t) + \mathbf{r}^{(m)}(t) G_{p\chi}^{(n-m)} \right\} \tag{22}$$

where, $\chi = \mathbf{r}(t)$ or $\mathbf{r}^{(1)}(t)$, $\psi_{\chi}^{(m)} = \frac{\partial \mathbf{r}^{(m)}(t)}{\partial \chi}$, $F_{\chi}^{(m)} = \frac{\partial f^{(m)}(t)}{\partial \chi}$ and $G_{\chi}^{(m)} = \frac{\partial g_p^{(m)}(t)}{\partial \chi}$.

4 Higher Order State Transition Tensors Derivation Using Analytic Continuation

In this section, an approach to expand the STM to the Higher Order State Transition Tensors is described. For the simplicity of computation, an index notation based approach is used. Using the index expression, the first order STM and second order State Transition Tensor can be presented as [17],

$$\phi_{ij}^{1} = \frac{\partial \chi_i}{\partial \chi_{0j}} \tag{23}$$

$$\phi_{ijk}^{2} = \frac{\partial^2 \chi_i}{\partial \chi_{0j} \partial \chi_{0k}} \tag{24}$$

where, the indices i, j, k are from 1 to 6 and χ_i is the i-th element of the state vector, $\boldsymbol{\chi} = [x, y, z, \dot{x}, \dot{y}, \dot{z}]^{\mathrm{T}}$ at time t, $\boldsymbol{\chi}_0$ is the initial state vector.

Taylor series expansion has been implemented on the elements of the second order State Transition Tensor as,

$$\phi_{ijk}^{2}(t + dT, t) = \frac{\partial^2 \chi_i(t + dT)}{\partial \chi_j(t) \partial \chi_k(t)} = \sum_{m=0}^{n} \frac{\partial^2 \chi_i^{(m)}(t)}{\partial \chi_j(t) \partial \chi_k(t)} \frac{dT^{(m)}}{m!} \tag{25}$$

The second order partial derivatives with respect to the states, of the higher order time derivatives of the variables are computed recursively as follows,

$$\frac{\partial^2 \mathbf{r}^{(n+2)}(t)}{\partial \chi_j \partial \chi_k} = -\mu \sum_{m=0}^{n} \binom{n}{m} \left\{ \psi_{jk}^{(m)} g_p^{(n-m)}(t) + \psi_j^{(m)} G_k^{(n-m)} + \psi_k^{(m)} G_j^{(n-m)} + \mathbf{r}^{(m)}(t) G_{jk}^{(n-m)} \right\} \quad (26)$$

$$\frac{\partial^2 f^{(n)}(t)}{\partial \chi_j \partial \chi_k} = \sum_{m=0}^{n} \binom{n}{m} \left\{ \psi_{jk}^{(m)} \mathbf{r}^{(n-m)}(t) + \psi_j^{(m)} \psi_k^{(n-m)} + \psi_k^{(m)} \psi_j^{(n-m)} + \mathbf{r}^{(m)}(t) \psi_{jk}^{(n-m)} \right\} \quad (27)$$

$$\begin{aligned} \frac{\partial^2 g_p^{(n+1)}(t)}{\partial \chi_j \partial \chi_k} = &-\frac{1}{f(t)} \Big[G_k^{(n+1)} F_j + g_p^{(n+1)}(t) F_{jk} + \frac{p}{2} \left(F_{jk}^{(1)} g_p^{(n)}(t) + F_j^{(1)} G_k^{(n)} + F_k^{(1)} G_j^{(n)} + f^{(1)}(t) G_{jk}^{(n)} \right) \\ &+ \sum_{m=1}^{n} \binom{n}{m} \Big\{ \frac{p}{2} \left(F_{jk}^{(m+1)} g_p^{(n-m)}(t) + F_j^{(m+1)} G_k^{(n-m)} + F_k^{(m+1)} G_j^{(n-m)} + f^{(m+1)}(t) G_{jk}^{(n-m)} \right) \\ &+ \left(F_{jk}^{(m)} g_p^{(n-m+1)}(t) + F_j^{(m)} G_k^{(n-m+1)} + F_k^{(m)} G_j^{(n-m+1)} + f^{(m)}(t) G_{jk}^{(n-m+1)} \right) \Big\} \Big] \end{aligned} \quad (28)$$

where, $\psi_j^{(m)} = \frac{\partial \mathbf{r}^{(m)}(t)}{\partial \chi_j}$, $\psi_{jk}^{(m)} = \frac{\partial^2 \mathbf{r}^{(m)}(t)}{\partial \chi_j \partial \chi_k}$, $F_j^{(m)} = \frac{\partial f^{(m)}(t)}{\partial \chi_j}$, $F_{jk}^{(m)} = \frac{\partial^2 f^{(m)}(t)}{\partial \chi_j \partial \chi_k}$, $G_j^{(m)} = \frac{\partial g_p^{(m)}(t)}{\partial \chi_j}$ and $G_{jk}^{(m)} = \frac{\partial^2 g_p^{(m)}(t)}{\partial \chi_j \partial \chi_k}$.

5 $J_2 - J_6$ Perturbation for Higher Order State Transition Tensors

In this section, Analytic Continuation technique has been implemented to compute $J_2 - J_6$ perturbation accelerations and their higher order partial derivatives with respect to the initial states and implemented on the STM and the second order State Transition Tensors. The $J_2 - J_6$ perturbation acceleration vectors are defined and simplified using Analytic Continuation as, [19, 27],

$$\mathbf{r}_{J_2}^{(2)}(t) = -\frac{3}{2} J_2 \left(\frac{\mu}{r^2}\right) \left(\frac{r_{eq}}{r}\right)^2 \begin{bmatrix} \left(1 - 5(\frac{z}{r})^2\right) \frac{x}{r} \\ \left(1 - 5(\frac{z}{r})^2\right) \frac{y}{r} \\ \left(3 - 5(\frac{z}{r})^2\right) \frac{z}{r} \end{bmatrix} = -\frac{3}{2} J_2 \mu r_{eq}^2 \begin{bmatrix} x g_5 - 5 x z^2 g_7 \\ y g_5 - 5 y z^2 g_7 \\ 3 z g_5 - 5 z^3 g_7 \end{bmatrix} \quad (29)$$

$$\mathbf{r}_{J_3}^{(2)}(t) = \frac{1}{2} J_3 \left(\frac{\mu}{r^2}\right) \left(\frac{r_{eq}}{r}\right)^3 \begin{bmatrix} 5\left(7(\frac{z}{r})^3 - 3(\frac{z}{r})\right) \frac{x}{r} \\ 5\left(7(\frac{z}{r})^3 - 3(\frac{z}{r})\right) \frac{y}{r} \\ 3\left(1 - 10(\frac{z}{r})^2 + \frac{35}{3}(\frac{z}{r})^4\right) \end{bmatrix} = \frac{1}{2} J_3 \mu r_{eq}^3 \begin{bmatrix} 5\left(7xz^3 g_9 - 3xz g_7\right) \\ 5\left(7yz^3 g_9 - 3yz g_7\right) \\ 3\left(g_5 - 10z^2 g_7 + \frac{35}{3} z^4 g_9\right) \end{bmatrix} \quad (30)$$

$$\mathbf{r}_{J_4}^{(2)}(t) = \frac{5}{8} J_4 \left(\frac{\mu}{r^2}\right)\left(\frac{r_{eq}}{r}\right)^4 \begin{bmatrix} (3 - 42(\frac{z}{r})^2 + 63(\frac{z}{r})^4)\frac{x}{r} \\ (3 - 42(\frac{z}{r})^2 + 63(\frac{z}{r})^4)\frac{y}{r} \\ (15 - 70(\frac{z}{r})^2 + 63(\frac{z}{r})^4)\frac{z}{r} \end{bmatrix} = \frac{5}{8} J_4 \mu r_{eq}^4 \begin{bmatrix} 3xg_7 - 42xz^2g_9 + 63xz^4g_{11} \\ 3yg_7 - 42yz^2g_9 + 63yz^4g_{11} \\ 15zg_7 - 70z^3g_9 + 63xz^5g_{11} \end{bmatrix} \tag{31}$$

$$\begin{aligned} \mathbf{r}_{J_5}^{(2)}(t) &= \frac{1}{8} J_5 \left(\frac{\mu}{r^2}\right)\left(\frac{r_{eq}}{r}\right)^5 \begin{bmatrix} 3\left(35(\frac{z}{r}) - 210(\frac{z}{r})^3 + 231(\frac{z}{r})^5\right)\frac{x}{r} \\ 3\left(35(\frac{z}{r}) - 210(\frac{z}{r})^3 + 231(\frac{z}{r})^5\right)\frac{y}{r} \\ \left(693(\frac{z}{r})^6 - 945(\frac{z}{r})^4 + 315(\frac{z}{r})^2 - 15\right) \end{bmatrix} \\ &= \frac{1}{8} J_5 \mu r_{eq}^4 \begin{bmatrix} 3\left(35xzg_9 - 210xz^3g_{11} + 231xz^5g_{13}\right) \\ 3\left(35yzg_9 - 210yz^3g_{11} + 231yz^5g_{13}\right) \\ 693z^6g_{13} - 945z^4g_{11} + 315z^2g_9 - 15g_7 \end{bmatrix} \end{aligned} \tag{32}$$

$$\begin{aligned} \mathbf{r}_{J_6}^{(2)}(t) &= -\frac{1}{16} J_6 \left(\frac{\mu}{r^2}\right)\left(\frac{r_{eq}}{r}\right)^6 \begin{bmatrix} \left(35 - 945(\frac{z}{r})^2 + 3465(\frac{z}{r})^4 - 3003(\frac{z}{r})^6\right)\frac{x}{r} \\ \left(35 - 945(\frac{z}{r})^2 + 3465(\frac{z}{r})^4 - 3003(\frac{z}{r})^6\right)\frac{y}{r} \\ \left(245 - 2205(\frac{z}{r})^2 + 4851(\frac{z}{r})^4 - 3003(\frac{z}{r})^6\right)\frac{z}{r} \end{bmatrix} \\ &= -\frac{1}{16} J_6 \mu r_{eq}^6 \begin{bmatrix} \left(35xg_9 - 945xz^2g_{11} + 3465xz^4g_{13} - 3003xz^6g_{15}\right) \\ \left(35yg_9 - 945yz^2g_{11} + 3465yz^4g_{13} - 3003yz^6g_{15}\right) \\ \left(245zg_9 - 2205z^3g_{11} + 4851z^5g_{13} - 3003z^7g_{15}\right) \end{bmatrix} \end{aligned} \tag{33}$$

where, r_{eq} is the equatorial radius of Earth and $J_2 = 1082.63 \times 10^{-6}$, $J_3 = -2.52 \times 10^{-6}$, $J_4 = -1.61 \times 10^{-6}$, $J_5 = -0.15 \times 10^{-6}$ and $J_6 = 0.57 \times 10^{-6}$ [27].

In order to simplify the derivation process of the higher time derivatives of the $J_2 - J_6$ perturbation accelerations and their higher order partials with respect to the states, 5 new constants ($C_{J_2}, C_{J_3}, C_{J_4}, C_{J_5}, C_{J_6}$) and two new variables (B_p and C_α) are defined as,

$$\begin{gathered} C_{J_2} = -\frac{3}{2} J_2 \mu r_{eq}^2 \quad C_{J_3} = \frac{1}{2} J_3 \mu r_{eq}^3 \quad C_{J_4} = \frac{5}{8} J_4 \mu r_{eq}^4 \\ C_{J_5} = \frac{1}{8} J_5 \mu r_{eq}^5 \quad C_{J_6} = -\frac{1}{16} J_6 \mu r_{eq}^6 \quad B_p = \mathbf{r} g_p \quad C_\alpha = z^\alpha \end{gathered} \tag{34}$$

Implementing these variables and constants, the $J_2 - J_6$ perturbation acceleration of Eqs. (29)–(33) has been simplified as,

$$\mathbf{r}_{J_2}^{(2)}(t) = C_{J_2} \left\{ \begin{bmatrix} 1 & 0 & 0 \\ 0 & 1 & 0 \\ 0 & 0 & 3 \end{bmatrix} B_5 - 5B_7C_2 \right\} \tag{35}$$

$$\mathbf{r}_{J_3}^{(2)} = C_{J_3} \left\{ -\begin{bmatrix} 15 & 0 & 0 \\ 0 & 15 & 0 \\ 0 & 0 & 30 \end{bmatrix} B_7 z + 35B_9C_3 + \begin{bmatrix} 0 \\ 0 \\ 3 \end{bmatrix} g_5 \right\} \tag{36}$$

$$\mathbf{r}_{J_4}^{(2)} = C_{J_4} \left\{ \begin{bmatrix} 3 & 0 & 0 \\ 0 & 3 & 0 \\ 0 & 0 & 15 \end{bmatrix} B_7 - \begin{bmatrix} 42 & 0 & 0 \\ 0 & 42 & 0 \\ 0 & 0 & 70 \end{bmatrix} B_9C_2 + 63B_{11}C_4 \right\} \tag{37}$$

$$\mathbf{r}_{J_5}^{(2)} = C_{J_5}\left\{\begin{bmatrix}105 & 0 & 0\\ 0 & 105 & 0\\ 0 & 0 & 315\end{bmatrix} B_9 z - \begin{bmatrix}630 & 0 & 0\\ 0 & 630 & 0\\ 0 & 0 & 945\end{bmatrix} B_{11}C_3 + 693B_{13}C_5 - \begin{bmatrix}0\\ 0\\ 15\end{bmatrix} g_7\right\} \tag{38}$$

$$\mathbf{r}_{J_6}^{(2)} = C_{J_6}\left\{\begin{bmatrix}35 & 0 & 0\\ 0 & 35 & 0\\ 0 & 0 & 245\end{bmatrix} B_9 - \begin{bmatrix}945 & 0 & 0\\ 0 & 945 & 0\\ 0 & 0 & 2205\end{bmatrix} B_{11}C_2 + \begin{bmatrix}3465 & 0 & 0\\ 0 & 3465 & 0\\ 0 & 0 & 4851\end{bmatrix} B_{13}C_4 - 3003B_{15}C_6\right\} \tag{39}$$

To compute the higher order time derivatives of the simplified $J_2 - J_6$ perturbation accelerations, Leibniz product rule is implemented to compute the higher order time derivatives of B_p and C_α as,

$$B_p^{(n)} = \sum_{m=0}^{n}\binom{n}{m}\mathbf{r}^{(m)} g_p^{(n-m)} \tag{40}$$

$$C_\alpha^{(n)} = \sum_{m=0}^{n}\binom{n}{m} C_{\alpha-1}^{(m)} z^{(n-m)} \tag{41}$$

Next, the higher order partial derivatives of $B_p^{(n)}$ and $C_\alpha^{(n)}$ are derived as,

$$\frac{\partial B_p^{(n)}}{\partial \chi_j} = \sum_{m=0}^{n}\binom{n}{m}\left(\frac{\partial \mathbf{r}^{(m)}}{\partial \chi_j} g_p^{(n-m)} + \mathbf{r}^{(m)}\frac{\partial g_p^{(n-m)}}{\partial \chi_j}\right) \tag{42}$$

$$\frac{\partial^2 B_p^{(n)}}{\partial \chi_j \chi_k} = \sum_{m=0}^{n}\binom{n}{m}\left(\frac{\partial^2 \mathbf{r}^{(m)}}{\partial \chi_j \chi_k} g_p^{(n-m)} + \frac{\partial \mathbf{r}^{(m)}}{\partial \chi_j}\frac{\partial g_p^{(n-m)}}{\partial \chi_k} + \frac{\partial \mathbf{r}^{(m)}}{\partial \chi_k}\frac{\partial g_p^{(n-m)}}{\partial \chi_j} + \mathbf{r}^{(m)}\frac{\partial^2 g_p^{(n-m)}}{\partial \chi_j \chi_k}\right) \tag{43}$$

$$\frac{\partial C_\alpha^{(n)}}{\partial \chi_j} = \sum_{m=0}^{n}\binom{n}{m}\left(\frac{\partial C_{\alpha-1}^{(m)}}{\partial \chi_j} z^{(n-m)} + C_{\alpha-1}^{(m)}\frac{\partial z^{(n-m)}}{\partial \chi_j}\right) \tag{44}$$

$$\frac{\partial^2 C_\alpha^{(n)}}{\partial \chi_j \partial \chi_k} = \sum_{m=0}^{n}\binom{n}{m}\left(\frac{\partial^2 C_{\alpha-1}^{(m)}}{\partial \chi_j \partial \chi_k} z^{(n-m)} + \frac{\partial C_{\alpha-1}^{(m)}}{\partial \chi_j}\frac{\partial z^{(n-m)}}{\partial \chi_k} + \frac{\partial C_{\alpha-1}^{(m)}}{\partial \chi_k}\frac{\partial z^{(n-m)}}{\partial \chi_j} + C_{\alpha-1}^{(m)}\frac{\partial^2 z^{(n-m)}}{\partial \chi_j \partial \chi_k}\right) \tag{45}$$

where, χ is $\mathbf{r}(t)$ or $\mathbf{r}^{(1)}(t)$.

Finally, the higher order time derivatives of B_p and C_α and their higher order partial derivatives are used in the following equations to recursively compute the higher order time derivatives of the J_2 perturbation acceleration and its higher order partials,

$$\mathbf{r}_{J_2}^{(n+2)}(t) = C_{J_2}\left\{\begin{bmatrix}1&0&0\\0&1&0\\0&0&3\end{bmatrix}B_5^{(n)} - 5\sum_{m=0}^{n}\binom{n}{m}B_7^{(m)}C_2^{(n-m)}\right\} \tag{46}$$

$$\frac{\partial \mathbf{r}_{J_2}^{(n+2)}(t)}{\partial \chi_j} = C_{J_2}\left\{\begin{bmatrix}1&0&0\\0&1&0\\0&0&3\end{bmatrix}\frac{\partial B_5^{(n)}}{\partial \chi_j} - 5\sum_{m=0}^{n}\binom{n}{m}\left(\frac{\partial B_7^{(m)}}{\partial \chi_j}C_2^{(n-m)} + B_7^{(m)}\frac{\partial C_2^{(n-m)}}{\partial \chi_j}\right)\right\} \tag{47}$$

$$\begin{aligned}\frac{\partial^2 \mathbf{r}_{J_2}^{(n+2)}(t)}{\partial \chi_j \partial \chi_k} = C_{J_2}\Bigg\{\begin{bmatrix}1&0&0\\0&1&0\\0&0&3\end{bmatrix}\frac{\partial^2 B_5^{(n)}}{\partial \chi_j \partial \chi_k} - 5\sum_{m=0}^{n}\binom{n}{m}\Bigg(&\frac{\partial^2 B_7^{(m)}}{\partial \chi_j \partial \chi_k}C_2^{(n-m)} + \frac{\partial B_7^{(m)}}{\partial \chi_j}\frac{\partial C_2^{(n-m)}}{\partial \chi_k}\\ &+\frac{\partial B_7^{(m)}}{\partial \chi_k}\frac{\partial C_2^{(n-m)}}{\partial \chi_j} + B_7^{(m)}\frac{\partial^2 C_2^{(n-m)}}{\partial \chi_j \partial \chi_k}\Bigg)\Bigg\}\end{aligned} \tag{48}$$

$\frac{\partial \mathbf{r}_{J_2}^{(n+2)}(t)}{\partial \chi(t)}$ and $\frac{\partial^2 \mathbf{r}_{J_2}^{(n+2)}(t)}{\partial \chi(t)\chi(t)}$ are added to Eqs. (23) and (24) to include J_2 perturbation in STM and second order STT. The higher order time derivatives of $J_3 - J_6$ perturbation accelerations and their higher order partials use the similar approach of derivation and these equations are shown in Appendix A.

6 Atmospheric Drag Perturbation for Higher Order State Transition Tensors

In this research work, an Exponential Atmospheric Model is used to incorporate the drag effect in Analytic Continuation method, [28, 29]. By definition, the perturbation acceleration due to atmospheric drag is given by,

$$a_{dr} = \mathbf{r}_{dr}^{(2)}(t) = -\frac{1}{2}\rho(r)\frac{C_D A}{m}v_{rel}^2\frac{\mathbf{v}_{rel}}{|\mathbf{v}_{rel}|} = -\tilde{\rho}||\mathbf{v}_{rel}||\mathbf{v}_{rel} \tag{49}$$

where, $\rho(r)$ is the atmospheric density from a distance r from the center of the Earth, C_D is the drag coefficient, A is the exposed cross-sectional area, and $\mathbf{v}_{rel}$ is the relative velocity of the satellite with respect to the Earth. In general, $\frac{m}{C_D A}$ is called the ballistic coefficient and in the presented simulation results it has been considered as 1.

The relative velocity of the satellite with respect to the Earth is defined as,

$$\mathbf{v}_{rel} = \mathbf{r}^{(1)} - \omega_{\oplus} \times \mathbf{r} = \begin{bmatrix} x^{(1)} + y\omega_{\oplus} \\ y^{(1)} - x\omega_{\oplus} \\ z^{(1)} \end{bmatrix} \tag{50}$$

where, $\omega_{\oplus}$ is the angular velocity of the Earth. The higher order time derivatives of $\mathbf{v_{rel}}$ is calculated using,

$$\mathbf{v}_{rel}^{(n)} = \begin{bmatrix} x^{(n+1)} + y^{(n)}\omega_{\oplus} \\ y^{(n+1)} - x^{(n)}\omega_{\oplus} \\ z^{(n+1)} \end{bmatrix} \tag{51}$$

The higher order time derivatives of the relative velocity magnitude, $||\mathbf{v}_{rel}||^{(n)}$, is computed with the help of two new variables, f_{vrel} and g_{vrel} defined as,

$$\begin{aligned} f_{vrel} &= \mathbf{v}_{rel}.\mathbf{v}_{rel} \\ g_{vrel} &= f_{vrel}^{(1/2)} = ||\mathbf{v}_{rel}|| \end{aligned} \tag{52}$$

Applying Leibniz product rule, the higher order time derivatives of the variables in Eq. (52) are expressed as,

$$f_{vrel}^{(n)} = \sum_{m=0}^{n} \binom{n}{m} \mathbf{v}_{rel}^{(m)}.\mathbf{v}_{rel}^{(n-m)} \tag{53}$$

$$||\mathbf{v}_{rel}||^{(n+1)} = g_{vrel}^{(n+1)} = \frac{1}{f_{\ vrel}} \left\{ \frac{1}{2} f_{vrel}^{(1)} g_{vrel}^{(n)} + \sum_{m=1}^{n} \binom{n}{m} \left(\frac{1}{2} f_{vrel}^{(m+1)} g_{vrel}^{(n-m)} - f_{vrel}^{(m)} g_{vrel}^{(n-m+1)} \right) \right\} \tag{54}$$

According to the Exponential Model of atmospheric density, $\rho(r)$ is defined as [29]

$$\rho(r) = \rho_0 e^{-\frac{h_{ellp} - h_0}{H}} = \rho_0 e^{\frac{R_{eq} + h_0}{H}} e^{-\frac{|\mathbf{r}|}{H}} \tag{55}$$

where, ρ_0 is the reference density at the reference altitude h_0, R_{eq} is the equatorial radius of Earth, h_{ellp} is the actual altitude above the ellipsoid, and H is the scale height. From Eq. (49) to (55), the equation of $\tilde{\rho}$ is given by,

$$\tilde{\rho} = \frac{1}{2} \frac{C_d A}{m} \rho_0 e^{\frac{R_{eq} + h_0}{H}} e^{-\frac{|\mathbf{r}|}{H}} = \beta e^{-\frac{|\mathbf{r}|}{H}} \tag{56}$$

where, β is a constant. The first order time derivative of $\tilde{\rho}$ is,

$$\tilde{\rho}^{(1)} = -\frac{1}{H} \tilde{\rho} |\mathbf{r}|^{(1)} \tag{57}$$

Leibniz product rule is implemented on Eq. (57) to get the higher order time derivative of $\tilde{\rho}$:

$$\tilde{\rho}^{(n+1)} = -\frac{1}{H} \sum_{m=0}^{n} \binom{n}{m} \tilde{\rho}^{(m)} |\mathbf{r}|^{(n-m+1)} \tag{58}$$

The higher order time derivatives of the drag perturbation acceleration is calculated using [28]

$$\mathbf{a}_{dr}^{(n)} = \mathbf{r}_{dr}^{(n+2)}(t) = -\sum_{i=0}^{n}\sum_{j=0}^{n-i}\frac{n!}{i!j!(n-i-j)!}\rho^{(i)}||\mathbf{v}_{rel}||^{(j)}\mathbf{v}_{rel}^{(n-i-j)} \tag{59}$$

Finally, the higher order partial derivatives of the higher order time derivatives of the drag perturbation acceleration with respect to the states are computed recursively using Eq. (60) and Eq. (61),

$$\frac{\partial \mathbf{r}_{dr}^{(n+2)}(t)}{\partial \chi_j} = -\sum_{m_1=0}^{n}\sum_{m_2=0}^{n-m_1}\frac{n!}{m_1!m_2!(n-m_1-m_2)!}\left(\frac{\partial \rho^{(m_1)}}{\partial \chi_j}||\mathbf{v}_{rel}||^{(m_2)}\mathbf{v}_{rel}^{(n-m_1-m_2)} + \rho^{(m_1)}\frac{\partial ||\mathbf{v}_{rel}||^{(m_2)}}{\partial \chi_j}\mathbf{v}_{rel}^{(n-m_1-m_2)} + \rho^{(m_1)}||\mathbf{v}_{rel}||^{(m_2)}\frac{\partial \mathbf{v}_{rel}^{(n-m_1-m_2)}}{\partial \chi_j}\right) \tag{60}$$

$$\begin{aligned}\frac{\partial^2 \mathbf{r}_{dr}^{(n+2)}(t)}{\partial \chi_j \partial \chi_k} = -\sum_{m_1=0}^{n}\sum_{m_2=0}^{n-m_1}\frac{n!}{m_1!m_2!(n-m_1-m_2)!}\Bigg\{&\left(\frac{\partial^2 \rho^{(m_1)}}{\partial \chi_j \partial \chi_k}||\mathbf{v}_{rel}||^{(m_2)}\mathbf{v}_{rel}^{(n-m_1-m_2)} + \frac{\partial \rho^{(m_1)}}{\partial \chi_j}\frac{\partial ||\mathbf{v}_{rel}||^{(m_2)}}{\partial \chi_k}\mathbf{v}_{rel}^{(n-m_1-m_2)} + \frac{\partial \rho^{(m_1)}}{\partial \chi_j}||\mathbf{v}_{rel}||^{(m_2)}\frac{\partial \mathbf{v}_{rel}^{(n-m_1-m_2)}}{\partial \chi_k}\right) + \\ &\left(\frac{\partial \rho^{(m_1)}}{\partial \chi_k}\frac{\partial ||\mathbf{v}_{rel}||^{(m_2)}}{\partial \chi_j}\mathbf{v}_{rel}^{(n-m_1-m_2)} + \rho^{(m_1)}\frac{\partial^2 ||\mathbf{v}_{rel}||^{(m_2)}}{\partial \chi_j \partial \chi_k}\mathbf{v}_{rel}^{(n-m_1-m_2)} + \rho^{(m_1)}\frac{\partial ||\mathbf{v}_{rel}||^{(m_2)}}{\partial \chi_j}\frac{\mathbf{v}_{rel}^{(n-m_1-m_2)}}{\partial \chi_k}\right) + \\ &\left(\frac{\partial \rho^{(m_1)}}{\partial \chi_k}||\mathbf{v}_{rel}||^{(m_2)}\frac{\partial \mathbf{v}_{rel}^{(n-m_1-m_2)}}{\partial \chi_j} + \rho^{(m_1)}\frac{\partial ||\mathbf{v}_{rel}||^{(m_2)}}{\partial \chi_k}\frac{\partial \mathbf{v}_{rel}^{(n-m_1-m_2)}}{\partial \chi_j} + \rho^{(m_1)}||\mathbf{v}_{rel}||^{(m_2)}\frac{\partial^2 \mathbf{v}_{rel}^{(n-m_1-m_2)}}{\partial \chi_j \chi_k}\right)\Bigg\}\end{aligned} \tag{61}$$

$\frac{\partial \mathbf{r}_{dr}^{(n+2)}(t)}{\partial \chi(t)}$ and $\frac{\partial^2 \mathbf{r}_{dr}^{(n+2)}(t)}{\partial \chi(t)\chi(t)}$ are added to Eqs. (23) and (24) to include drag perturbation in STM and second order STT. The complete procedure for computing $J_2 - J_6$ and drag perturbed STM, and second order State Transition Tensor are shown as an algorithm in Appendix B.

7 Numerical Results

Numerical results for unperturbed, $J_2 - J_6$ and drag perturbed STM, and second order State Transition Tensors derived using Analytic Continuation method are presented in this section of the paper. Four different types of orbits, as shown in Table 1, are selected for the simulation. The codes are written and compiled using MATLAB R2019a and Canonical unit system is used for numerical stability [29].

Three methods are used for presenting the accuracy of the method- Root Mean Square (RMS) error, Symplectic error check and Error Propagation Prediction.

The RMS error of the elements of the unperturbed STMs is computed from the difference between the elements of the unperturbed closed form solution using Battin's method, [4, 27], and the Analytic Continuation method for 10 orbit periods as shown below:

$$E_{ij} = \sqrt{\sum_{k=1}^{n} \frac{(M_{ijk} - L_{ijk})^2}{n}} \tag{62}$$

where, n is the total number of steps, M_{ijk} and L_{ijk} are the (i, j)-th terms of the STMs at k-th time period from Battin's method and Analytic Continuation method, respectively, and E_{ij} is the RMS error of the (i, j)-th term of the STM. The results of the RMS error are shown in Tables 2, 3, 4 and 5 for LEO, MEO, GTO and HEO respectively. It is evident from the results that in all elements of the unperturbed STM double precision accuracy is achieved .

Next, symplectic nature unperturbed as well as $J_2 - J_6$ gravity perturbed STMs and second order State Transition Tensors is used to compare the accuracy of the method. By definition, STM (ϕ^1) is symplectic if it satisfies Eq. (63) [27],

$$\phi^{1T}[J]\phi^1 = [J] \tag{63}$$

Table 1 Orbits used for numerical simulation

Orbit Type	a, m	e	f, deg	i, deg	ω, deg	Ω, deg	t_p, s
LEO	7.3090×10^6	0.1	0	60	30	45	6.2187×10^3
MEO	1.0964×10^7	0.4	0	60	30	45	1.1425×10^4
GTO	2.6353×10^7	0.6	0	60	30	45	4.2574×10^4
HEO	2.6999×10^7	0.7	0	60	30	45	4.4153×10^4

Table 2 RMS error of the elements of the unperturbed STM of LEO orbit

1.3942×10^{-14}	1.4502×10^{-14}	1.5149×10^{-14}	2.8722×10^{-15}	1.9173×10^{-15}	1.9126×10^{-15}
1.4524×10^{-14}	1.3981×10^{-14}	1.5425×10^{-14}	1.9381×10^{-15}	3.2506×10^{-15}	1.9576×10^{-15}
1.5159×10^{-14}	1.5443×10^{-14}	1.6973×10^{-14}	1.9479×10^{-15}	1.9667×10^{-15}	2.9844×10^{-15}
7.3583×10^{-14}	7.5392×10^{-14}	8.1742×10^{-14}	1.3858×10^{-14}	1.4425×10^{-14}	1.5458×10^{-14}
7.5938×10^{-14}	7.3192×10^{-14}	8.0913×10^{-14}	1.4448×10^{-14}	1.4288×10^{-14}	1.5357×10^{-14}
8.1056×10^{-14}	8.1186×10^{-14}	8.7673×10^{-14}	1.5471×10^{-14}	1.5377×10^{-14}	1.6897×10^{-14}

Table 3 RMS error of the elements of the unperturbed STM of MEO orbit

5.4882×10^{-15}	5.9018×10^{-15}	5.6524×10^{-15}	3.8847×10^{-15}	4.7179×10^{-16}	4.6054×10^{-16}
5.9069×10^{-15}	5.6210×10^{-15}	6.4514×10^{-15}	4.7309×10^{-16}	3.8636×10^{-15}	5.2706×10^{-16}
5.6614×10^{-15}	6.4561×10^{-15}	6.2877×10^{-15}	4.6095×10^{-16}	5.2853×10^{-16}	3.8625×10^{-15}
4.6376×10^{-14}	4.9715×10^{-14}	4.6324×10^{-14}	5.5678×10^{-15}	5.8725×10^{-15}	5.6240×10^{-15}
4.9802×10^{-14}	4.6210×10^{-14}	5.4206×10^{-14}	5.8759×10^{-15}	5.5534×10^{-15}	6.5757×10^{-15}
4.6436×10^{-14}	5.4273×10^{-14}	5.3116×10^{-14}	5.6329×10^{-15}	6.5815×10^{-15}	6.2836×10^{-15}

Table 4 RMS error of the elements of the unperturbed STM of GTO orbit

9.0156×10^{-14}	8.0121×10^{-14}	9.8163×10^{-14}	1.8060×10^{-14}	1.0671×10^{-14}	1.3122×10^{-14}
8.0140×10^{-14}	8.5378×10^{-14}	9.8094×10^{-14}	1.0485×10^{-14}	1.8265×10^{-14}	1.3622×10^{-14}
9.8144×10^{-14}	9.8059×10^{-14}	9.3233×10^{-14}	1.3247×10^{-14}	1.3913×10^{-14}	1.8845×10^{-14}
4.5602×10^{-13}	4.2441×10^{-13}	4.8815×10^{-13}	9.3278×10^{-14}	8.3577×10^{-14}	9.8080×10^{-14}
4.2700×10^{-13}	4.1972×10^{-13}	4.8327×10^{-13}	8.3600×10^{-14}	8.4944×10^{-14}	1.0054×10^{-13}
4.8649×10^{-13}	4.7965×10^{-13}	4.7750×10^{-13}	9.8062×10^{-14}	1.0050×10^{-13}	9.7554×10^{-14}

Table 5 RMS error of the elements of the unperturbed STM of HEO orbit

3.1151×10^{-12}	3.0525×10^{-12}	3.5452×10^{-12}	8.5518×10^{-13}	9.4291×10^{-13}	7.4181×10^{-13}
3.0536×10^{-12}	3.0748×10^{-12}	3.2955×10^{-12}	9.7523×10^{-13}	7.4920×10^{-13}	8.9767×10^{-13}
3.5437×10^{-12}	3.2934×10^{-12}	3.4803×10^{-12}	7.8115×10^{-13}	8.9159×10^{-13}	1.1442×10^{-12}
1.9463×10^{-11}	2.0158×10^{-11}	2.1143×10^{-11}	3.4252×10^{-12}	3.3592×10^{-12}	3.8517×10^{-12}
2.0477×10^{-11}	1.7508×10^{-11}	1.9179×10^{-11}	3.3607×10^{-12}	3.3031×10^{-12}	3.5563×10^{-12}
2.0899×10^{-11}	1.8752×10^{-11}	2.0422×10^{-11}	3.8501×10^{-12}	3.5538×10^{-12}	3.8493×10^{-12}

where, $[J]$ is defined by,

$$[J] = \begin{bmatrix} 0_{3\times3} & I_{3\times3} \\ -I_{3\times3} & 0_{3\times3} \end{bmatrix} \tag{64}$$

For presenting the error in the symplectic nature of the STMs, $[E^1_{sym}]$ is calculated using Eq. (65), [22, 30],

$$[E^1_{sym}] = \phi^{1T}[J]\phi^1 - [J] \tag{65}$$

In order to generate the necessary conditions for checking the symplectic nature of the second order State Transition Tensor, Eq. (65) is differentiated with respect to the initial states [17],

$$\frac{\partial}{\partial \mathbf{x}_0}\{\phi^{1T}[J]\phi^1 = [J]\} \Rightarrow \phi^{2T}[J]\phi^1 + \phi^{1T}[J]\phi^2 = [0_{6\times6}] \tag{66}$$

The error in the symplectic nature second order State Transition Tensor, $[E^2_{sym}]$, is defined as,

$$[E^2_{sym}] = \phi^{2T}[J]\phi^1 + \phi^{1T}[J]\phi^2 \tag{67}$$

The elements of $[E^1_{sym}]$ and $[E^2_{sym}]$ of every time step for unperturbed and $J_2 - J_6$ gravity perturbed STMs and second order State Transition Tensors are plotted against corresponding time steps in Figs. 1, 2, 3, 4, 5, 6, 7 and 8 for 10 orbit periods. Again, the results are showing double precision level of accuracy in the symplectic nature of the STMs and second order State Transition Tensors.

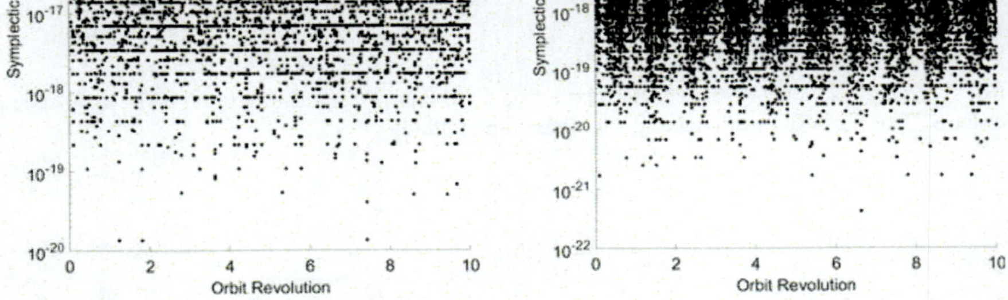

Fig. 1 Symplectic check versus orbit revolution for the unperturbed first order STM and second order STT of the LEO using analytic continuation method

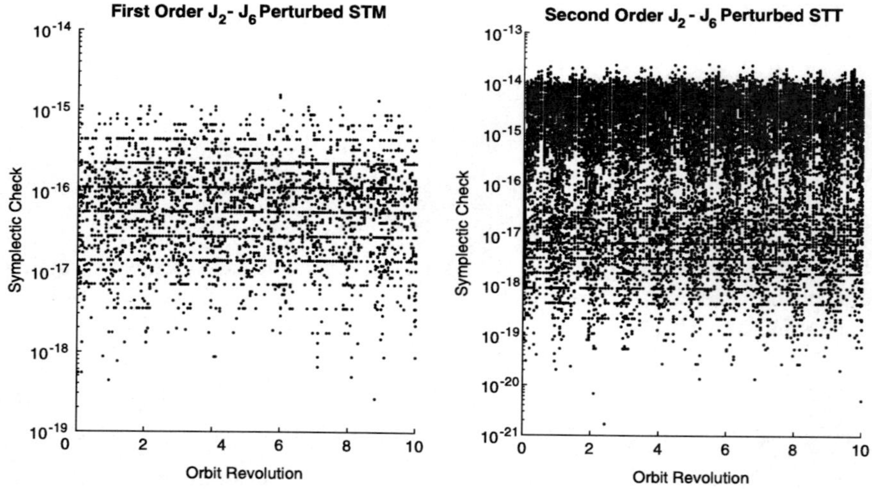

Fig. 2 Symplectic check versus orbit revolution for the $J_2 - J_6$ perturbed first order STM and second order STT of the LEO using analytic continuation method

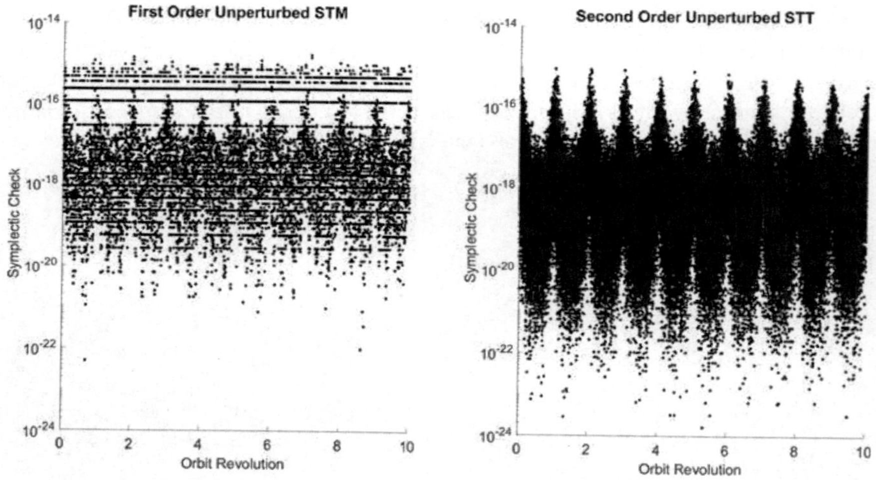

Fig. 3 Symplectic check versus orbit revolution for the unperturbed first order STM and second order STT of the MEO using analytic continuation method

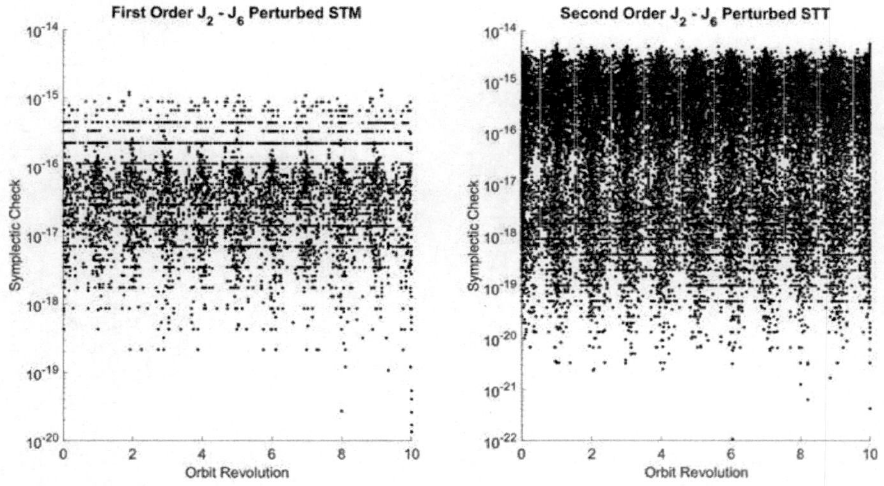

Fig. 4 Symplectic check versus orbit revolution for the $J_2 - J_6$ perturbed first order STM and second order STT of the MEO using analytic continuation method

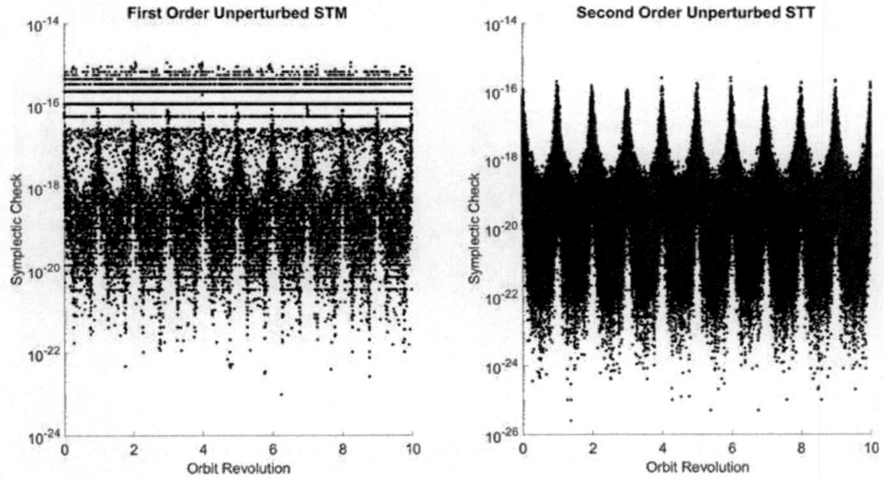

Fig. 5 Symplectic check versus orbit revolution for the unperturbed first order STM and second order STT of the GTO using analytic continuation method

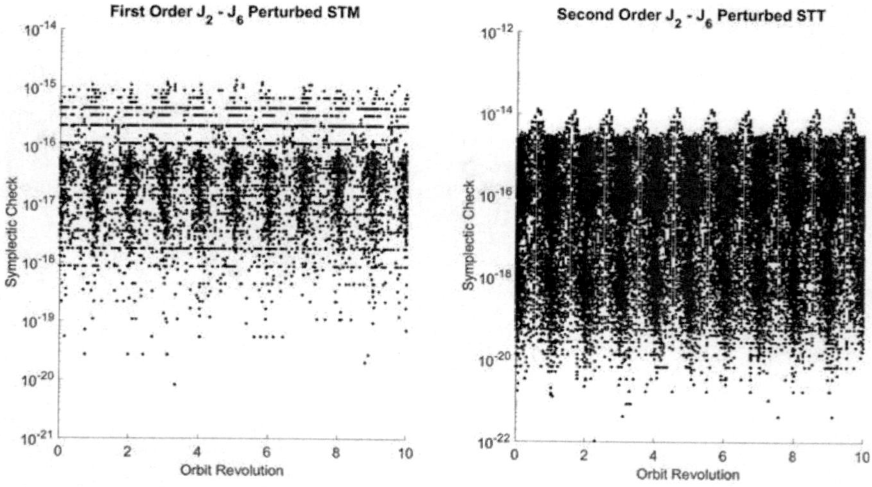

Fig. 6 Symplectic check versus orbit revolution for the $J_2 - J_6$ perturbed first order STM and second order STT of the GTO using analytic continuation method

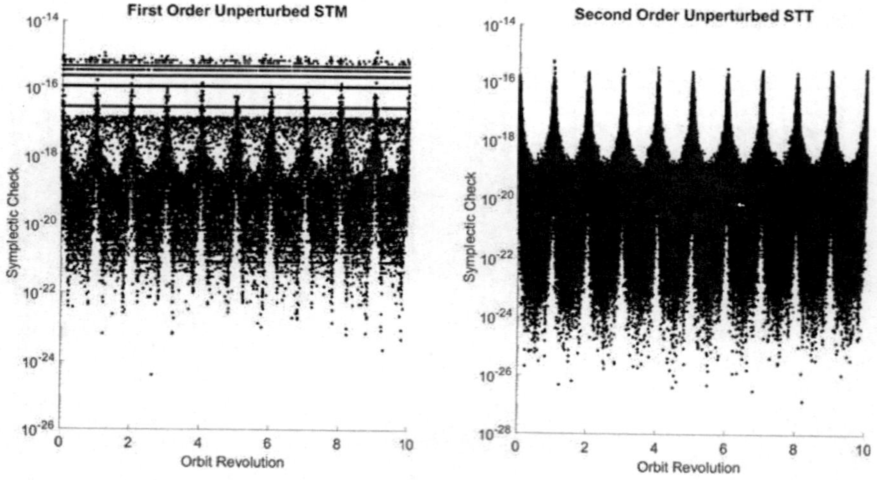

Fig. 7 Symplectic check versus orbit revolution for the unperturbed first order STM and second order STT of the HEO using analytic continuation method

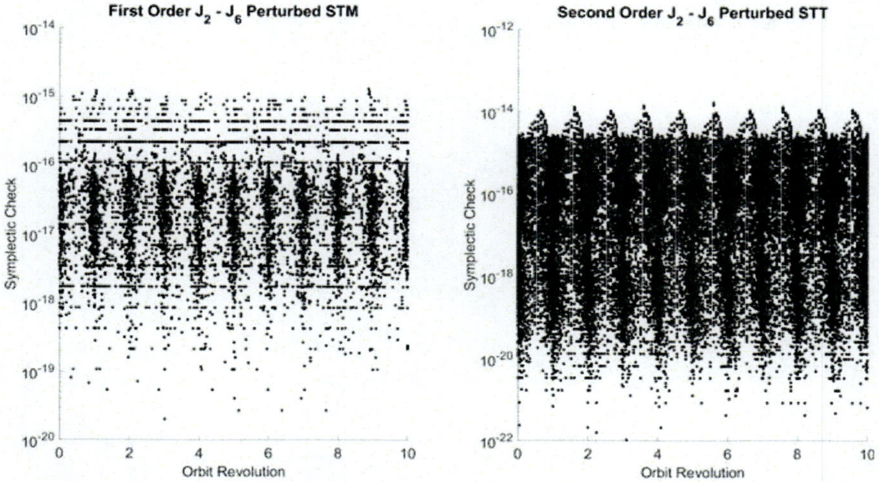

Fig. 8 Symplectic check versus orbit revolution for the $J_2 - J_6$ perturbed first order STM and second order STT of the HEO using analytic continuation method

To check the accuracy of the error propagation of the states using gravity and drag perturbed STM and second order State Transition Tensors, the orbits presented in Table 1 are generated using their initial states to get the nominal trajectories, $\mathbf{x}$. Then initial errors of $10^{-6} \times \mathbf{r}(t_0)$ and $10^{-7} \times \mathbf{r}^{(1)}(t_0)$ are added to the initial position and velocity of the orbits and propagated for one orbit period using the same time steps as the nominal trajectory to generate the gravity and drag perturbed "True" trajectories with initial errors, $\mathbf{x}_{\otimes}$. Next, the associated error to the nominal trajectory, $\mathbf{x}$, at each time step due to the initial error is propagated using only STM as shown in Eq. (68) and using both STM and second order State Transition Tensor as shown in Eq. (69) and added to the states of the nominal trajectory at the corresponding time steps to get $\hat{\mathbf{x}}^1$ and $\hat{\mathbf{x}}^2$. Finally the prediction error at every time step, $Error^1$ and $Error^2$ are computed by taking the difference between $\hat{\mathbf{x}}^1$ and $\mathbf{x}_{\otimes}$, and $\hat{\mathbf{x}}^2$ and $\mathbf{x}_{\otimes}$ respectively as shown in Eq. (71).

$$\delta\mathbf{x}^1(t_i) = \phi^1 \delta\mathbf{x}(t_{i-1}) \tag{68}$$

$$\delta\mathbf{x}^2(t_i) = \phi^1 \delta\mathbf{x}(t_{i-1}) + \frac{1}{2!}\phi^2 \delta\mathbf{x}(t_{i-1})\delta\mathbf{x}(t_{i-1}) \tag{69}$$

$$\begin{aligned} \hat{\mathbf{x}}^1(t_i) &= \mathbf{x}(t_i) + \delta\mathbf{x}^1(t_i) \\ \hat{\mathbf{x}}^2(t_i) &= \mathbf{x}(t_i) + \delta\mathbf{x}^2(t_i) \end{aligned} \tag{70}$$

$$\begin{aligned} Error^1(t_i) &= \mathbf{x}_{\otimes}(t_i) - \hat{\mathbf{x}}^1(t_i) \\ Error^2(t_i) &= \mathbf{x}_{\otimes}(t_i) - \hat{\mathbf{x}}^2(t_i) \end{aligned} \tag{71}$$

Table 6 Initial states and initial errors of the orbits

Orbit Type	Initial states		Initial errors	
LEO	P_0, m	$[2.8654\ \ 5.1911\ \ 2.8484] \times 10^6$	δP_0, m	$[2.8654\ \ 5.1911\ \ 2.8484]$
	V_0, m/s	$[-5.3862\ \ -0.3867\ \ 6.1232] \times 10^3$	δV_0, m/s	$[-0.5386\ \ -0.0387\ \ 0.6123] \times 10^{-3}$
MEO	P_0, m	$[2.8654\ \ 5.1911\ \ 2.8484] \times 10^6$	δP_0, m	$[2.8654\ \ 5.1911\ \ 2.8484]$
	V_0, m/s	$[-6.0765\ \ -0.4363\ \ 6.9078] \times 10^3$	δV_0, m/s	$[-0.6077\ \ -0.0436\ \ 0.6908] \times 10^{-3}$
GTO	P_0, m	$[4.5916\ \ 8.3184\ \ 4.5644] \times 10^6$	δP_0, m	$[4.5916\ \ 8.3184\ \ 4.5644]$
	V_0, m/s	$[-5.1317\ \ -0.3684\ \ 5.8338] \times 10^3$	δV_0, m/s	$[-0.5132\ \ -0.0368\ \ 0.5834] \times 10^{-3}$
HEO	P_0, m	$[3.5283\ \ 6.3921\ \ 3.5074] \times 10^6$	δP_0, m	$[3.5283\ \ 6.3921\ \ 3.5074]$
	V_0, m/s	$[-6.0343\ \ -0.4332\ \ 6.8598] \times 10^3$	δV_0, m/s	$[-0.6034\ \ -0.0433\ \ 0.6859] \times 10^{-3}$

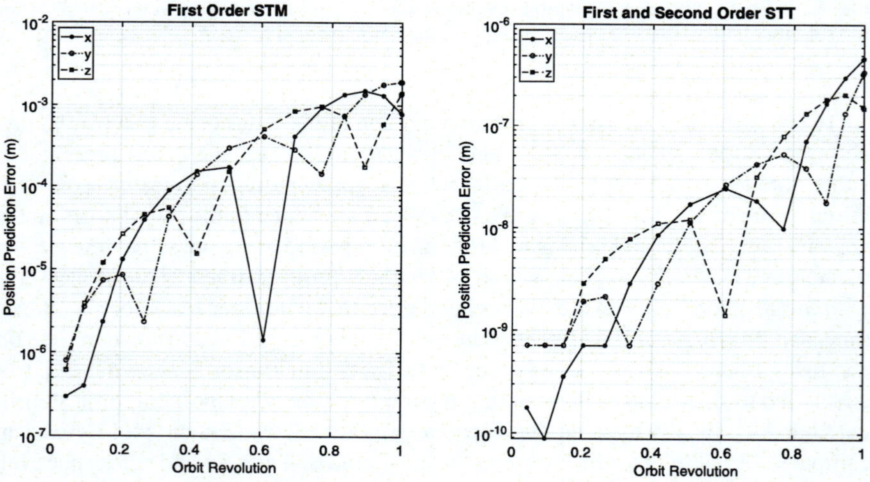

Fig. 9 Position prediction error versus orbit revolution using the $J_2 - J_6$ and drag perturbed first order STM and first and second order STT of the LEO orbit using analytic continuation method

The initial states of the simulated LEO, MEO, GTO and HEO and their corresponding initial errors are presented in Table 6.

The results of prediction error using $J_2 - J_6$ gravity and drag perturbed STMs and second order State Transition Tensors for the LEO are shown in Figs. 9 and 10, for the MEO are shown in Figs. 11 and 12, for the GTO are shown in Figs. 13 and 14 and for the HEO are shown in Figs. 15 and 16.

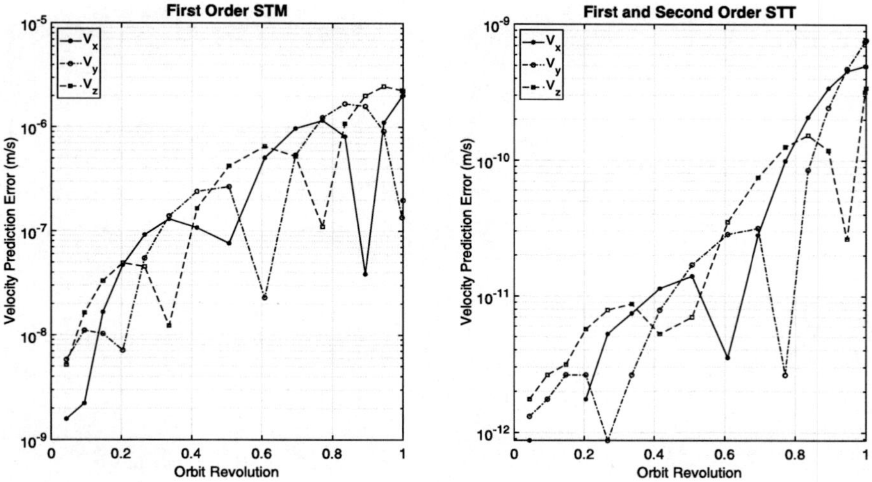

Fig. 10 Velocity prediction error versus orbit revolution using the $J_2 - J_6$ and drag perturbed first order STM and first and second order STT of the LEO orbit using analytic continuation method

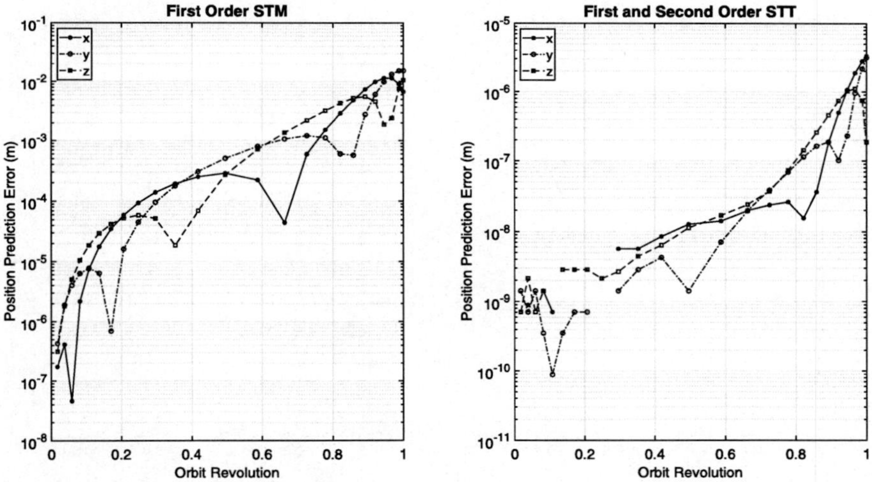

Fig. 11 Position prediction error verusu orbit revolution using the $J_2 - J_6$ and drag perturbed first order STM and first and second order STT of the MEO orbit using analytic continuation method

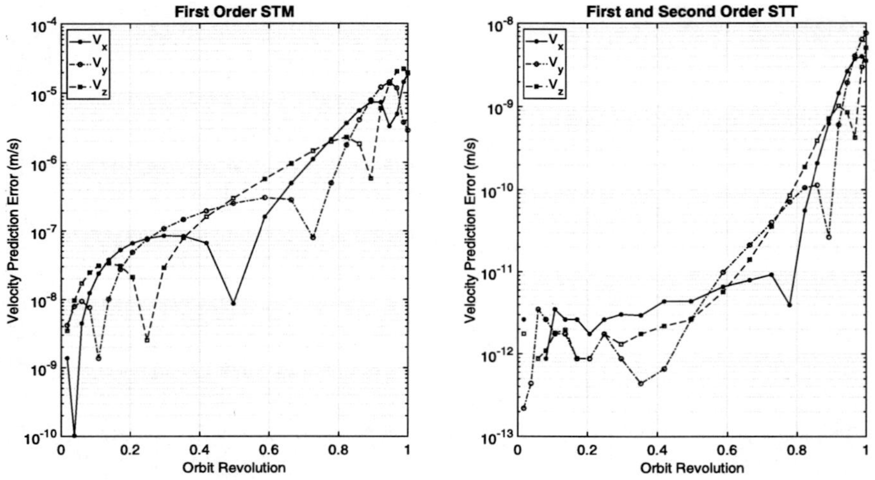

Fig. 12 Velocity prediction error versus orbit revolution using the $J_2 - J_6$ and drag perturbed first order STM and first and second order STT of the MEO orbit using analytic continuation method

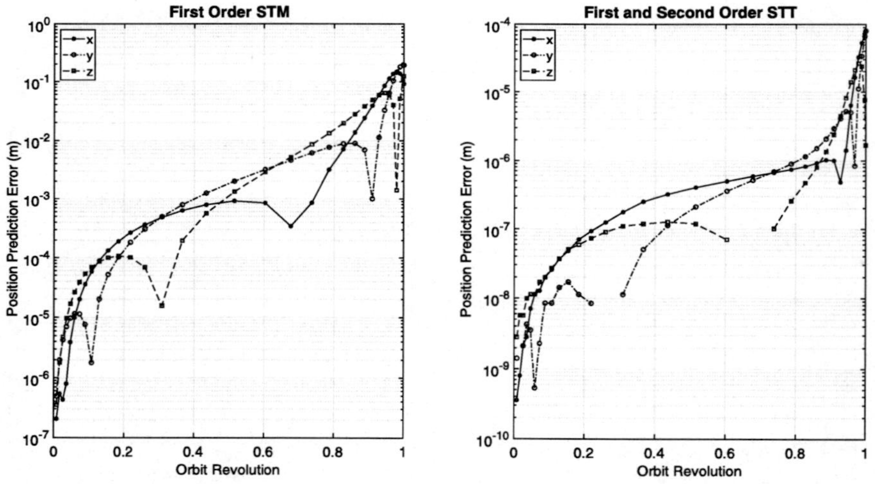

Fig. 13 Position prediction error versus orbit revolution using the $J_2 - J_6$ and drag perturbed first order STM and first and second order STT of the GTO orbit using analytic continuation method

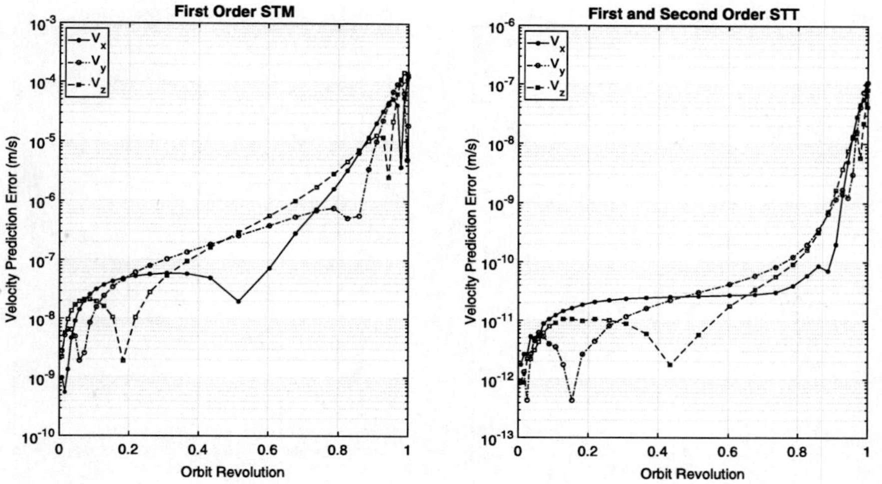

Fig. 14 Velocity prediction error versus orbit revolution using the $J_2 - J_6$ and drag perturbed first order STM and first and second order STT of the GTO orbit using analytic continuation method

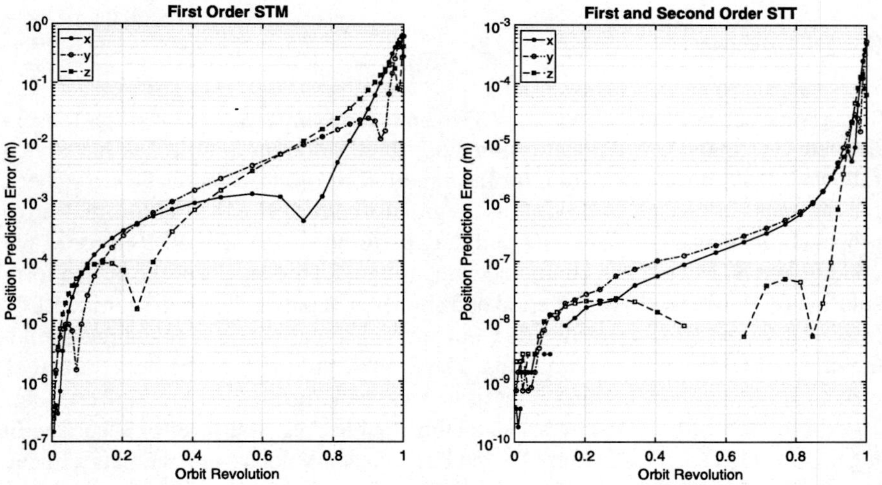

Fig. 15 Position prediction error vs orbit revolution using the $J_2 - J_6$ and drag perturbed first order STM and first and second order STT of the HEO orbit using analytic continuation method

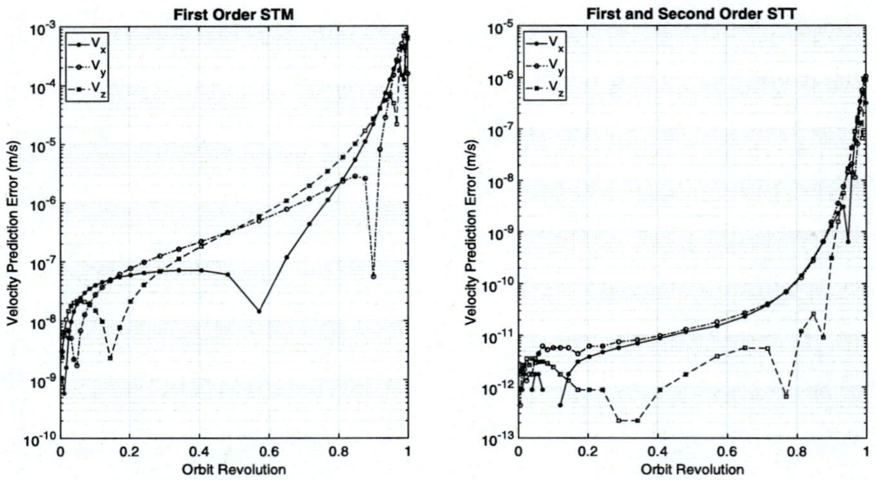

Fig. 16 Velocity prediction error versus orbit revolution using the $J_2 - J_6$ and drag perturbed first order STM and first and second order STT of the HEO orbit using analytic continuation method

8 Discussion

To show the accuracy of the computed STM and second order State Transition Tensors derived via Analytic Continuation method, three different approaches have been followed: RMS error calculation for the unperturbed orbits, Symplectic error check for the unperturbed as well as $J_2 - J_6$ gravity perturbed cases, and initial error propagation using $J_2 - J_6$ gravity and drag perturbed cases. The RMS error of the unperturbed STM has been computed comparing with the results of the closed form solution by Battin for 10 orbit period and shown in Tables 2, 3, 4 and 5 for the LEO, MEO, GTO and HEO respectively. Double precision accuracy is maintained in all the cases of RMS error computations. The Symplectic Error of the unperturbed as well as $J_2 - J_6$ gravity perturbed cases are shown in Figs. 1, 2, 3, 4, 5, 6, 7 and 8 for 10 orbit period of time and again maintains double precision level of accuracy. In Figs. 9, 10, 11, 12, 13, 14, 15 and 16, the Position and Velocity prediction error have been shown for the $J_2 - J_6$ gravity and drag perturbed cases, showing at least 3 to 4 digits of accuracy improvement when incorporating second order State Transition Tensors for error propagation.

9 Conclusion

In this paper, the procedure for deriving $J_2 - J_6$ gravity and drag perturbed Two-body problem STM and second order State Transition tensors are presented using the Analytic Continuation technique. The recursive nature of computing the higher order partials is a lucrative approach for expanding the method further. Four different types of orbits: LEO, MEO, GTO and HEO are simulated to show the versatility of the method. To present the accuracy of the method, at first RMS errors for 10 orbit period are presented comparing to the closed form solutions of Battin, which show double precision accuracy of the unperturbed STMs. Second, the symplectic nature of the unperturbed as well as $J_2 - J_6$ gravity perturbed STM and second order State Transition Tensors are presented, showing again machine precision level of accuracy. Finally, prediction error of initial error propagation is presented for one orbit period. In every case, at least 3 to 4 digits of accuracy has been improved using gravity and drag perturbed second order State Transition Tensors for one orbit revolution. The future expansion of this research work will be to implement the results for solving perturbed Multi-revolution Lambert's problem and to quantify uncertainty propagation of the states.

Acknowledgements This publication is part of a research supported by the Center of Excellence for the Commercial Space Transportation (COE-CST) of Federal Aviation Administration (Award Number FAA-15-C-CST-UCF-011). The authors express their gratitude for the support.

Appendix A

The higher order time derivatives of $J_3 - J_6$ perturbation accelerations and their higher order partials are shown in the following equations:

$$\mathbf{r}_{J_3}^{(n+2)} = C_{J_3}\left\{-\begin{bmatrix}15 & 0 & 0\\ 0 & 15 & 0\\ 0 & 0 & 30\end{bmatrix}\sum_{m=0}^{n}\binom{n}{m}B_7^{(m)}z^{(n-m)} + 35\sum_{m=0}^{n}\binom{n}{m}B_9^{(m)}C_3^{(n-m)} + \begin{bmatrix}0\\0\\3\end{bmatrix}g_5^{(n)}\right\} \tag{72}$$

$$\begin{aligned}\frac{\partial \mathbf{r}_{J_3}^{(n+2)}}{\partial \chi_j} = C_{J_3}\Bigg\{&-\begin{bmatrix}15 & 0 & 0\\ 0 & 15 & 0\\ 0 & 0 & 30\end{bmatrix}\sum_{m=0}^{n}\binom{n}{m}\left(\frac{\partial B_7^{(m)}}{\partial \chi_j}z^{(n-m)} + B_7^{(m)}\frac{\partial z^{(n-m)}}{\partial \chi_j}\right)\\ &+35\sum_{m=0}^{n}\binom{n}{m}\left(\frac{\partial B_9^{(m)}}{\partial \chi_j}C_3^{(n-m)} + B_9^{(m)}\frac{\partial C_3^{(n-m)}}{\partial \chi_j}\right) + \begin{bmatrix}0\\0\\3\end{bmatrix}\frac{\partial g_5^{(n)}}{\partial \chi_j}\Bigg\}\end{aligned} \tag{73}$$

$$\begin{aligned}\frac{\partial \mathbf{r}_{J_3}^{(n+2)}}{\partial \chi_j \partial \chi_k} = C_{J_3}\Bigg\{&-\begin{bmatrix}15 & 0 & 0\\ 0 & 15 & 0\\ 0 & 0 & 30\end{bmatrix}\sum_{m=0}^{n}\binom{n}{m}\left(\frac{\partial^2 B_7^{(m)}}{\partial \chi_j \partial \chi_k}z^{(n-m)} + \frac{\partial B_7^{(m)}}{\partial \chi_j}\frac{\partial z^{(n-m)}}{\partial \chi_k} + \frac{\partial B_7^{(m)}}{\partial \chi_k}\frac{\partial z^{(n-m)}}{\partial \chi_j} + B_7^{(m)}\frac{\partial^2 z^{(n-m)}}{\partial \chi_j \partial \chi_k}\right)\\ &+35\sum_{m=0}^{n}\binom{n}{m}\left(\frac{\partial^2 B_9^{(m)}}{\partial \chi_j \chi_k}C_3^{(n-m)} + \frac{\partial B_9^{(m)}}{\partial \chi_j}\frac{\partial C_3^{(n-m)}}{\partial \chi_k} + \frac{\partial B_9^{(m)}}{\partial \chi_k}\frac{\partial C_3^{(n-m)}}{\partial \chi_j} + B_9^{(m)}\frac{\partial^2 C_3^{(n-m)}}{\partial \chi_j \partial \chi_k}\right) + \begin{bmatrix}0\\0\\3\end{bmatrix}\frac{\partial^2 g_5^{(n)}}{\partial \chi_j \chi_k}\Bigg\}\end{aligned} \tag{74}$$

$$\mathbf{r}_{J_4}^{(n+2)} = C_{J_4}\left\{\begin{bmatrix}3&0&0\\0&3&0\\0&0&15\end{bmatrix}B_7^{(n)} - \begin{bmatrix}42&0&0\\0&42&0\\0&0&70\end{bmatrix}\sum_{m=0}^{n}\binom{n}{m}B_9^{(m)}C_2^{(n-m)} + 63\sum_{m=0}^{n}\binom{n}{m}B_{11}^{(m)}C_4^{(n-m)}\right\} \tag{75}$$

$$\frac{\partial \mathbf{r}_{J_4}^{(n+2)}}{\partial \chi_j} = C_{J_4}\left\{\begin{bmatrix}3&0&0\\0&3&0\\0&0&15\end{bmatrix}\frac{\partial B_7^{(n)}}{\partial \chi_j} - \begin{bmatrix}42&0&0\\0&42&0\\0&0&70\end{bmatrix}\sum_{m=0}^{n}\binom{n}{m}\left(\frac{\partial B_9^{(m)}}{\partial \chi_j}C_2^{(n-m)} + B_9^{(m)}\frac{\partial C_2^{(n-m)}}{\partial \chi_j}\right) + 63\sum_{m=0}^{n}\binom{n}{m}\left(\frac{\partial B_{11}^{(m)}}{\partial \chi_j}C_4^{(n-m)} + B_{11}^{(m)}\frac{\partial C_4^{(n-m)}}{\partial \chi_j}\right)\right\} \tag{76}$$

$$\frac{\partial^2 \mathbf{r}_{J_4}^{(n+2)}}{\partial \chi_j \partial \chi_k} = C_{J_4}\left\{\begin{bmatrix}3&0&0\\0&3&0\\0&0&15\end{bmatrix}\frac{\partial^2 B_7^{(n)}}{\partial \chi_j \partial \chi_k} - \begin{bmatrix}42&0&0\\0&42&0\\0&0&70\end{bmatrix}\sum_{m=0}^{n}\binom{n}{m}\left(\frac{\partial^2 B_9^{(m)}}{\partial \chi_j \partial \chi_k}C_2^{(n-m)} + \frac{\partial B_9^{(m)}}{\partial \chi_j}\frac{\partial C_2^{(n-m)}}{\partial \chi_k} + \frac{\partial B_9^{(m)}}{\partial \chi_k}\frac{\partial C_2^{(n-m)}}{\partial \chi_j} + B_9^{(m)}\frac{\partial^2 C_2^{(n-m)}}{\partial \chi_j \partial \chi_k}\right) + 63\sum_{m=0}^{n}\binom{n}{m}\left(\frac{\partial^2 B_{11}^{(m)}}{\partial \chi_j \partial \chi_k}C_4^{(n-m)} + \frac{\partial B_{11}^{(m)}}{\partial \chi_j}\frac{\partial C_4^{(n-m)}}{\partial \chi_k} + \frac{\partial B_{11}^{(m)}}{\partial \chi_k}\frac{\partial C_4^{(n-m)}}{\partial \chi_j} + B_{11}^{(m)}\frac{\partial^2 C_4^{(n-m)}}{\partial \chi_j \partial \chi_k}\right)\right\} \tag{77}$$

$$\mathbf{r}_{J_5}^{(n+2)} = C_{J_5}\left\{\begin{bmatrix}105&0&0\\0&105&0\\0&0&315\end{bmatrix}\sum_{m=0}^{n}\binom{n}{m}B_9^{(m)}z^{(n-m)} - \begin{bmatrix}630&0&0\\0&630&0\\0&0&945\end{bmatrix}\sum_{m=0}^{n}\binom{n}{m}B_{11}^{(m)}C_3^{(n-m)} + 693\sum_{m=0}^{n}\binom{n}{m}B_{13}^{(m)}C_5^{(n-m)} - \begin{bmatrix}0\\0\\15\end{bmatrix}g_7^{(n)}\right\} \tag{78}$$

$$\frac{\partial \mathbf{r}_{J_5}^{(n+2)}}{\partial \chi_j} = C_{J_5}\left\{\begin{bmatrix}105&0&0\\0&105&0\\0&0&315\end{bmatrix}\sum_{m=0}^{n}\binom{n}{m}\left(\frac{\partial B_9^{(m)}}{\partial \chi_j}z^{(n-m)} + B_9^{(m)}\frac{\partial z^{(n-m)}}{\partial \chi_j}\right) - \begin{bmatrix}630&0&0\\0&630&0\\0&0&945\end{bmatrix}\sum_{m=0}^{n}\binom{n}{m}\left(\frac{\partial B_{11}^{(m)}}{\partial \chi_j}C_3^{(n-m)} + B_{11}^{(m)}\frac{\partial C_3^{(n-m)}}{\partial \chi_j}\right) + 693\sum_{m=0}^{n}\binom{n}{m}\left(\frac{\partial B_{13}^{(m)}}{\partial \chi_j}C_5^{(n-m)} + B_{13}^{(m)}\frac{\partial C_5^{(n-m)}}{\partial \chi_j}\right) - \begin{bmatrix}0\\0\\15\end{bmatrix}\frac{\partial g_7^{(n)}}{\partial \chi_j}\right\} \tag{79}$$

$$\frac{\partial^2 \mathbf{r}_{J_5}^{(n+2)}}{\partial \chi_j \partial \chi_k} = C_{J_5}\left\{\begin{bmatrix}105 & 0 & 0\\ 0 & 105 & 0\\ 0 & 0 & 315\end{bmatrix}\sum_{m=0}^{n}\binom{n}{m}\left(\frac{\partial^2 B_9^{(m)}}{\partial \chi_j \partial \chi_k}z^{(n-m)} + \frac{\partial B_9^{(m)}}{\partial \chi_j}\frac{\partial z^{(n-m)}}{\partial \chi_k} + \frac{\partial B_9^{(m)}}{\partial \chi_k}\frac{\partial z^{(n-m)}}{\partial \chi_j} + B_9^{(m)}\frac{\partial^2 z^{(n-m)}}{\partial \chi_j \partial \chi_k}\right)\right.$$
$$-\begin{bmatrix}630 & 0 & 0\\ 0 & 630 & 0\\ 0 & 0 & 945\end{bmatrix}\sum_{m=0}^{n}\binom{n}{m}\left(\frac{\partial^2 B_{11}^{(m)}}{\partial \chi_j \partial \chi_k}C_3^{(n-m)} + \frac{\partial B_{11}^{(m)}}{\partial \chi_j}\frac{\partial C_3^{(n-m)}}{\partial \chi_k} + \frac{\partial B_{11}^{(m)}}{\partial \chi_k}\frac{\partial C_3^{(n-m)}}{\partial \chi_j} + B_{11}^{(m)}\frac{\partial^2 C_3^{(n-m)}}{\partial \chi_j \partial \chi_k}\right)$$
$$\left.+693\sum_{m=0}^{n}\binom{n}{m}\left(\frac{\partial^2 B_{13}^{(m)}}{\partial \chi_j \partial \chi_k}C_5^{(n-m)} + \frac{\partial B_{13}^{(m)}}{\partial \chi_j}\frac{\partial C_5^{(n-m)}}{\partial \chi_k} + \frac{\partial B_{13}^{(m)}}{\partial \chi_k}\frac{\partial C_5^{(n-m)}}{\partial \chi_j} + B_{13}^{(m)}\frac{\partial^2 C_5^{(n-m)}}{\partial \chi_j \partial \chi_k}\right) - \begin{bmatrix}0\\ 0\\ 15\end{bmatrix}\frac{\partial^2 g_7^{(n)}}{\partial \chi_j \partial \chi_k}\right\} \tag{80}$$

$$\mathbf{r}_{J_6}^{(n+2)} = C_{J_6}\left\{\begin{bmatrix}35 & 0 & 0\\ 0 & 35 & 0\\ 0 & 0 & 245\end{bmatrix}B_9^{(n)} - \begin{bmatrix}945 & 0 & 0\\ 0 & 945 & 0\\ 0 & 0 & 2205\end{bmatrix}\sum_{m=0}^{n}\binom{n}{m}B_{11}^{(m)}C_2^{(n-m)}\right.$$
$$\left.+\begin{bmatrix}3465 & 0 & 0\\ 0 & 3465 & 0\\ 0 & 0 & 4851\end{bmatrix}\sum_{m=0}^{n}\binom{n}{m}B_{13}^{(m)}C_4^{(n-m)} - 3003\sum_{m=0}^{n}\binom{n}{m}B_{15}^{(m)}C_6^{(n-m)}\right\} \tag{81}$$

$$\frac{\partial \mathbf{r}_{J_6}^{(n+2)}}{\partial \chi_j} = C_{J_6}\left\{\begin{bmatrix}35 & 0 & 0\\ 0 & 35 & 0\\ 0 & 0 & 245\end{bmatrix}\frac{\partial B_9^{(n)}}{\partial \chi_j} - \begin{bmatrix}945 & 0 & 0\\ 0 & 945 & 0\\ 0 & 0 & 2205\end{bmatrix}\sum_{m=0}^{n}\binom{n}{m}\left(\frac{\partial B_{11}^{(m)}}{\partial \chi_j}C_2^{(n-m)} + B_{11}^{(m)}\frac{\partial C_2^{(n-m)}}{\partial \chi_j}\right)\right.$$
$$\left.+\begin{bmatrix}3465 & 0 & 0\\ 0 & 3465 & 0\\ 0 & 0 & 4851\end{bmatrix}\sum_{m=0}^{n}\binom{n}{m}\left(\frac{\partial B_{13}^{(m)}}{\partial \chi_j}C_4^{(n-m)} + B_{13}^{(m)}\frac{\partial C_4^{(n-m)}}{\partial \chi_j}\right) - 3003\sum_{m=0}^{n}\binom{n}{m}\left(\frac{\partial B_{15}^{(m)}}{\partial \chi_j}C_6^{(n-m)} + B_{15}^{(m)}\frac{\partial C_6^{(n-m)}}{\partial \chi_j}\right)\right\} \tag{82}$$

$$\frac{\partial^2 \mathbf{r}_{J_6}^{(n+2)}}{\partial \chi_j \partial \chi_k} = C_{J_6}\left\{\begin{bmatrix}35 & 0 & 0\\ 0 & 35 & 0\\ 0 & 0 & 245\end{bmatrix}\frac{\partial^2 B_9^{(n)}}{\partial \chi_j \partial \chi_k} - \begin{bmatrix}945 & 0 & 0\\ 0 & 945 & 0\\ 0 & 0 & 2205\end{bmatrix}\sum_{m=0}^{n}\binom{n}{m}\left(\frac{\partial^2 B_{11}^{(m)}}{\partial \chi_j \partial \chi_k}C_2^{(n-m)} + \frac{\partial B_{11}^{(m)}}{\partial \chi_j}\frac{\partial C_2^{(n-m)}}{\partial \chi_k}\right.\right.$$
$$\left.+\frac{\partial B_{11}^{(m)}}{\partial \chi_k}\frac{\partial C_2^{(n-m)}}{\partial \chi_j} + B_{11}^{(m)}\frac{\partial^2 C_2^{(n-m)}}{\partial \chi_j \partial \chi_k}\right)$$
$$+\begin{bmatrix}3465 & 0 & 0\\ 0 & 3465 & 0\\ 0 & 0 & 4851\end{bmatrix}\sum_{m=0}^{n}\binom{n}{m}\left(\frac{\partial^2 B_{13}^{(m)}}{\partial \chi_j \partial \chi_k}C_4^{(n-m)} + \frac{\partial B_{13}^{(m)}}{\partial \chi_j}\frac{\partial C_4^{(n-m)}}{\partial \chi_k} + \frac{\partial B_{13}^{(m)}}{\partial \chi_k}\frac{\partial C_4^{(n-m)}}{\partial \chi_j} + B_{13}^{(m)}\frac{\partial^2 C_4^{(n-m)}}{\partial \chi_j \partial \chi_k}\right)$$
$$\left.-3003\sum_{m=0}^{n}\binom{n}{m}\left(\frac{\partial^2 B_{15}^{(m)}}{\partial \chi_j \partial \chi_k}C_6^{(n-m)} + \frac{\partial B_{15}^{(m)}}{\partial \chi_j}\frac{\partial C_6^{(n-m)}}{\partial \chi_k} + \frac{\partial B_{15}^{(m)}}{\partial \chi_k}\frac{\partial C_6^{(n-m)}}{\partial \chi_j} + B_{15}^{(m)}\frac{\partial^2 C_6^{(n-m)}}{\partial \chi_j \partial \chi_k}\right)\right\} \tag{83}$$

Appendix B

In this section, the algorithm to derive the $J_2 - J_6$ and drag perturbed STM and second order State Transition Tensors is presented. For brevity, the notation $\mathbf{r}$ is used instead of $\mathbf{r}(t)$.

Algorithm 1: Algorithm for J_2 Perturbed First and Second Order State Transition Tensor Computation using Analytic Continuation Method

Initialize $\mathbf{r}(t_0)$, $\mathbf{r}^{(1)}(t_0)$ and T, where T is the orbital period;
Set value for $\delta_a, \delta_r, \delta_{fac}$ and k_{inc};
Get N ; // ▷ Eq. (9)
for $k = 1 \rightarrow i_{max}$ **do**
$f = \mathbf{r}.\mathbf{r}$; $f^{(1)} = 2\mathbf{r}.\mathbf{r}^{(1)}$; $f^{(2)} = 2\mathbf{r}^{(1)}.\mathbf{r}^{(1)} + 2\mathbf{r}.\mathbf{r}^{(2)}$; $g_p = f^{-\frac{p}{2}}$; $g_p^{(1)} = -\frac{p}{2} f^{(1)} \frac{g_p}{f}$;
$g_p^{(2)} = -\frac{1}{f}\left(\frac{p}{2} f^{(1)} g_p^{(1)} + \frac{p}{2} f^{(2)} g_p + f^{(1)} g_p^{(1)}\right)$; $B_p = \mathbf{r} g_p$; $B_p^{(1)} = \mathbf{r}^{(1)} g_p + \mathbf{r} g_p^{(1)}$; $C_2 = zz$;
$C_2^{(1)} = z^{(1)} z + z z^{(1)}$; $C_\alpha = C_{\alpha-1} z$; $C_\alpha^{(1)} = C_{\alpha-1}^{(1)} z + C_{\alpha-1} z^{(1)}$;
Get $\mathbf{r}_{J_2}^{(2)}, \mathbf{r}_{J_3}^{(2)}, \mathbf{r}_{J_4}^{(2)}, \mathbf{r}_{J_5}^{(2)}$ and $\mathbf{r}_{J_6}^{(2)}$; // ▷ Eq. (35) - (39)
Get $\mathbf{r}_{dr}^{(2)}$; // ▷ Eq. (49)
$\mathbf{r}_{per}^{(2)}(t) = -\mu \frac{\mathbf{r}(t)}{(\mathbf{r}(t).\mathbf{r}(t))^{3/2}} + \mathbf{r}_{J_2}^{(2)} + \mathbf{r}_{J_3}^{(2)} + \mathbf{r}_{J_4}^{(2)} + \mathbf{r}_{J_5}^{(2)} + \mathbf{r}_{J_6}^{(2)} + \mathbf{r}_{dr}^{(2)}$;
$B_p^{(2)} = \mathbf{r}^{(2)} g_p + 2\mathbf{r}^{(1)} g_p^{(1)} + \mathbf{r} g_p^{(2)}$;
$C_2^{(2)} = z^{(2)} z + 2z^{(1)} z^{(1)} + z z^{(2)}$; $C_\alpha^{(2)} = C_{\alpha-1}^{(2)} z + 2C_{\alpha-1}^{(1)} z^{(1)} + C_{\alpha-1} z^{(2)}$;
where, p is 3, 5, 7, 9, 11, 13 and 15 and α is 3, 4, 5 and 6.
Get $\mathbf{r}_{J_2}^{(3)}, \mathbf{r}_{J_3}^{(3)}, \mathbf{r}_{J_4}^{(3)}, \mathbf{r}_{J_5}^{(3)}, \mathbf{r}_{J_6}^{(3)}$ and $\mathbf{r}_{J_{dr}}^{(3)}$; // ▷ Eq. (46), (72), (75), (78), (81), (59)
for $n = 3 \rightarrow N$ **do**
$\mathbf{r}_{per}^{(n)} = -\mu \sum_{m=0}^{n} \binom{n}{m} \mathbf{r}^{(m)} g_3^{(n-m)} + \mathbf{r}_{J_2}^{(n)} + \mathbf{r}_{J_3}^{(n)} + \mathbf{r}_{J_4}^{(n)} + \mathbf{r}_{J_5}^{(n)} + \mathbf{r}_{J_6}^{(n)} + \mathbf{r}_{dr}^{(n)}$;
Get $f^{(n)}, g_p^{(n)}, B_p^{(n)}$; // ▷ Eq. (4), (5), (41)
$C_2^{(n)} = \sum_{m=0}^{n} z^{(m)} z^{(n-m)}$;
Get $C_\alpha^{(n)}$; // ▷ Eq. (41)
Get $\mathbf{r}_{J_2}^{(n+1)}, \mathbf{r}_{J_3}^{(n+1)}, \mathbf{r}_{J_4}^{(n+1)}, \mathbf{r}_{J_5}^{(n+1)}, \mathbf{r}_{J_6}^{(n+1)}, \mathbf{r}_{J_{dr}}^{(n+1)}$; // ▷ Eq. (46), (72), (75), (78), (81), (59)
end
Get dT ; // ▷ Eq. (12)
$\mathbf{r}_{per}(t + dT) = \sum_{m=0}^{n} (\mathbf{r}^{(m)}(t) + \mathbf{r}_{J_2}^{(m)}(t) + \mathbf{r}_{J_3}^{(m)}(t) + \mathbf{r}_{J_4}^{(m)}(t) + \mathbf{r}_{J_5}^{(m)}(t) + \mathbf{r}_{J_6}^{(m)}(t) + \mathbf{r}_{J_{dr}}^{(m)}(t))$;
$\mathbf{r}_{per}^{(1)}(t + dT) = \sum_{m=1}^{n} (\mathbf{r}^{(m)}(t) + \mathbf{r}_{J_2}^{(m)}(t) + \mathbf{r}_{J_3}^{(m)}(t) + \mathbf{r}_{J_4}^{(m)}(t) + \mathbf{r}_{J_5}^{(m)}(t) + \mathbf{r}_{J_6}^{(m)}(t) + \mathbf{r}_{J_{dr}}^{(m)}(t))$; Get $\frac{\partial \mathbf{r}}{\partial \mathbf{r}}, \frac{\partial \mathbf{r}^{(1)}}{\partial \mathbf{r}^{(1)}}, \frac{\partial \mathbf{r}^{(1)}}{\partial \mathbf{r}}, \frac{\partial \mathbf{r}}{\partial \mathbf{r}^{(1)}}, \frac{\partial f(t)}{\partial \mathbf{r}(t)}, \frac{\partial g_p(t)}{\partial \mathbf{r}(t)}, \frac{\partial f(t)}{\partial \mathbf{r}^{(1)}(t)}, \frac{\partial g_p(t)}{\partial \mathbf{r}^{(1)}(t)}$; // ▷ Eq. (19)
Get $\frac{\partial B_p}{\partial \mathbf{r}}$ and $\frac{\partial B_p}{\partial \mathbf{r}^{(1)}}$; // ▷ Eq. (42)
$\frac{\partial C_2}{\partial \mathbf{r}} = \frac{\partial z}{\partial \mathbf{r}} z + z \frac{\partial z}{\partial \mathbf{r}}$; $\frac{\partial C_2}{\partial \mathbf{r}^{(1)}} = \frac{\partial z}{\partial \mathbf{r}^{(1)}} z + z \frac{\partial z}{\partial \mathbf{r}^{(1)}}$;
Get $\frac{\partial C_\alpha}{\partial \mathbf{r}}, \frac{\partial C_\alpha}{\partial \mathbf{r}^{(1)}}, \frac{\partial \mathbf{r}_{J_2}^{(2)}}{\partial \mathbf{r}}, \frac{\partial \mathbf{r}_{J_2}^{(2)}}{\partial \mathbf{r}^{(1)}}$; // ▷ Eq. (44), (53)
for $n = 0 \rightarrow N$ **do**
Get $\frac{\partial \mathbf{r}^{(n+2)}}{\partial \mathbf{r}}, \frac{\partial \mathbf{r}^{(n+2)}}{\partial \mathbf{r}^{(1)}}$; // ▷ Eq. (22)
Get $\frac{\partial \mathbf{r}_{J_2-J_6}^{(n+2)}}{\partial \mathbf{r}}, \frac{\partial \mathbf{r}_{J_2-J_6}^{(n+2)}}{\partial \mathbf{r}^{(1)}}, \frac{\partial \mathbf{r}_{dr}^{(n+2)}}{\partial \mathbf{r}}, \frac{\partial \mathbf{r}_{dr}^{(n+2)}}{\partial \mathbf{r}^{(1)}}$; // ▷ Eq. (47), (73), (76), (79), (82), (60)
$\frac{\partial \mathbf{r}_{per}^{(n+2)}}{\partial \mathbf{r}} = \frac{\partial \mathbf{r}^{(n+2)}}{\partial \mathbf{r}} + \frac{\partial \mathbf{r}_{J_2-J_6}^{(n+2)}}{\partial \mathbf{r}} + \frac{\partial \mathbf{r}_{dr}^{(n+2)}}{\partial \mathbf{r}}$; $\frac{\partial \mathbf{r}_{per}^{(n+2)}}{\partial \mathbf{r}^{(1)}} = \frac{\partial \mathbf{r}^{(n+2)}}{\partial \mathbf{r}^{(1)}} + \frac{\partial \mathbf{r}_{J_2-J_6}^{(n+2)}}{\partial \mathbf{r}^{(1)}} + \frac{\partial \mathbf{r}_{dr}^{(n+2)}}{\partial \mathbf{r}^{(1)}}$;
Get $\frac{\partial f^{(n+1)}}{\partial \mathbf{r}}, \frac{\partial f^{(n+1)}}{\partial \mathbf{r}^{(1)}}, \frac{\partial g_p^{(n+1)}}{\partial \mathbf{r}}, \frac{\partial g_p^{(n+1)}}{\partial \mathbf{r}^{(1)}}, \frac{\partial B_p^{(n+1)}}{\partial \mathbf{r}}, \frac{\partial B_p^{(n+1)}}{\partial \mathbf{r}^{(1)}}, \frac{\partial C_\alpha^{(n+1)}}{\partial \mathbf{r}}, \frac{\partial C_\alpha^{(n+1)}}{\partial \mathbf{r}^{(1)}}$; // ▷ Eq. (20), (21), (42), (44)
end
Get $\phi_{11}^1(t + dT, t), \phi_{12}^1(t + dT, t), \phi_{21}^1(t + dT, t), \phi_{22}^1(t + dT, t)$; // ▷ Eq. (15), (16), (17), (18)
$\phi_{11per}^1 = \phi_{11}^1 + \sum_{m=0}^{n} \left(\frac{\partial \mathbf{r}_{J_2-J_6}^{(m)}(t)}{\partial \mathbf{r}(t)} + \frac{\partial \mathbf{r}_{dr}^{(m)}(t)}{\partial \mathbf{r}(t)} \right) \frac{dT^{(m)}}{m!}$;
$\phi_{12per}^1 = \phi_{12}^1 + \sum_{m=0}^{n} \left(\frac{\partial \mathbf{r}_{J_2-J_6}^{(m)}(t)}{\partial \mathbf{r}^{(1)}(t)} + \frac{\partial \mathbf{r}_{dr}^{(m)}(t)}{\partial \mathbf{r}(t)} \right) \frac{dT^{(m)}}{m!}$;

$\phi^1_{21per} = \phi^1_{21} + \sum_{m=1}^{n}\left(\frac{\partial \mathbf{r}^{(m)}_{J_2-J_6}(t)}{\partial \mathbf{r}(t)} + \frac{\partial \mathbf{r}^{(m)}_{dr}(t)}{\partial \mathbf{r}(t)}\right)\frac{dT^{(m-1)}}{(m-1)!}$; $\phi^1_{22per} = \phi^1_{22} + \sum_{m=1}^{n}\left(\frac{\partial \mathbf{r}^{(m)}_{J_2-J_6}(t)}{\partial \mathbf{r}^{(1)}(t)} + \frac{\partial \mathbf{r}^{(m)}_{dr}(t)}{\partial \mathbf{r}(t)}\right)\frac{dT^{(m-1)}}{(m-1)!}$;

for $n = 0 \rightarrow N$ **do**

Get $\frac{\partial^2 \mathbf{r}^{(n+2)}}{\partial \mathbf{r}\partial \mathbf{r}}$, $\frac{\partial^2 \mathbf{r}^{(n+2)}}{\partial \mathbf{r}\partial \mathbf{r}^{(1)}}$, $\frac{\partial^2 \mathbf{r}^{(n+2)}}{\partial \mathbf{r}^{(1)}\partial \mathbf{r}}$, $\frac{\partial^2 \mathbf{r}^{(n+2)}}{\partial \mathbf{r}^{(1)}\partial \mathbf{r}^{(1)}}$; // ▷ Eq. (26)

Get $\frac{\partial^2 \mathbf{r}^{(n+2)}_{J_2-J_6}}{\partial \mathbf{r}\partial \mathbf{r}}$, $\frac{\partial^2 \mathbf{r}^{(n+2)}_{J_2-J_6}}{\partial \mathbf{r}\partial \mathbf{r}^{(1)}}$, $\frac{\partial^2 \mathbf{r}^{(n+2)}_{J_2-J_6}}{\partial \mathbf{r}^{(1)}\partial \mathbf{r}}$, $\frac{\partial^2 \mathbf{r}^{(n+2)}_{J_2-J_6}}{\partial \mathbf{r}^{(1)}\partial \mathbf{r}^{(1)}}$; // ▷ Eq. (48), (74), (77), (80), (83)

Get $\frac{\partial^2 \mathbf{r}^{(n+2)}_{dr}}{\partial \mathbf{r}\partial \mathbf{r}}$, $\frac{\partial^2 \mathbf{r}^{(n+2)}_{dr}}{\partial \mathbf{r}\partial \mathbf{r}^{(1)}}$, $\frac{\partial^2 \mathbf{r}^{(n+2)}_{dr}}{\partial \mathbf{r}^{(1)}\partial \mathbf{r}}$, $\frac{\partial^2 \mathbf{r}^{(n+2)}_{dr}}{\partial \mathbf{r}^{(1)}\partial \mathbf{r}^{(1)}}$; // ▷ Eq. (61)

$\frac{\partial^2 \mathbf{r}^{(n+2)}_{per}}{\partial \mathbf{r}\partial \mathbf{r}} = \frac{\partial^2 \mathbf{r}^{(n+2)}}{\partial \mathbf{r}\partial \mathbf{r}} + \frac{\partial^2 \mathbf{r}^{(n+2)}_{J_2-J_6}}{\partial \mathbf{r}\partial \mathbf{r}} + \frac{\partial^2 \mathbf{r}^{(n+2)}_{dr}}{\partial \mathbf{r}\partial \mathbf{r}}$; $\frac{\partial^2 \mathbf{r}^{(n+2)}_{per}}{\partial \mathbf{r}\partial \mathbf{r}^{(1)}} = \frac{\partial^2 \mathbf{r}^{(n+2)}}{\partial \mathbf{r}\partial \mathbf{r}^{(1)}} + \frac{\partial^2 \mathbf{r}^{(n+2)}_{J_2-J_6}}{\partial \mathbf{r}\partial \mathbf{r}^{(1)}} + \frac{\partial^2 \mathbf{r}^{(n+2)}_{dr}}{\partial \mathbf{r}\partial \mathbf{r}^{(1)}}$;

$\frac{\partial^2 \mathbf{r}^{(n+2)}_{per}}{\partial \mathbf{r}^{(1)}\partial \mathbf{r}} = \frac{\partial^2 \mathbf{r}^{(n+2)}}{\partial \mathbf{r}^{(1)}\partial \mathbf{r}} + \frac{\partial^2 \mathbf{r}^{(n+2)}_{J_2-J_6}}{\partial \mathbf{r}^{(1)}\partial \mathbf{r}} + \frac{\partial^2 \mathbf{r}^{(n+2)}_{dr}}{\partial \mathbf{r}^{(1)}\partial \mathbf{r}}$; $\frac{\partial^2 \mathbf{r}^{(n+2)}_{per}}{\partial \mathbf{r}^{(1)}\partial \mathbf{r}^{(1)}} = \frac{\partial^2 \mathbf{r}^{(n+2)}}{\partial \mathbf{r}^{(1)}\partial \mathbf{r}^{(1)}} + \frac{\partial^2 \mathbf{r}^{(n+2)}_{J_2-J_6}}{\partial \mathbf{r}^{(1)}\partial \mathbf{r}^{(1)}} + \frac{\partial^2 \mathbf{r}^{(n+2)}_{dr}}{\partial \mathbf{r}^{(1)}\partial \mathbf{r}^{(1)}}$;

Get $\frac{\partial^2 f^{(n+1)}}{\partial \mathbf{r}\partial \mathbf{r}}$, $\frac{\partial^2 f^{(n+1)}}{\partial \mathbf{r}\partial \mathbf{r}^{(1)}}$, $\frac{\partial^2 f^{(n+1)}}{\partial \mathbf{r}^{(1)}\partial \mathbf{r}}$, $\frac{\partial^2 f^{(n+1)}}{\partial \mathbf{r}^{(1)}\partial \mathbf{r}^{(1)}}$; // ▷ Eq. (27)

Get $\frac{\partial^2 g_p^{(n+1)}}{\partial \mathbf{r}\partial \mathbf{r}}$, $\frac{\partial^2 g_p^{(n+1)}}{\partial \mathbf{r}\partial \mathbf{r}^{(1)}}$, $\frac{\partial^2 g_p^{(n+1)}}{\partial \mathbf{r}^{(1)}\partial \mathbf{r}}$, $\frac{\partial^2 g_p^{(n+1)}}{\partial \mathbf{r}^{(1)}\partial \mathbf{r}^{(1)}}$; // ▷ Eq. (28)

Get $\frac{\partial^2 B_p^{(n+1)}}{\partial \mathbf{r}\partial \mathbf{r}}$, $\frac{\partial^2 B_p^{(n+1)}}{\partial \mathbf{r}\partial \mathbf{r}^{(1)}}$, $\frac{\partial^2 B_p^{(n+1)}}{\partial \mathbf{r}^{(1)}\partial \mathbf{r}}$, $\frac{\partial^2 B_p^{(n+1)}}{\partial \mathbf{r}^{(1)}\partial \mathbf{r}^{(1)}}$; // ▷ Eq. (43)

Get $\frac{\partial^2 C_\alpha^{(n+1)}}{\partial \mathbf{r}\partial \mathbf{r}}$, $\frac{\partial^2 C_\alpha^{(n+1)}}{\partial \mathbf{r}\partial \mathbf{r}^{(1)}}$, $\frac{\partial^2 C_\alpha^{(n+1)}}{\partial \mathbf{r}^{(1)}\partial \mathbf{r}}$, $\frac{\partial^2 C_\alpha^{(n+1)}}{\partial \mathbf{r}^{(1)}\partial \mathbf{r}^{(1)}}$; // ▷ Eq. (45)

Get $\phi^2_{111}(t+dT,t)$, $\phi^2_{112}(t+dT,t)$, $\phi^2_{121}(t+dT,t)$, $\phi^2_{122}(t+dT,t)$, $\phi^2_{211}(t+dT,t)$, $\phi^2_{212}(t+dT,t)$, $\phi^2_{221}(t+dT,t)$, $\phi^2_{222}(t+dT,t)$; // ▷ Eq. (24)

$\phi^2_{111per}(t+dT,t) = \phi^2_{111}(t+dT,t) + \sum_{m=0}^{n}\left(\frac{\partial^2 \mathbf{r}^{(m)}_{J_2-J_6}(t)}{\partial \mathbf{r}(t)\partial \mathbf{r}(t)} + \frac{\partial^2 \mathbf{r}^{(m)}_{dr}(t)}{\partial \mathbf{r}(t)\partial \mathbf{r}(t)}\right)\frac{dT^{(m)}}{m!}$;

$\phi^2_{112per}(t+dT,t) = \phi^2_{112}(t+dT,t) + \sum_{m=0}^{n}\left(\frac{\partial^2 \mathbf{r}^{(m)}_{J_2-J_6}(t)}{\partial \mathbf{r}(t)\partial \mathbf{r}^{(1)}(t)} + \frac{\partial^2 \mathbf{r}^{(m)}_{dr}(t)}{\partial \mathbf{r}(t)\partial \mathbf{r}^{(1)}(t)}\right)\frac{dT^{(m)}}{m!}$;

$\phi^2_{121per}(t+dT,t) = \phi^2_{121}(t+dT,t) + \sum_{m=0}^{n}\left(\frac{\partial^2 \mathbf{r}^{(m)}_{J_2-J_6}(t)}{\partial \mathbf{r}^{(1)}(t)\partial \mathbf{r}(t)} + \frac{\partial^2 \mathbf{r}^{(m)}_{dr}(t)}{\partial \mathbf{r}^{(1)}(t)\partial \mathbf{r}(t)}\right)\frac{dT^{(m)}}{m!}$;

$\phi^2_{122per}(t+dT,t) = \phi^2_{122}(t+dT,t) + \sum_{m=0}^{n}\left(\frac{\partial^2 \mathbf{r}^{(m)}_{J_2-J_6}(t)}{\partial \mathbf{r}^{(1)}(t)\partial \mathbf{r}^{(1)}(t)} + \frac{\partial^2 \mathbf{r}^{(m)}_{dr}(t)}{\partial \mathbf{r}^{(1)}(t)\partial \mathbf{r}^{(1)}(t)}\right)\frac{dT^{(m)}}{m!}$;

$\phi^2_{211per}(t+dT,t) = \phi^2_{211}(t+dT,t) + \sum_{m=1}^{n}\left(\frac{\partial^2 \mathbf{r}^{(m)}_{J_2-J_6}(t)}{\partial \mathbf{r}(t)\partial \mathbf{r}(t)} + \frac{\partial^2 \mathbf{r}^{(m)}_{dr}(t)}{\partial \mathbf{r}(t)\partial \mathbf{r}(t)}\right)\frac{dT^{(m-1)}}{(m-1)!}$;

$\phi^2_{212per}(t+dT,t) = \phi^2_{212}(t+dT,t) + \sum_{m=1}^{n}\left(\frac{\partial^2 \mathbf{r}^{(m)}_{J_2-J_6}(t)}{\partial \mathbf{r}(t)\partial \mathbf{r}^{(1)}(t)} + \frac{\partial^2 \mathbf{r}^{(m)}_{dr}(t)}{\partial \mathbf{r}(t)\partial \mathbf{r}^{(1)}(t)}\right)\frac{dT^{(m-1)}}{(m-1)!}$;

$\phi^2_{221per}(t+dT,t) = \phi^2_{221}(t+dT,t) + \sum_{m=1}^{n}\left(\frac{\partial^2 \mathbf{r}^{(m)}_{J_2-J_6}(t)}{\partial \mathbf{r}^{(1)}(t)\partial \mathbf{r}(t)} + \frac{\partial^2 \mathbf{r}^{(m)}_{dr}(t)}{\partial \mathbf{r}^{(1)}(t)\partial \mathbf{r}(t)}\right)\frac{dT^{(m-1)}}{(m-1)!}$;

$\phi^2_{222per}(t+dT,t) = \phi^2_{222}(t+dT,t) + \sum_{m=1}^{n}\left(\frac{\partial^2 \mathbf{r}^{(m)}_{J_2-J_6}(t)}{\partial \mathbf{r}^{(1)}(t)\partial \mathbf{r}^{(1)}(t)} + \frac{\partial^2 \mathbf{r}^{(m)}_{dr}(t)}{\partial \mathbf{r}^{(1)}(t)\partial \mathbf{r}^{(1)}(t)}\right)\frac{dT^{(m-1)}}{(m-1)!}$;

if $t + dT = T$ **then**
  break;
else if $t + dT > T$ **then**
  dT = T - t ;
  continue;
else
  Get N ; // ▷ Eq. (9)
  continue;

References

1. Kessler, D.J., Johnson, N.L., Liou, J., Matney, M.: The kessler syndrome: implications to future space operations. Adv. Astronaut. Sci. **137**(8), 2010 (2010)
2. Probe, A., Elgohary, T.A., Junkins, J.L.: A new method for space objects probability of collision. In: AIAA/AAS Astrodynamics Specialist Conference, p. 5653 (2016)
3. Probe, A., Elgohary, T.A., Junkins, J.L.: Orbital probability of collision using orthogonal polynomial approximations. In: 1st IAA Conference on Space Situational Awareness (ICSSA) (2017)
4. Battin, R.H.: An introduction to the mathematics and methods of astrodynamics, revised edition. American Institute of Aeronautics and Astronautics (1999)
5. Geller, D.K.: Linear covariance techniques for orbital rendezvous analysis and autonomous onboard mission planning. J. Guid. Control Dyn. **29**(6), 1404–1414 (2006)
6. Sabol, C., Hill, K., Alfriend, K., Sukut, T.: Nonlinear effects in the correlation of tracks and covariance propagation. Acta Astronaut. **84**, 69–80 (2013)
7. Vittaldev, V., Russell, R.P.: Space object collision probability via Monte Carlo on the graphics processing unit. J. Astronaut. Sci. **64**(3), 285–309 (2017)
8. Yang, Z., Luo, Y.-Z., Zhang, J.: Nonlinear semi-analytical uncertainty propagation of trajectory under impulsive maneuvers. Astrodynamics **3**(1), 61–77 (2019)
9. Vittaldev, V., Russell, R.P., Linares, R.: Spacecraft uncertainty propagation using gaussian mixture models and polynomial chaos expansions. J. Guid. Control Dyn. 2615–2626 (2016)
10. Giza, D., Singla, P., Jah, M.: An approach for nonlinear uncertainty propagation: Application to orbital mechanics. In: AIAA Guidance, Navigation, and Control Conference, p. 6082 (2009)
11. Junkins, J.L., Majji, M., Turner, J.D.: High order keplerian state transition tensors. In: Proceedings of the F. Landis Markley Astronautics Symposium, AAS Cambridge, Maryland, pp. 169–186 (2008)
12. Elgohary, T.A., Turner, J.D.: State transition tensor models for the uncertainty propagation of the two-body problem. Adv. Astronaut. Sci.: AAS/AIAA Astrodyn. Conf. **150**, 1171–1194 (2014)
13. Lantoine, G., Russell, R.P.: A fast second-order algorithm for preliminary design of low-thrust trajectories. In: 59th International Astronautical Congress, Glasgow, Scotland, vol. 29 (2008)
14. Younes, A.B., Turner, J., Majji, M., Junkins, J.: High-order uncertainty propagation using state transition tensor series. In: Jer-Nan Juang Astrodynamics Symposium, Univelt, Inc., San Diego, CA, No. AAS, pp. 12–636 (2012)
15. Turner, J.: OCEA user manual. Amdyn System (2006)
16. Alhulayil, M., Younes, A.B., Turner, J.D.: Higher order algorithm for solving lambert's problem. J. Astronaut. Sci. **65**(4), 400–422 (2018)
17. Bani Younes, A.: Exact computation of high-order state transition tensors for perturbed orbital motion. J. Guid. Control Dyn. **42**(6), 1365–1371 (2019)
18. Turner, J., Elgohary, T., Majji, M., Junkins, J.: High accuracy trajectory and uncertainty propagation algorithm for long-term asteroid motion prediction. In: K. Alfriend, M. Akella, J. Hurtado, J. Turner (eds.) Adventures on the Interface of Mechanics and Control, pp. 15–34 (2012)
19. Hernandez, K., Read, J.L., Elgohary, T.A., Turner, J.D., Junkins, J.L.: Analytic Power Series Solutions for Two-body and J2–J6 Trajectories and State Transition Models. In: Advances in Astronautical Sciences: AAS/AIAA Astrodynamics Specialist Conference (2015)
20. Hernandez, K., Elgohary, T.A., Turner, J.D., Junkins, J.L.: Analytic continuation power series solution for the two-body problem with atmospheric drag. In: Advances in Astronautical Sciences Spaceflight Mechanics, vol. 158 (2016a)
21. Hernandez, K., Elgohary, T.A., Turner, J.D., Junkins, J.L.: A novel analytic continuation power series solution for the perturbed two-body problem. Celest. Mech. Dyn. Astron. **131**(10), 48 (2019)

22. Tasif, T.H., Elgohary, T.A.: A high order analytic continuation technique for the perturbed two-body problem state transition matrix. In: Advances in Astronautical Sciences: AAS/AIAA Space Flight Mechanics Meeting (2019)
23. Tasif, T.H., Elgohary, T.A.: An adaptive analytic continuation method for computing the perturbed two-body problem state transition matrix. J. Astronaut. Sci. **67**(4), 1412–1444 (2020)
24. Tasif, T.H., Elgohary, T.A.: An adaptive analytic continuation technique for the computation of the higher order state transition tensors for the perturbed two-body problem. AIAA Scitech 2020 Forum, p. 0958 (2020b)
25. Tasif, T.H., Hippelheuser, J.E., Elgohary, T.A.: Analytic continuation extended kalman filter framework for space-based inertial orbit estimation via a network of observers. In: IAA 7th Annual Space Traffic Management Conference (2021)
26. Abad, A., Barrio, R., Blesa, F., Rodríguez, M.: Algorithm 924: TIDES, a Taylor series integrator for differential equations. ACM Trans. Math. Softw. (TOMS) **39**(1), 5 (2012)
27. Schaub, H., Junkins, J.L.: Analytical mechanics of space systems. American Institute of Aeronautics and Astronautics (2005)
28. Hernandez, K., Elgohary, T.A., Turner, J.D., Junkins, J.L.: Analytic continuation power series solution for the two-body problem with atmospheric drag. In: Advances in Astronautical Sciences: AAS/AIAA Space Flight Mechanics Meeting, pp. 2605–2614 (2016b)
29. David, A.V., McClain, W.: Fundamentals of astrodynamics and applications. The Space Technology Library, California (2013)
30. Read, J.L., Younes, A.B., Macomber, B., Turner, J., Junkins, J.L.: State transition matrix for perturbed orbital motion using modified Chebyshev Picard iteration. J. Astronaut. Sci. **62**(2), 148–167 (2015)

Research on Dynamic Pressure Sensor Based on ZigBee Technology

Tao Li, Ying Wu, and Yanxi Yu

Abstract With the rapid development of information technology, sensor technology has played a positive role in daily life and social production. Aiming at the research of safety performance and human resource cost saving in some specific places, combined with the intelligent and industrial development steps in modern society, wireless pressure sensor has gradually become a brand-new technology product. Therefore, on the basis of understanding the status of ZigBee technology development, according to the ZigBee technology as the core of the dynamic pressure sensor system composition, respectively from the hardware design and software design to study how to achieve, in order to lay a foundation for the development of technology in various fields.

Keywords ZigBee technology · Dynamic pressure sensor · Hardware · Software · Communication protocol · Node

1 Introduction

To put it simply, ZigBee technology is a technical standard developed with 802.15.4 wireless standard approved by IEEE as the core. It is mainly used in devices with low power consumption, low cost and low propagation rate, and can use one point to multipoint for fast networking. From a practical point of view, the use of ZigBee technology has strong compatibility and security. In IEEE802.15.4 wireless network, devices can be divided into two types according to their communication capabilities. One is a full-function device (FFD), and the other is a thin function device (RFD). The former can communicate with all types of functional devices, while the latter can only communicate with FFD [1–3]. Therefore, a compact functional device can only be used for simple control applications. It transmits less information and data and occupies few types of resources. It is usually used as a communication terminal in the network structure. ZigBee technology protocol suite is simple and compact, during

T. Li (✉) · Y. Wu · Y. Yu
Chongqing Academy of Metrology and Quality Inspection, Chongqing, China
e-mail: ltxiaocong@163.com

H. Dai (ed.), *Computational and Experimental Simulations in Engineering*, Mechanisms and Machine Science 119,
https://doi.org/10.1007/978-3-031-02097-1_24

the implementation of the proposed hardware requirements are low, only need to use eight-bit microprocessor can meet the basic requirements, full function protocol software needs to use 32 K bytes of ROM, minimum power protocol software to use 4 K bytes of ROM. ZigBee technical equipment has the characteristics of low power consumption, the practice of communication distance is mainly controlled within 100 m, mainly with two kinds of capabilities, one is the link quality indication ability, the other is the energy detection ability. Using the detection results obtained by these two capabilities, ZigBee technology equipment can autonomously adjust the transmitted power, and effectively control the equipment energy loss on the basis of meeting the communication link quality requirements. From the point of view of networking performance, technology can build point-to-point or star network structure, which has great capacity in practice [4–6].

2 Methods

The design structure of the dynamic pressure sensor with ZigBee technology as the core is mainly divided into two modules. On the one hand, it refers to the field acquisition transmission module, on the other hand, it refers to the monitoring center, and the communication area needs to use ZigBee wireless network protocol for design and implementation. Generally speaking, the monitoring system is installed in the monitoring dispatch department, which includes the ZigBee network central node and the running computer system. The pressure sensor is used to detect the pressure information in the area to be measured, and the working condition in the field is transformed into a specific value of voltage or current, which is then transmitted to the wireless data collector through the ZigBee communication module. At the bottom of the pressure sensor should focus on the pressure signal, when collecting information and the data acquisition system is to use RF chip CC2430, collecting and transmission circuit design, and transfer data to obtain the information out, collect data results show that the presented in the LCD screen, also can use RS-485 communication interface and PC connection communication. At the same time, the data acquisition module will use the converter ADC to quantize and encode the acquired field signals, convert them into digital signals, and then transmit them to the microprocessor. The main control unit mainly receives and processes the relevant data. The specific structure is shown in Fig. 1 as follows.

2.1 Hardware

When designing the hardware, the RF chip CC2430 should be regarded as the core device. It is mainly packaged with 7 mm × 7mm QLP, which contains 48 pins. All pins are divided into three types: one refers to the I/O port, one refers to the

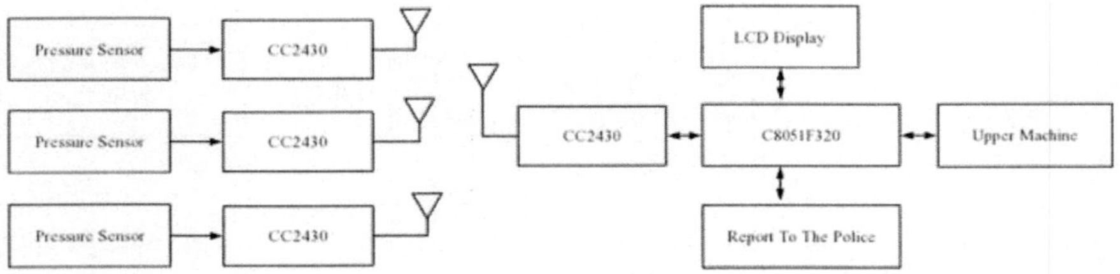

Fig. 1 System structure

power cord, and the last refers to the control line. CC2430, as a new ZigBee wireless microcontroller series chip, not only contains RF transceiver, but also installed enhanced 8051 microcontroller. In practice, it is mainly in the 2.4 GHz frequency band, and the power supply voltage is mainly controlled between 2.0 and 3.6 V, and the current consumption in standby state is only 0.2 UA. Battery life can reach more than six months, the highest transfer rate can reach 250 kbit/s. In practice, CC2430 only needs to add a few peripheral components to achieve the hardware design of ZigBee wireless communication function. From the point of view of data acquisition and transmission hardware, it contains several modules, such as ZigBee data analysis, data transmission, power supply, pressure sensor, etc. The pressure sensor will transmit the pressure signal collected in the field to the ZigBee circuit, and the ZigBee circuit will conduct in-depth analysis and processing of the relevant data and information. It should be noted that the ZigBee transmission module will convert the pressure signal into an electrical signal and the base station to complete the communication, and the power supply needs to use two 1.5 V alkaline batteries. The circuit principle of the transmitting module for data acquisition is shown in Fig. 2.

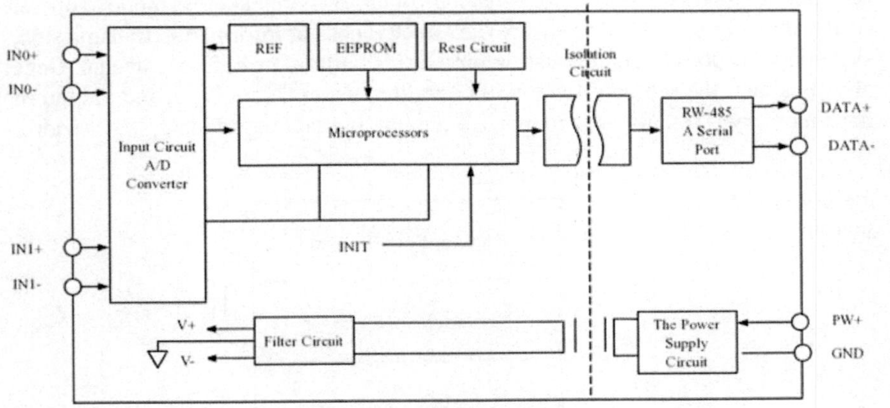

Fig. 2 Transmission module circuit principle of data acquisition

Fig. 3 Node frame diagram

2.2 *Software*

First, node design. Firstly, the RF chip CC2430 is studied by using the functions of the physical layer and MAC layer of IEEE802.15.4 wireless standardization. Second, the upper layer protocol should use the mature Z-stack protocol Stack; Thirdly, during the design of user programs, we should use the standard ZigBee technical specifications and API functions provided by various manufacturers to realize the system functions, and thus complete the networking development; Finally, design the TinyOS system. The specific framework is shown in Fig. 3.

Second, protocol stack. The systematic ZigBee protocol station is divided into four layers: the first is the physical layer, the second is the MAC layer, the second is the network layer and the security layer, and the last is the application layer, the specific structure is shown in Fig. 4.

According to the proposed IEEE802.15.4 wireless standardization, the protocols of the physical layer and MAC layer are defined, while the network layer and the security city are made by the ZigBee Alliance, and the application layer needs to be developed and utilized according to the user needs. In the development of wireless communication technology, in order to solve the problems between radio carriers, designers use the conflict-free multi-carrier channel access mode to operate. At the same time, in order to ensure the effectiveness of information transmission, a systematic response communication protocol is built in the software design. Generally speaking, the protocol stack will be written using C language, and the internal flash memory is mainly used to manage the binding table, grid table, MAC address,

Fig. 4 Protocol stack structure

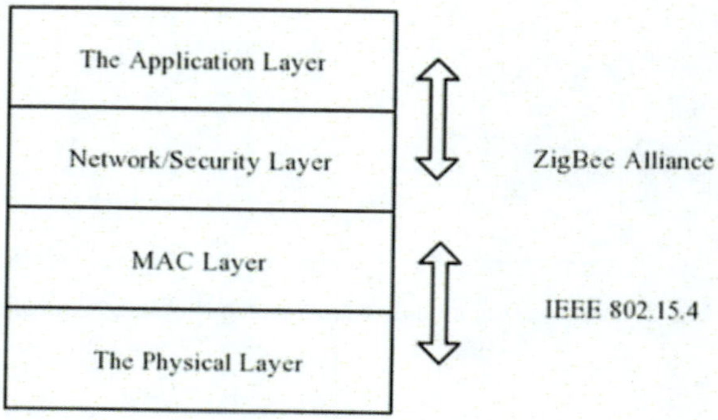

so during the design to use programmable flash memory microcontroller. In combination with the stack structure diagram shown in Fig. 3, the logic should be divided into multiple layers in combination with the canonical ZigBee definition, and the code for each layer should be in a separate source file, while the services and application interfaces need to be defined in header files. In addition, to ensure that the system is modular and abstract, the top-level design of the C header files define all the APL functions supported. The user application and the support sub-layer (APS) and application layer (APL) interact with each other. Representative applications will always be connected to the application layer and its interfaces. The APL module will provide advanced protocol stack management functions for the system operation, while the user application, using this module, can carry out scientific control of related functions. The application support sublayer mainly provides the endpoint interface. This layer can be used to open or close multiple ports during the application, to obtain or transmit relevant data information, and to provide original statements for key values and packet data transmission. When the binding table is in an empty state during the first programming to the coordinator, the main application must adjust the appropriate binding API functions to build new binding items. APS also includes the indirect send buffer RAM, which is used to store and manage indirect frames until the target receiver requests the relevant frames. As the time of the number of node requests increases, the time for data packets to be stored in the indirect sending buffer will also increase. The longer the time, the larger the buffer space required [7–9].

Third, communication protocol. ZigBee wireless network can use a variety of types of network configuration, the system uses the star network for communication connection, the network configuration includes a coordinator and a number of terminal nodes, the relevant terminal devices can only communicate with the coordinator. In order to achieve this goal, the coordinator must master all acquisition node network addresses, and each node in the network address before adding network can provide a coordinator, the coordinator will store data into the system, and thus build a address table, so convenient for the user in the process of collecting data in accordance with the address table to collect information. The maximum length of A MAC packet is 127 bytes. All data contains the header byte and 16 CRC values. During transmission, the response data transmission mechanism needs to be used for processing, and the frame with ACK bit of 1 should be designed to be answered by the receiver. Assuming no response is obtained within a certain range, then it represents a problem with the collection node. The operation flow chart of THE CC2430 software of the main control unit is as follows (Fig. 5).

In the case of the coordinator to obtain information, according to the first identification character of the data to determine the type of data analysis, one is the sensor bucket network address, the other is the sensor to obtain data. Assuming the sensor's network address then that means the content will be stored in the address table. Assuming that it is the data obtained by the sensor, it should be uploaded to C0851F320 after completing the pre-processing, and wait for all the sensors in the monitoring area to collect the data. On the basis of effective fusion of internal data, the final result will be presented on the LCD screen. After the system user uses the

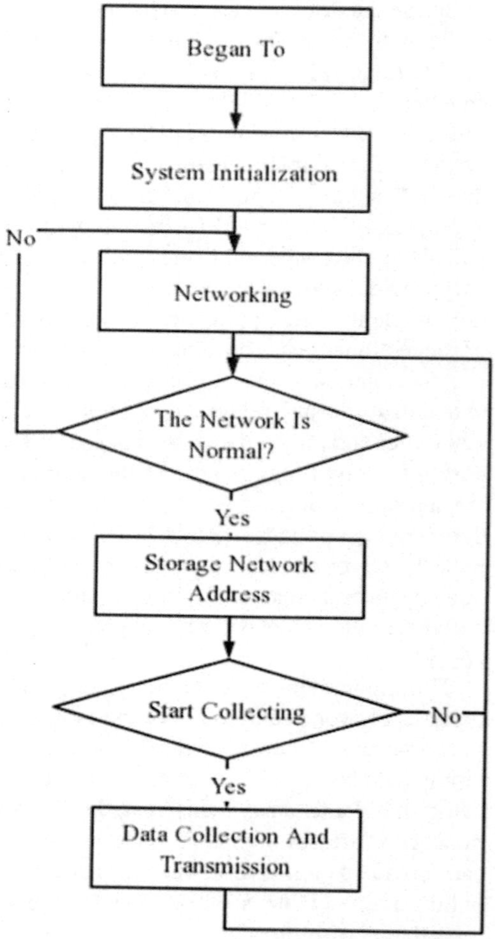

Fig. 5 Operation flow chart of CC2430 software of main control unit

upper computer monitoring system to put forward the communication request, the single chip microcomputer will transfer the effective data to the upper computer in the window, and the user can use the upper computer to collect and draw charts and statistical analysis and other basic work.

3 Result Analysis

From a practical point of view, ZigBee technology, as a new form of short-range wireless communication application, has strong security in practice, and can effectively control power consumption and cost expenditure, which has a positive role in the development of current information technology. By using ZigBee technology to build a wireless network, and thus get a systematic infinite dynamic pressure sensor, not only can solve the previous industrial field wiring design problems, but also can accelerate the pace of communication technology research and development. Science and technology personnel in the practice of research and integration of advanced ZigBee protocol stack and tool kit at home and abroad as a reference design, not only put forward a representative ZigBee solution, but also laid a foundation for the development of high performance and low consumption wireless products. Nowadays, as enterprises strengthen their own technology research and development efforts at the same time, more and more people realize the value of ZigBee technology application, and the relevant ZigBee alliance standards have been standardized and improved, the corresponding chip system price will gradually drop. I believe that with the wide use of ZigBee technology products, each field in the rapid development of more high-quality innovative ideas can be put forward. At present, the dynamic pressure sensor with ZigBee technology as the core is mainly used in wireless induction network, industrial control, wireless tracking and other fields, and has obtained excellent results in practice and exploration [10–13].

4 Conclusion

To sum up, in this paper, we study the ZigBee technology as the core system of wireless dynamic pressure sensor, need to use short transmission system node implementations, and will obtain the high precision numerical simulation data into pressure, for the corresponding numerical in conversion chip, and then use wireless communication technology to transform data values into wireless data receiver node, Finally, the design of wireless pressure sensor is completed. Compared with other short-range wireless transmission technologies, ZigBee technology can accelerate the transmission speed and expand the transmission range at the same time. The practical operation is very simple and effective.

Acknowledgments Preparation of calibration specification for fire hydrant hydraulic test machine and development of calibration device cstc2020jxjl120011.

References

1. Jiang, S., Xu, H., Zhang, L.: Research on wireless pressure sensor based on ZigBee technology. Instrum. Tech. Sens. (0Z1), 212–214 (2009)
2. Yang, Y., Wu, X., Cen, R.: Development of wearable pulse wave detection module based on Zigbee technology. Chin. J. Sens. Sens. **2009**(11), 1538–1541 (2009)
3. Zhou, J., Peng, Y., Chang, P., et al.: Research on pressure sensor calibration system based on ZIGBEE. Electron. Des. Eng. **2012**(11), 61–63 (2012)

4. Kuang, X.-B., Song, C.-Z., Xiong, Z.-Y., et al.: Design and implementation of wireless pressure sensor system based on ZigBee. Digit. Technol. Appl. **2017**(8), 147–148 (2017)
5. Xu, Z., Ren, Z.: Application research of ZigBee pressure sensor based on CC2530. Electron. Technol. **046**(009), 35–36, 34 (2017)
6. Yang, Y., Wu, X., Cen, R.: Development of a wearable pulse monitoring module based on Zigbee technology. Chin. J. Sens. Actuat. **022**(011), 1538–1541 (2009)
7. Cui, J., Li, X.: Research and application of Zigbee wireless technology in oilfield water injection well pressure monitoring. (2012)
8. Zeng, X.: Research on coal mine gas monitoring system based on ZigBee technology. East China Sci. Technol. (Comprehensive) (012), 380–380 (2018)
9. Zhao, M., Cheng, Y.: Simulation experimental Platform for Monitoring mine Anchor cables stress Based on ZigBee. J. Exp. Technol. Manag. **2014**(8), 159–163 (2014)
10. Zhang, L., Sun, X.: Design and research of internet of things intelligent coal mine safety monitoring system based on Zigbee. J. Hebei Inst. Civil Eng. Arch. **036**(002), 124–127 (2018)
11. Ren, B., Li, T., Li, X.: Research on dynamic inertial estimation technology for deck deformation of large ships. Sensors **19**(19), 4167 (2019)
12. Lin, Y., Yong, Z., Yingjin, L., et al.: Research on dynamic pricing of supply chain products based on channel advantages. Kybernetes **41**(9), 1377–1385(9) (2012)
13. Alemdar, H., Ersoy, C.: Wireless sensor networks for healthcare: a survey. Comput. Netw. **54**(15), 2688–2710 (2010)

Application of POA Algorithm in Optimal Operation of Reservoir Flood Control and Water Storage

Wenlong Dua and Hengfei An

Abstract By understanding step by step optimization algorithm and the reservoir flood control optimization scheduling model, in addressing the problem of local optimal acquired experience, combined with practical computing needs water balance as the core of the improved algorithm is put forward, thus preventing the total decision-making sequence in overly complex constraint conditions is divided into multiple subsequence of the role of each other is not, to ensure application algorithm is more rapid, It is not easy to fall into local optimization problems. Based on understanding the current research status on the basis of reservoir flood operations, according to the actual building model and the improved algorithm, to verify this practice, the application of the case analysis, the final result shows step by step optimization algorithm can guarantee the water balance at the same time, using piecewise compensation as the core of the improved algorithm under complex conditions, accurate calculation of reservoir operation, and the results.

Keywords Stepwise optimization algorithm · Water storage · Reservoir flood control · Optimal operation

1 Introduction

Reservoir flood control in the development of city construction of complex multi-stage and multi-objective decision-making process, the normalized operation mainly using the theory and experience, according to different situations of protection objects, the flood control level, the library chooses appropriate scheduling, such as the flood change rule is simple and intuitive operation, but still can't show the maximum effect of reservoir operation, Moreover, it is difficult to deal with the

W. Dua (✉)
Yellow River Engineering Consulting Co., Ltd., Zhengzhou 450003, China
e-mail: duanwenl@163.com

H. An
Key Laboratory of Water Management and Water Security for Yellow River Basin, Ministry of Water Resources (Under Construction), Zhengzhou 450003, China

H. Dai (ed.), *Computational and Experimental Simulations in Engineering*, Mechanisms and Machine Science 119,
https://doi.org/10.1007/978-3-031-02097-1_25

problem of reservoir group flood control dispatching. In practice, researchers put forward the optimal scheduling scheme, that is, carry out dynamic programming and master the corresponding improved algorithm [1–3]. It should be noted that when the number of variables in dynamic planning continues to rise, the actual calculation will be hindered by the dimension, resulting in an increase in the actual calculation number. At the same time, when dynamic variables are required to conform to no after-effect in dynamic planning, such algorithms cannot be used in case of flood budget. According to the advantages and disadvantages of dynamic programming, H.R. Johnson and others proposed a stepwise optimization algorithm, also known as POA algorithm, in the mid-1970s. By transforming the multi-period problems into multiple two-period problems, the optimization process is repeated until the convergence conditions are met. This kind of algorithm has been widely used in reservoir optimal operation because of its small storage space, relatively stable overall model, high practical application efficiency and convenient programming. But from a practical point of view, POA algorithm is often affected by the initial solution and cannot converge to global optimization. In practice, Chinese researchers have proposed that if the decision objective function cannot satisfy strict convexity and has first order continuous partial derivative, the result will converge to local optimal. In view of this phenomenon, some researchers put forward effective improvement schemes, such as the initial solution based on the optimal results obtained by the piecewise algorithm for subsequent calculation and analysis, so as to avoid unnecessary influence of the initial solution on the final result; The penalty function can also be referenced in THE POA algorithm to calculate and solve in the optimal scheduling model to improve the convergence level and search efficiency of the algorithm. It can also be processed according to the algorithm of progressive difference and variable stage optimization improvement strategy. Compared with the traditional POA algorithm, it shows stronger global search ability. At present, the proposed improved algorithm is difficult to ensure the actual computational efficiency when dealing with high-risk complex problems, and there are few research topics and results for reservoir flood control optimization scheduling. Therefore, this paper starts from the main reason why POA algorithm falls into local optimal, and proposes an improved algorithm based on subsection compensation according to water balance. Thus, its application performance in practical reservoir flood control optimization problem is verified [4–6].

2 Method

2.1 Analysis of Optimal Operation Model for Reservoir Flood Control

In order to ensure the optimal operation of reservoir flood control can obtain the best benefit, it is necessary to study the optimal criterion first. By understanding the current state of the reservoir basin, the selected optimal criteria will inevitably affect the actual operation results. In this paper, when studying regional flood, the optimal criterion is the maximum peak-cutting criterion, in other words, the reservoir regulation and storage function is used to ensure the uniformity of water discharge from the reservoir, so as to guide the maximum value of discharge under the reservoir to be minimized. The corresponding objective function is as follows [7]:

First, in the case of reservoir cross-section and no interval inflow, the objective function is:

$$\min z = \int_{T0}^{tD} q_t^2 dt$$

Second, in the case of reservoir cross-section and interval inflow, the objective function is:

$$\min z = \int_{t0}^{tD} (q_t + q_{\text{area,t}})^2 dt$$

Third, in the case of downstream flood control section, the objective function is:

$$\min z = \int_{t0}^{tD} (q_{\text{prevent,t}})^2 dt$$

2.2 Improved POA Algorithm

In the scheduling process for two-time optimization, the decision variables must meet the optimization requirements of the objective function in continuous adjustment. According to the constraint analysis of the water balance, it is assumed that the decision variable represents the downstream discharge of the reservoir. In order to ensure the consistency of the overall downstream discharge, the change of the downstream discharge in this period must be compensated according to the downstream discharge in other periods. Therefore, in order to apply the step-by-step optimization algorithm reasonably in the optimal operation of reservoir flood control and not be affected by the initial value, the following rules should be proposed for the compensation of the downstream discharge by using the improved method of subsection compensation:

Taking the downstream discharge {QTN} as the decision sequence, the following formula can be obtained:

$$\Delta q = q_l^{k'} - q_l^{k-1}$$

In the above formula, QTK-1 represents the downstream discharge value after k-1 optimization at the moment; QTK represents the corresponding adjustment value in the KTH optimization. The optimization flow values after k times and K-1 times are uniformly marked as QT, so the lower discharge flow needs to be scientifically adjusted according to the following formula:

$$q_t \Leftarrow q_t - \frac{\Delta q}{n_{\text{modulate}}}(t \in D)$$

In the above formula, n_{modulate}. On behalf of the meet $t \in D$. The number of qt traffic values under this condition.

For interval D, the following conditions should be met:

First, $(l-1, l, l+1) \notin D$;

Secondly, it is assumed that the water level in all periods conforms to the maximum and minimum water level limits, so three aspects can be considered:

When the operation criterion represents the maximum peak clipping of the reservoir section and there is no interval inflow, the following formula can be obtained:

$$\begin{cases} q_1 > \overline{q_t} + (q_{\min} - \overline{q_t})\alpha(\Delta q > 0) \\ q_1 < \overline{q_t} + (q_{\max} - \overline{q_t})\alpha(\Delta q < 0) \end{cases}$$

In the above formula, $\overline{q_t}, q_{\min}, q_{\max}$. Represents the average value and the maximum and minimum value of the sequence, α represents the scaling coefficient, and the value range is mainly controlled between 0 and 1, which can be set to 0 if the constraint conditions are too loose. Assuming that the calculation results converge to the local optimum, the scaling coefficient can be appropriately increased.

When the operation criterion belongs to the maximum peak clipping of reservoir section and there is an interval inflow, the following formula can be obtained:

$$\begin{cases} q_t + q_{\text{area,t}} > \left(\overline{q_t + q_{\text{area,t}}}\right) + \left[\left(q_t + q_{\text{area,t}}\right)_{\min} - \left(\overline{q_t + q_{\text{area,t}}}\right)\right]\alpha(\Delta q > 0) \\ q_t + q_{\text{area,t}} < \left(\overline{q_t + q_{\text{area,t}}}\right) + \left[\left(q_t + q_{\text{area,t}}\right)_{\max} - \left(\overline{q_t + q_{\text{area,t}}}\right)\right]\alpha(\Delta q < 0) \end{cases}$$

It is assumed that the regulation criterion represents the maximum peak clipping of the downstream flood control section and there is a certain range of inflow. In this case, the flood change should be calculated and the influence of τ time on the numerical composition of the downstream flood control section under the condition that the discharge under the reservoir has the aftereffect is analyzed. The specific formula is as follows:

$$\begin{cases} q_{t+T} + q_{\text{area,t+T}} > \left(\overline{q_{t+T} + q_{\text{area,t+T}}}\right) + \left[\left(q_{t+T} + q_{\text{area,t+T}}\right)_{\min} - \left(\overline{q_{t+T} + q_{\text{area,t+T}}}\right)\right]\alpha(\Delta q > 0) \\ q_{t+T} + q_{\text{area,t+T}} < \left(\overline{q_{t+T} + q_{\text{area,t+T}}}\right) + \left[\left(q_{t+T} + q_{\text{area,t+T}}\right)_{\max} - \left(\overline{q_{t+T} + q_{\text{area,t+T}}}\right)\right]\alpha(\Delta q < 0) \end{cases}$$

Assuming that the water level in a certain period cannot meet the requirements of constraint conditions, the following requirements should be met in interval D besides the above conditions:

When the water level exceeds the constraint condition of maximum water level, the following equation can be obtained:

$$\begin{cases} IFl \in [t_0, t_{\max}] \; and \; \Delta q < 0 \; then \; t \in [t_0, t_{\max}] \\ IFl \in [t_{\max}, t_D] \; and \; \Delta q > 0 \; then \; t \in [t_{\max}, t_D] \end{cases}$$

In the case that the water level is lower than the constraint condition of the lowest water level, it can be obtained:

$$\begin{cases} IFl \in [t_0, t_{\min}] \; and \; \Delta q > 0 \; then \; t \in [t_{\min}, t_D] \\ IFl \in [t_{\min}, t_D] \; and \; \Delta q < 0 \; then \; t \in [t_0, t_{\min}] \end{cases}$$

In the formula above, $t_{\max}$. Represents the time when the reservoir reaches its highest water level, $t_{\min}$. Represents the time when the lowest water level of the reservoir appears.

3 Result Analysis

In order to verify that the above algorithm in the application of optimal scheduling model of reservoir flood control effectiveness, this paper carried out according to the typical flood process of certain place a large reservoir simulation scheduling analysis, follow the principles of maximum peak clipping is standard, the actual decision variable refers to {QTN}, decision-making sequence length reached $122, comprehensive consideration of various constraints, By comparing the traditional stepwise optimization algorithm and the improved optimization algorithm, the optimal training of the discharge flow is obtained. The conditions followed during the calculation are divided into the following points: first, the water level of the lifting reaches 310 m, and the water level of the final stage reaches 315 m; Second, the highest control water level reached 319.3 m, the lowest control water level reached 306.5 m; Third, the maximum discharge is uniformly marked as 10,800 m^3/s, and the output is comprehensively studied to ensure that the discharge will not be lower than 700 m^3/s. Fourth, outbound flow can control the range of change at about 500 m^3/s; Fifth, the initial solution is the process of reservoir inflow. The corresponding basic parameters of the reservoir are shown in the following Table 1.

From a practical point of view, reservoir flood control can be divided into three cases: first combined with Fig. 1 analysis shows that when there is no interval inflow in the reservoir section, at this time of the flood control constraint condition is very loose, whether it is a step by step optimization algorithm in the traditional sense or the results obtained the improved step by step optimization algorithm have consistency,

Table 1 Parameter design

Dead water level	Limited water level by	Normal storage level	Flood high water level	Designed flood level	Check the flood level
299	310	315	319.3	320.9	322.1

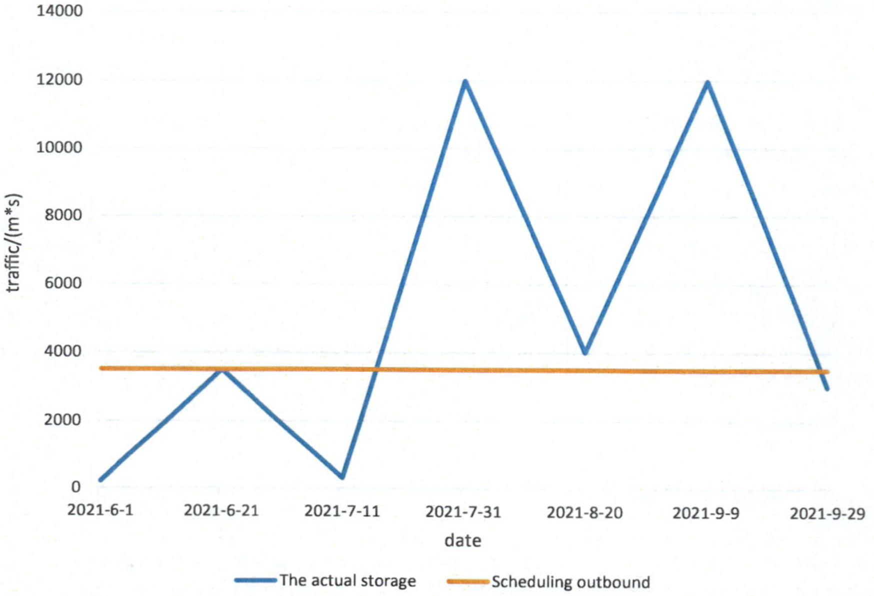

Fig. 1 Based on the analysis of calculation results without interval inflow

it is proved that both can guarantee the adjustment of the outbound traffic, has a balance, The highest position of the reservoir can reach 315.7 m, which meets the requirements of the objective function [8–10].

Secondly, combined with the comparison results shown in Figs. 2 and 3, the calculation under the condition of interval inflow of reservoir cross-section is studied. At this point, within the interval flood area with tighter constraints conditions, using the traditional optimization algorithm processing step by step, the top of the reservoir can reach 315.8 m, the actual overall decision sequence will be near the peak as a dividing line, thus obtaining more child sequence optimization processing, divided according to the actual result form to determine the sequence of state, In other words, it is more affected. By using the improved stepwise optimization algorithm, the highest water level of the reservoir can reach 315.5 m, which is better than the traditional algorithm. The outflow of the reservoir can obviously compensate the flood peak in the interval and guarantee the balance of the discharge under the non-flood peak area.

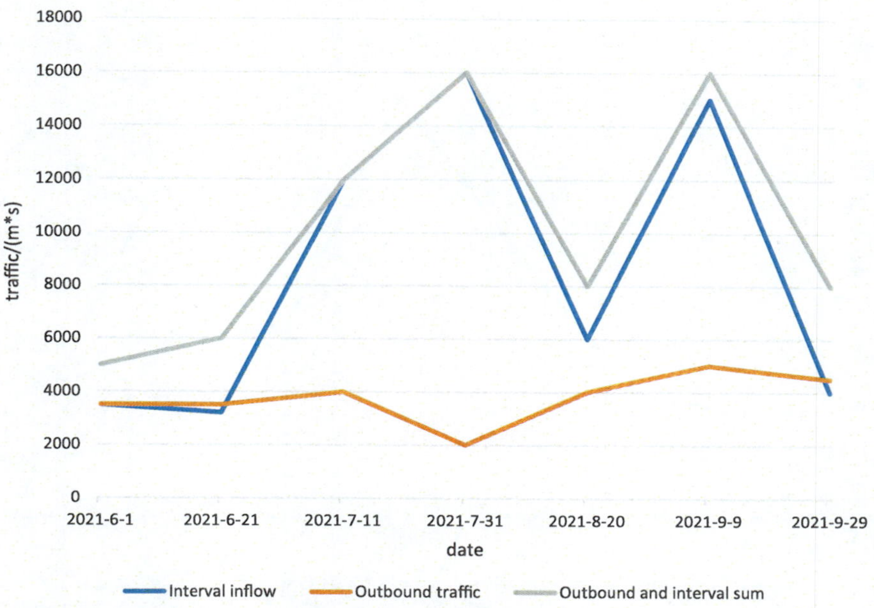

Fig. 2 Interval inflow calculation results of traditional POS algorithm

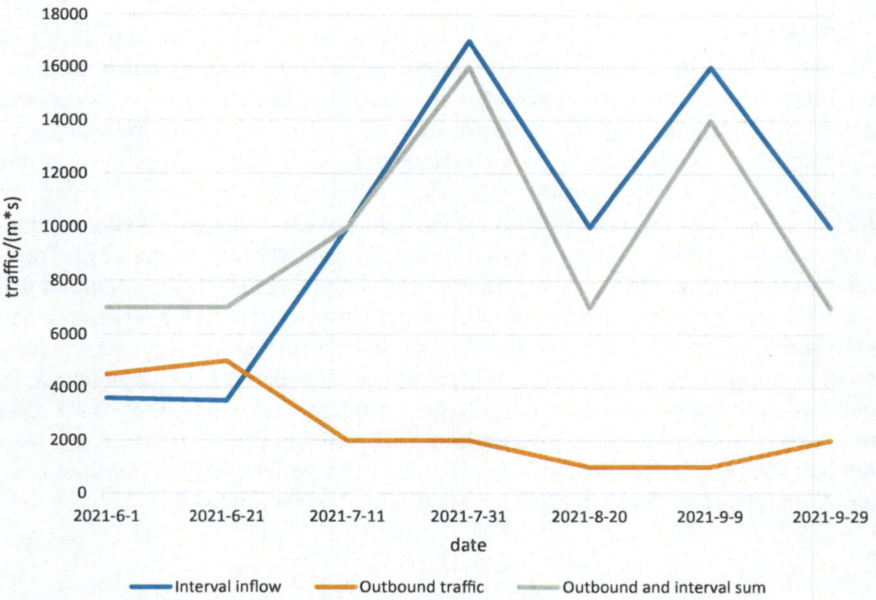

Fig. 3 Interval inflow calculation results of the improved POS algorithm

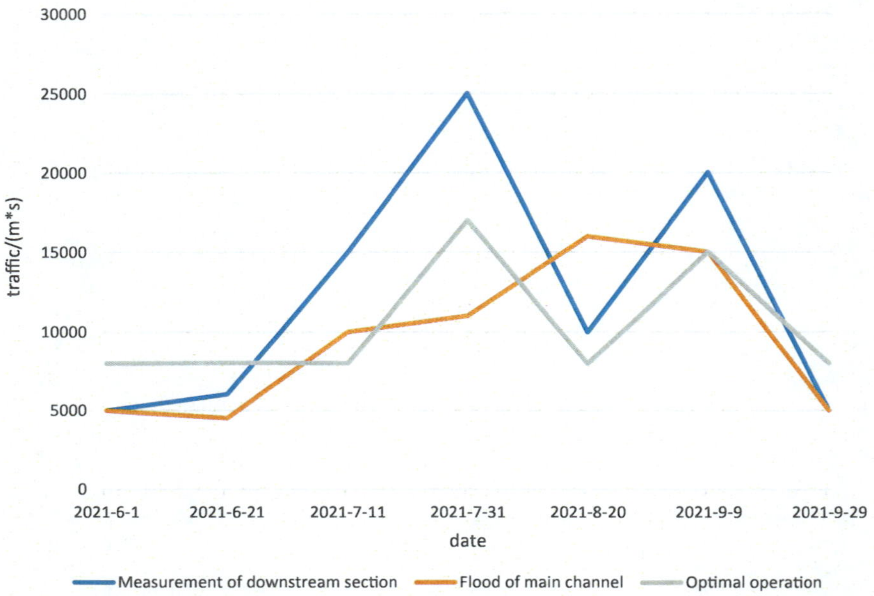

Fig. 4 Calculation results of interval inflow flood control section of traditional POS algorithm

Finally, the calculation of downstream flood control section should be analyzed based on Figs. 4 and 5 below. Because it is a representative double-branch flood, not only the flood peak is too concentrated, but also has a certain backside, which will have a negative impact on the downstream flood control. Similar to the scheduling situation of reservoir section, both the traditional optimization algorithm and the improved optimization algorithm can effectively control the flood peak, but the former is difficult to meet the requirements of the objective function, so it is difficult to ensure the balance of the flow of the section. By using the improved stepwise optimization algorithm, the reservoir will control the discharge under the condition that the main stream has flood peak, so as to prevent the superposition of flood peak. When the main flow is too low, the outflow flow of the actual reservoir will be increased, so as to control the flood pressure of the reservoir. Compared with the flood results obtained from actual measurement, the peak clipping rate can rise to 30.2% after optimized dispatching, and the water level will be reduced by 1.76 m. In this way, not only the application effect of peak clipping can be effectively demonstrated, but also the pressure of downstream flood control can be controlled.

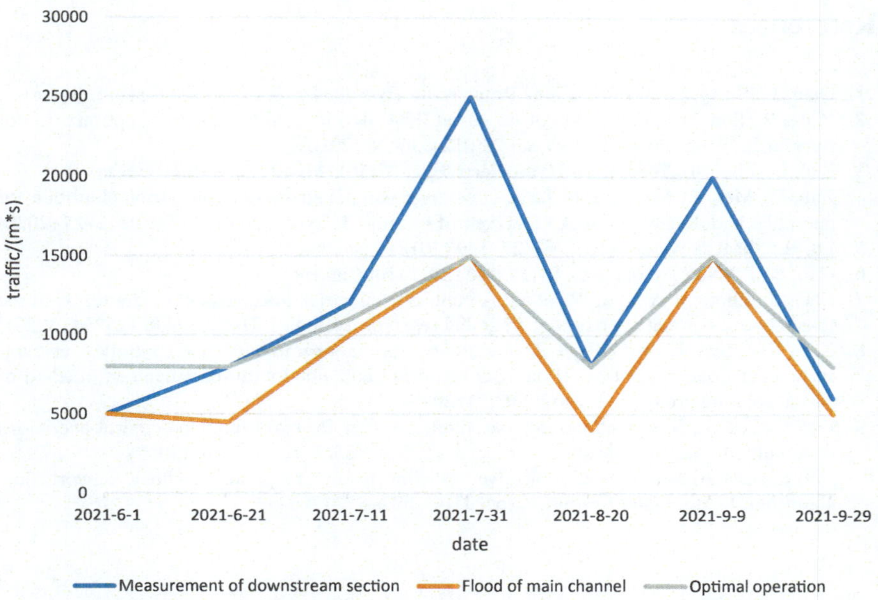

Fig. 5 Calculation results of interval inflow flood control section based on improved POS algorithm

4 Conclusion

In summary, according to the experimental comparison and analysis above, the traditional stepwise optimization algorithm is more dependent on the initial solution to deal with the problem, while the improved stepwise optimization algorithm puts forward improved countermeasures of segmental compensation while guaranteeing the water balance. In practical application, the latter can reasonably deal with the optimization scheduling of reservoir flood control and storage under too complicated circumstances, and show a strong role of flood blocking and peak cutting to ensure the balance of the discharge in the non-flood peak area, so as to control the influence of the initial solution on the application of the algorithm. It should be noted that the processing results of the optimal scheduling scheme are ideal, and it is difficult to meet the real-time flood control change requirements. Compared with the conventional scheduling scheme, more constraints need to be designed for practical analysis in order to improve the effectiveness of the actual operation.

References

1. Qian, J., Zhang, S., Xia, M.: China Rural Water Hydropower **8**, 22–25 (2014) (in Chinese)
2. Yang, B., Sun, W.: Application of improved POA algorithm in flood control optimization of river basin. Water Power Energy Sci **28**(012), 36–38 (2010)
3. Luo, C., Zhou, J., Yuan, L.: J. Hydroelectr. Eng. **37**(10), 41–49 (2018) (in Chinese)
4. Zhu, D., Mei, Y., Xu, X., Liu, Z.: A three-layer parallel stepwise optimization algorithm for optimal scheduling of complex flood control systems. J. Hydraulic Eng. **51**(10), 15–27 (2020)
5. Lu, H.: Shaanxi Water Resour. **6**, 127–129 (2018)
6. Guo, S.: J. Water Resour. Res. **01**(1), 1–6 (2012) (in Chinese)
7. Chen, J., Zhong, P.A., Liu, W., et al.: A multi-objective risk management model for real-time flood control optimal operation of a parallel reservoir system. J. Hydrol. **590**, 125264 (2020)
8. Kim, Y.G., Sun, B., Kim, P., et al.: A study on optimal operation of gate-controlled reservoir system for flood control based on PSO algorithm combined with rearrangement method of partial solution groups. J. Hydrol. **593**(7) (2020)
9. Wei, C.C., Hsu, N.S.: Optimal tree-based release rules for real-time flood control operations on a multipurpose multireservoir system. J. Hydrol. **365**(3–4), 213–224 (2009)
10. Brage, B., Barbosa, P.: Multiobjective real-time reservoir operation with a network flow algorithm. J. Am. Water Resour. Assoc. **37**(4), 837–852 (2001)

Research on Manipulator Control Based on RGB-D Sensor

Xiyuan Wan, Qingdong Luo, Yunhan Li, Jingjing Lou, and Pengfei Zheng

Abstract At present, the most mature way of teaching control of industrial robot manipulator is based on the control of teaching pendant joystick. With the development of human–computer interaction technology, more and more interaction technologies can be studied, such as voice control, gesture control and so on. This paper is devoted to the realization of new interaction methods and control logic, so that the manipulator can be controlled by gesture without contact. The implementation process of this method is as follows: firstly, get depth data by Kinect sensor [1], and use filter to reduce data jitter. Secondly, the user in the depth data is separated from the background to reduce the interference and the amount of calculation. Third, match depth data coordinate system and RGB image coordinate system. Use relative pixel coordinate in RGB image to detect the gesture. Fourth, the obtained coordinate data is transmitted to the rapid program of industrial robot through socket. Finally, the proposed method was tested and the results showed that: users can control manipulator to move and grasp objects by gesture smoothly. Through experiment, we can evaluate the performance and disadvantages of our system. The gesture recognition rate is more than 80% (only one gesture is below 90%. The system can lock the user as an operator target. After practice, the recognition rate of a freshman operator can be higher than 75%. What's more, this system has stability when program runs long time. The performance degradation is less than 5% after 30 min of operation. Though it doesn't have a universal protocol in the field of gesture recognition, we still believe the technology of HRI will impact our life in the future, and be expected to extend widely. We develop this system for extensibility. Our method is suitable for remote control and application scenarios under special working conditions, and has a good market application prospect.

Keywords Machine vision · Intelligent control · Human–computer interaction · Industrial robot

X. Wan · Q. Luo · Y. Li · J. Lou · P. Zheng (✉)
Yiwu Industrial & Commercial College, Yiwu 322000, China
e-mail: pfzheng@126.com

P. Zheng
East China University of Science and Technology, Shanghai 200237, China

H. Dai (ed.), *Computational and Experimental Simulations in Engineering*, Mechanisms and Machine Science 119,
https://doi.org/10.1007/978-3-031-02097-1_26

Fig. 1 Flow char of the system

1 Introduction

In recent years, major economies around the world are vigorously promoting the revival of manufacturing. Under the upsurge of industry 4.0, industrial Internet, Internet of things and cloud computing, many excellent manufacturing enterprises and scientific research institutes around the world have carried out the practice and research of intelligent control. Among them, human–computer interaction technology is a hot topic. Many scholars have studied this field [2–14].

Human computer interaction is a technology to study human and system and their interaction. Its research purpose is to use all possible information for human–computer communication and improve the naturalness and efficiency of interaction. Nowadays, manipulator has been widely used in factories and life, such as automobile assembly, handling and sorting, service fields and so on. Therefore, the research on human–computer interaction of manipulator is of great significance. At present, the way to control the manipulator is mainly teaching pendant. We hope to reduce the cognitive burden of learners and control the robot in a more natural way. We propose a manipulator control scheme based on gesture interaction, which mainly completes the following work: using Kinect 2.0 sensor to obtain human body depth data, separate human body and background, and design a set of gesture recognition logic after Gaussian filtering. The gesture coordinate data is transmitted to the robot through socket, and the manipulator motion is controlled through relative coordinate data. Besides, we develop virtual switch to reduce misoperation effectively (Fig. 1).

2 Hardware and Software

2.1 ABB Industrial Robots

In the exercise test, we used ABB IRB 120 (Fig. 2) from the industrial robot laboratory of Yiwu Industrial & Commercial College. The IRB 120 robot is the latest addition to ABB's new fourthgeneration of robotic technology with superior control and path accuracy. It weighs 25 kg and can load 3 kg.

Fig. 2 Industrial robot platform

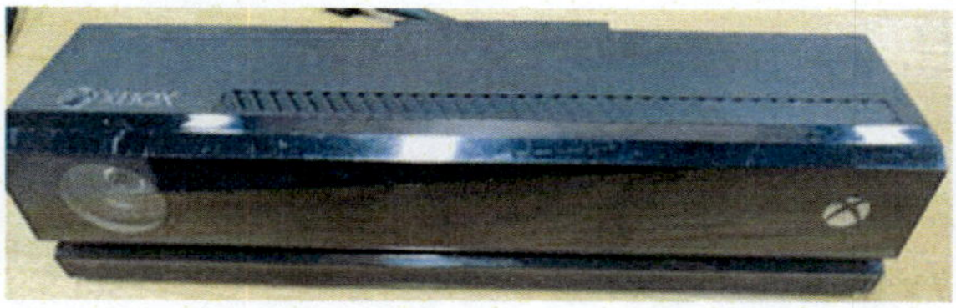

Fig. 3 Kinect sensor

2.2 *Kinect V2 Sensor*

We attempt to use Kinect as the sensor (Fig. 3), which is produced by Microsoft Corporation. Different from normal camera, Kinect can acquire depth image by infrared radiation projector and camera. Therefore, Kinect can build three dimensional data of environment or objects in captured images. In other words, Kinect is also a kind of 3D image camera. Kinect builds on range camera technology by Israeli developer PrimeSense [15], which developed a system that can interpret specific gestures, making completely hands-free control of electronic devices possible by using an infrared projector and camera and a special microchip to track the movement of objects and individuals in three dimensions. This 3D scanner system is called Light Coding what employs a variant of image-based 3D reconstruction. Thus, Kinect can make up for the limitation of normal camera when using for development. Due to the good performance on this aspect, we decided to use Kinect to achieve our project.

2.3 *Software Specifications*

Kinect for Windows SDK

The Architecture of Kinect SDK (Fig. 4) is based on "pipeline". From the SDK, we can get the raw data include image stream, depth stream, audio stream. Besides, the

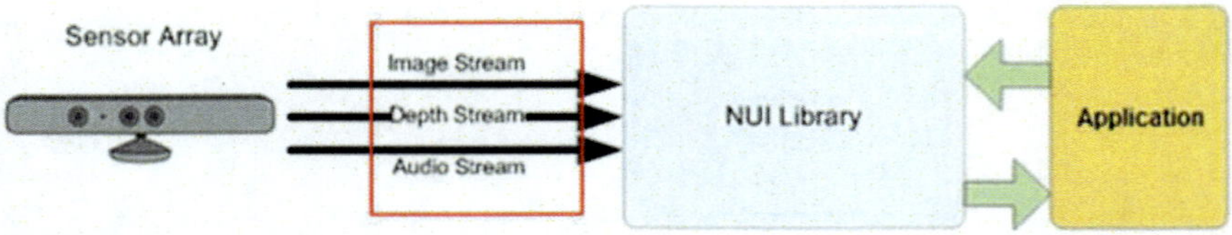

Fig. 4 Kinect for Windows SDK

SDK also provide the NUI library for developer to call. In this paper, we mainly use the raw data of image stream and depth stream, which is shown in the Fig. 4 red block.

Robot studio

Robot studio is a robot virtual simulation software provided by ABB. It can program industrial robots with rapid language. It also provides online function to communicate with external programs through socket.

3 Data Collection and Process

In this section, mainly discuss how to collect the depth data and the process of gesture recognition. At first, we introduce the feature of Kinect sensor's depth data. Second part is filtering, in order to move the jitter and make data smoother. Thirdly, we separate user's depth data with background, in order to reduce the noise of background. Fourthly, match the depth coordinate system and RGB camera's coordinate system. At last, we design a gesture recognition algorithm. The whole process is shown in Fig. 5.

Fig. 5 Process of gesture recognition

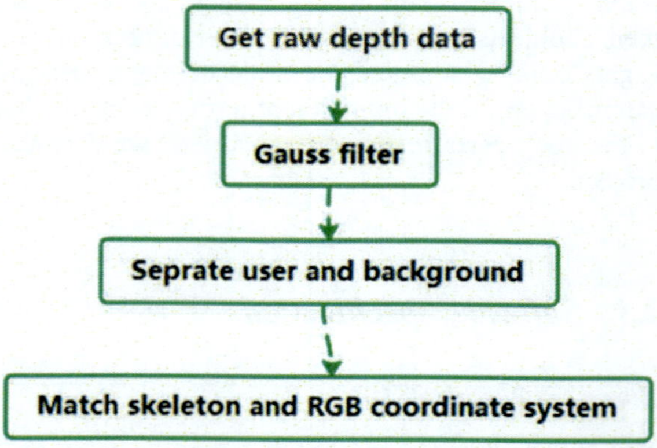

3.1 *Kinect Sensor's Depth Data*

Kinect sensor's depth data stream is made up by image frame (Fig. 6). In each depth image frame, every pixels have specific depth data, which is the depth from Kinect to object, the unit is millimeter. Each pixel has 16 bits data. The higher 13 bits are depth data from Kinect to object and the lower 3 bits are user index data. So the theoretical depth which can be measured by Kinect is 0–213 (mm).

As we can see in the Fig. 7, "Pixel a" is a pixel from a user, the lower three bits is "001", which means it's a human index. "Pixel b" is a pixel from a object, so the lower tree bits is "000".

The next target is how to calculate the real distance from object to Kinect. For example, consider in a particular frame, one of the pixel values is 20952. Figure 8 explains how the depth in millimeters is calculated from a particular pixel value by applying bit shifting.

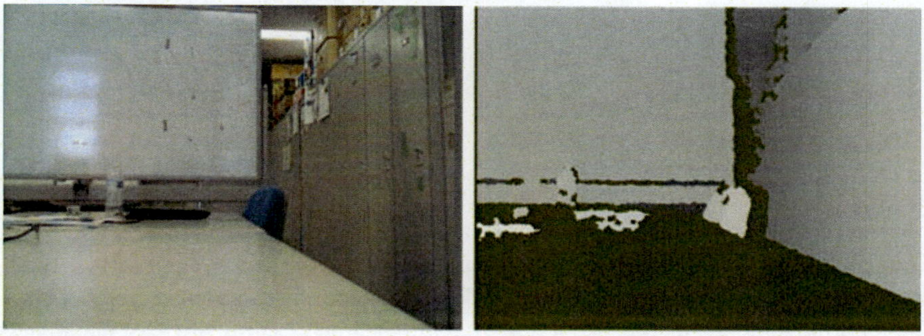

a. RGB image b. Depth image

Fig. 6 RGB image and depth image

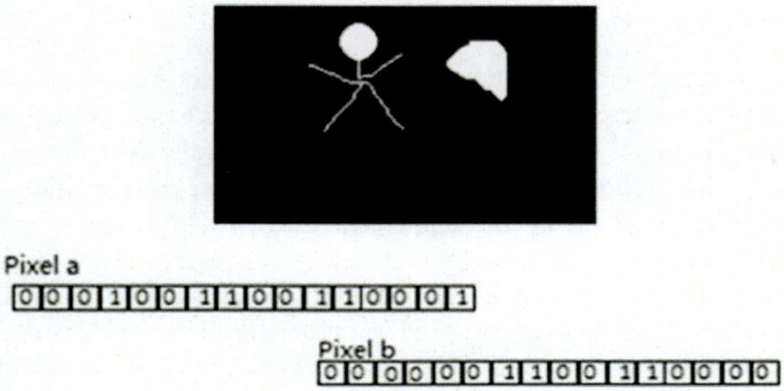

Fig. 7 Pixel data

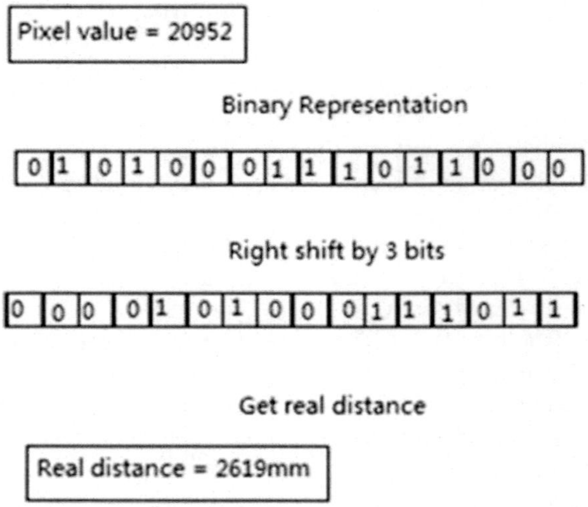

Fig. 8 Get distance from pixel data

3.2 *Filter*

Reasons cause jitter

Thought the depth data jitter could be caused by the application performance due to both software and hardware, there are several possible reasons for depth data jittering. One of the main reasons is processing large amounts of data over period of time. Because of the processing of large data, it's very difficult to calculate the accuracy of the joint movement. As to my function of depth data processing, the least calculated amount per minute is larger than 10 million. An important step before consuming depth data is to use a noise reduction filter to remove as much noise as possible from the data. Such filters are called smoothing filters because they result in smoother positions over time.

Balance between smooth and delay

An ideal filter would remove all unwanted noise and jitters from the data resulting in smooth position data over time. It would also follow the movements of the joint without any lag or delay. Unfortunately, there is a tradeoff between these two objectives in practice, and choosing a filtering technique that aggressively smoothes out the data would result in higher filtering delay, which would increase the perceived latency. As an intuitive explanation for this concept, consider a case where a person is standing still and therefore the input to the filter is mostly a constant position along with some noise. In order to produce a smooth output, the filter should not be sensitive to the changes in input due to noise. Now suppose the person starts moving his/her hand. In order to be responsive to these movements, the filter should be designed to be sensitive to changes due to movement, which is an opposite of the requirement for noise removal. In practice, most filters take some time to see enough movement

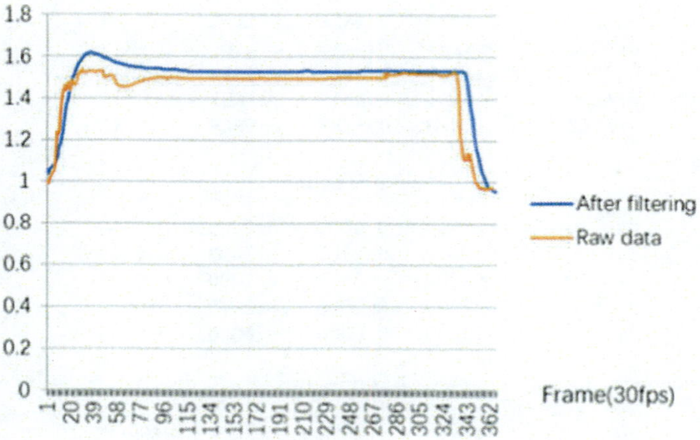

Fig. 9 Comparison of filter and without filter

before they start following these changes in output, and therefore their output lags behind the changes in input.

Result of filter

When we reach out to Kinect and hold it for a period of time, the detected depth data is shown in the figure below. The red line is the original data and the blue line is the filtered data. As we can see in the Fig. 9 the output data after filtering is smoother than the raw data, while it has some delay.

3.3 Separate User and Background

After some test, we can know some background data will cause little noise, hence the recognition result will make error sometimes. So, separating user and background is necessary for recognizing.

The method of data process is shown at Fig. 10. The matrix a stands for an example of a simplified depth image's 16 bits data matrix. It is include a user and other background. We left shift 13 bits of every pixel of the matrix a, then we can get matrix b. From matrix b, it's easy for us to separate which data stands for user and which is background. Thus, we can get the users' data from matrix a, which is shown in matrix c (Fig. 11).

0xAAA8	0xAA28	0xAA58	0xABA8	0xFAA8
0xFFA8	0xAAB1	0xAAA9	0xAAC1	0xA3A8
0xCCA8	0xAAA9	0xAAB1	0xAAA9	0x93A8
0xAAD8	0xAAC1	0xAAA9	0xAAB1	0x9538
0xFAA8	0xA4A8	0x87A8	0x98A8	0x12A8

Matrix a

0x0000	0x0000	0x0000	0x0000	0x0000
0x0000	0x2000	0x2000	0x2000	0x0000
0x0000	0x2000	0x2000	0x2000	0x0000
0x0000	0x2000	0x2000	0x2000	0x0000
0x0000	0x0000	0x0000	0x0000	0x0000

Matrix b

0x0000	0x0000	0x0000	0x0000	0x0000
0x0000	0xAAB1	0xAAA9	0xAAC1	0x0000
0x0000	0xAAA9	0xAAB1	0xAAA9	0x0000
0x0000	0xAAC1	0xAAA9	0xAAB1	0x0000
0x0000	0x0000	0x0000	0x0000	0x0000

Matrix c

Fig. 10 Process of separating user and background

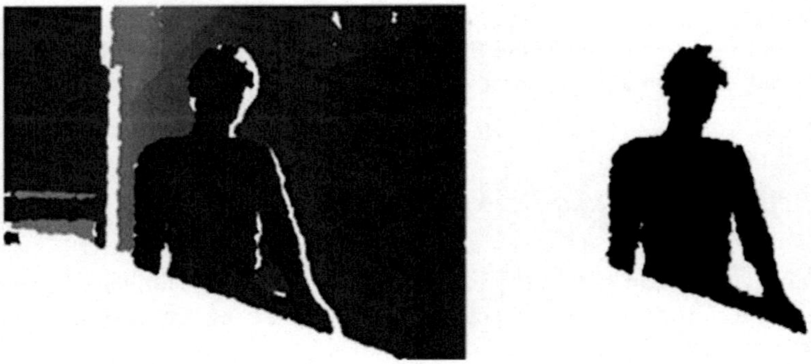

Fig. 11 Result of separating user and background

4 Experiment and Evaluation

4.1 Control Method

At present, we have designed seven control strategies (Table 1).

Table 1 Control method

Gesture	Manipulator action
Raise hands above head	Virtual switch
Right hand from left to right	X axis coordinates increase
Right hand right left to left	X axis coordinate reduction
Right hand forward	Y axis coordinates increase
Right hand back	Y axis coordinate reduction
Right hand up	Z axis coordinates increase
Right hand down	Z axis coordinate reduction

4.2 Evaluation

Experiment is an important part of developing, which is used to verify whether the software and hardware conform to the design. Through a whole experiment, we can evaluate the efficiency, correctness, robustness, etc. of the system. We test each unit of this system; find the errors or bugs in each module; set some groups, test the environment interference to the system, such as passerby, run time, etc (Table 2).

Passerby test

When a passer-by appears behind the operator, the program will only track the original operator. From the table (Recognition accuracy), we can come to a conclusion: there is no influence by passerby (Table 3).

Run time test

The purpose of this test is to evaluate the time robust of this system. We set two groups to compare with the control group. In group 1, we test the recognition rate after the program running 30 min; in group 2, we test the recognition rate after the program running 60 min. The result of this test is showed in Table 4.

Table 2 Test plan

Test project	Purpose
Control group	The reference of the experiment result
Group 1: have pass by	Record the recognition rate; evaluate the influence of pass by
Group 2: run program long time	Record the recognition rate; evaluate the influence when program run long time
Group 3: other users (Fresh)	Record the recognition rate of other user; compare the result of fresh user and after practicing
Group 4: other users (After practice)	

Table 3 With/without passerby test

	Have passby	Control group
Virtual switch	48/50 (96%)	48/50 (96%)
X axis coordinates increase	48/50 (96%)	49/50 (98%)
X axis coordinate reduction	48/50 (96%)	49/50 (98%)
Y axis coordinates increase	48/50 (96%)	48/50(96%)
Y axis coordinate reduction	42/50 (84%)	43/50 (86%)
Z axis coordinates increase	50/50 (100%)	50/50 (100%)
Z axis coordinate reduction	50/50 (100%)	50/50 (100%)

Table 4 Run time test

	Run 30 min	Run 60 min	Control group
Virtual switch	46/50 (92%)	41/50 (82%)	48/50 (96%)
X axis coordinates increase	47/50 (94%)	44/50 (88%)	49/50 (98%)
X axis coordinate reduction	49/50 (98%)	44/50 (88%)	49/50 (98%)
Y axis coordinates increase	47/50 (94%)	40/50 (80%)	48/50(96%)
Y axis coordinate reduction	44/50 (88%)	39/50 (78%)	48/50(96%)
Z axis coordinates increase	49/50 (98%)	46/50 (92%)	50/50 (100%)
Z axis coordinate reduction	46/50 (92%)	46/50 (92%)	50/50 (100%)

Operability test

In this section, we plan to test the operability of this system. Firstly, we invite a volunteer to control this system and record the recognition rate. Secondly, we record the recognition rate after doing some exercise (Table 5).

4.3 Experiment Summary

We test the recognition rate by different users; evaluate the robustness when the program runs long time.

Through experiment, we can evaluate the performance of our system. We basically get the expected results. The system can lock the user as an operator target. After short practice, a freshman can use the system. The recognition rate of a freshman operator can be higher than 75%. What's more, this system has stability when program runs long time.

Table 5 Operability test

	Volunteer A		Author
	Freshman	After exercise	Control group
Virtual switch	35/50 (70%)	42/50 (84%)	48/50 (96%)
X axis coordinates increase	39/50 (78%)	47/50 (94%)	49/50 (98%)
X axis coordinate reduction	38/50 (76%)	45/50 (90%)	49/50 (98%)
Y axis coordinates increase	45/50 (90%)	46/50 (92%)	48/50(96%)
Y axis coordinate reduction	30/50 (60%)	38/50 (76%)	43/50 (86%)
Z axis coordinates increase	46/50 (92%)	48/50 (96%)	50/50 (100%)
Z axis coordinate reduction	47/50 (94%)	48/50 (96%)	50/50 (100%)

5 Conclusion

This paper mainly studies the natural gesture interaction technology, and applies the natural gesture interaction to the teaching of manipulator, which improves the interactive experience of instructors. At present, there are still some deficiencies in this research, such as the inability to control the speed of the manipulator in real time through gestures. In the future, we will continue to devote ourselves to research in this field.

Funding Statement This study was supported by the Domestic Visiting Engineers Project of Zhejiang Education Department in 2020, China (Grant No. FG2020196, FG2020197), the second batch of teaching reform research projects in the 13th Five-Year Plan of Zhejiang Higher Education, China (Grant No. jg20190878), and the public welfare science and technology research project of Jinhua, Zhejiang Province, China (Grant No. 2021–4-386).

Conflicts of Interest The authors declare that they have no conflicts of interest to report regarding the present study.

References

1. https://developer.microsoft.com/zh-cn/windows/kinect/
2. Kawarazaki, N., Diaz, A.I.B.: Gesture recognition system for wheelchair control using a depth sensor. In: IEEE Symposium on Computational Intelligence in Rehabilitation and Assistive Technologies (2013)

3. Wei, L., Hu, H.: Evaluating the performance of a face movement based. In: International Conference on Robotics and Biomimetics (2010)
4. Gleeson, B., Maclean, K., Haddadi, A., Croft, E., Alcazar, J.: Gestures for industry intuitive human-robot communication from human observation. In: ACM/IEEE International Conference on Human-Robot Interaction (2013)
5. Liu, H., Wang, L.: Gesture recognition for human-robot collaboration: a review. Int. J. Ind. Ergon. **68**, 355–367 (2017)
6. Croft, E.A., et al.: Cooperative gestures for industry: exploring the efficacy of robot hand configurations in expression of instructional gestures for human-robot interaction. Int. J. Robot. Res. **36**, 5–7 (2017)
7. Ge, L., Wang, H., Xing, J.: Maintenance robot motion control based on Kinect gesture recognition. In: 7th International Symposium on Test Automation and Instrumentation (ISTAI 2018) (2019)
8. Mocan, B., Fulea, M., Brad, S.: Designing a multimodal human-robot interaction interface for an industrial robot. Springer International Publishing (2016)
9. Laguillaumie, P., Laribi, M.A., Seguin, P., et al.: From human motion capture to industrial robot imitation. Springer International Publishing (2016)
10. Ganapathyraju, S.: Hand gesture recognition using convexity hull defects to control an industrial robot. In: International Conference on Instrumentation Control & Automation. IEEE (2014)
11. Veeriah, V., Swaminathan, P.L.: Robust hand gesture recognition algorithm for simple mouse control. J. Comput. Commun. Eng. **2**(2), 219–221 (2013)
12. Shirwalkar, S., Singh, A., Sharma, K.: Telemanipulation of an industrial robotic arm using gesture recognition with Kinect. In: IEEE 2013 International Conference on Control, Automation, Robotics and Embedded Systems (CARE) (2014)
13. Vasiljevic, G., Jagodin, N., Kovacic, Z.: Kinect-based robot teleoperation by velocities control in the joint/Cartesian frames. In: 10th IFAC Symposium on Robot Control (2012)
14. Popov, V., Ahmed, S., Shake, V.N.: Gesture-based Interface for real-time control of a Mitsubishi SCARA robot manipulator-sciencedirect. IFAC-PapersOnLine **52**(25), 180–185 (2019)
15. http://en.wikipedia.org/wiki/PrimeSense

Research on Cement Sheath Integrity Under High Temperature During In-Situ Development for Shale Oil Well

Xueli Guo, Fengzhong Qi, Yongjin Yu, Jianzhou Jin, Yuchao Guo, Hongfei Ji, Yongqin Cheng, and Zhengyang Zhao

Abstract The temperature is as high as 500 °C at the bottom of shale oil wells during in-situ development. During shale oil production, the fluids with high temperature are transmitted to the wellhead, and the wellbore temperature will rise extremely, which greatly affects the cement sheath integrity. In this paper, the linear expansion coefficient of cement stone under the temperature of 500 °C were measured. Based on the experimental data and the field data of shale oil well, the stage finite element modeling method was adopted to establish a casing-cement-formation (CCF) model considering the temperature and pressure coupling. The results revealed that the linear expansion coefficient of cement stone decreases linearly with the increase of temperature. When the temperature was less than 250 °C, the cement stone expanded. However, when the temperature exceeded 250 °C, it shrank. The maximum shrinkage ratio can be 25×10^{-6}/°C for the temperature of 500 °C. The circumferential stress of the inner wall of cement sheath first increased rapidly, then slowly decreased to be stable. The larger the cement stone shrink ratio was, the greater the circumferential stress was. The maximum circumferential stress can exceed 10 MPa for the shrink ratio of 25×10^{-6}/°C. It can be concluded that the circumferential stress appears in a tensile state under high temperature conditions, which can easily lead to the failure of the cement sheath seal. The strength of the cement stone should be considered when designing the cement slurry property. In addition to the decay performance, the expansion of the cement stone should also be designed to ensure that the cement stone does not shrink when subjected to high temperature conditions.

Keywords In-situ development of oil shale · Cement sheath seal failure · Ultra-high temperature · Cement shrink · Stage finite element modeling

X. Guo (✉) · F. Qi · Y. Yu · J. Jin · Y. Guo · H. Ji
CNPC Engineering Technology R&D Company Limited, Beijing, China
e-mail: guoxldr@cnpc.com.cn; clouder0713@163.com

Y. Cheng
Research Institute of Petroleum Exploration and Development, Beijing, China

Y. Cheng · Z. Zhao
China University of Petroleum, Beijing, China

H. Dai (ed.), *Computational and Experimental Simulations in Engineering*, Mechanisms and Machine Science 119,
https://doi.org/10.1007/978-3-031-02097-1_27

1 Instruction

The underground in-situ conversion of shale oil can be called an "underground refinery". The horizontal well with electric heating to lightweight technology was used to continuously heat the organic-rich shale interval with a buried depth of 300–3000 m, so that multiple types of organic matter can be light-weighted during the physical and chemical process. The bottom temperature of the shale oil in-situ conversion heating well can reach up to 500 °C, and the heat was transmitted along the casing. High temperature made the upper wellbore temperature rise, leading to the casing and cement expansion, and the integrity of the wellbore damage. Conventional cement cannot meet the requirements. Working conditions with large changes in wellbore temperature and pressure are becoming more and more common. In order to achieve long-term sealing, the cement sheath needs to be adapted to the changes of temperature and pressure during drilling and production. Therefore, quantitatively determining the stress state of the cement sheath at different stages of the life cycle of an oil and gas well is a prerequisite for effective countermeasures, so as to give an accurate judgment on how the cement sheath seal integrity will fail.

Researchers established the casing-cement sheath-formation model considering the whole life of a wellbore. Pattillo and Kristiansen [1] modeled the behavior the wellbore and surrounding region with a finite element method considering the pre-wellbore depletion, drilling the wellbore, installation of casing and cement, and subsequent draw down. Gray [2] proposed a staged-finite-element procedure during well construction, sequentially considering the stress state and displacements at and near the wellbore. The model replicated the complicated stress states that arise from the simulations action of far-field stresses, overburden pressure, cement hardening and shrinkage, debonding at the interfaces, and plastic flow of cement sheath and rock formation. Mackay and Fontoura [3] carried out a finite element model that focused on the drilling of the wellbore and on the hardening of the cement. Zhang [4] presented a finite element modeling approach to simulate the well construction processes and injection cycle considering the in-situ stress field, mud and cement slurry pressure, and the periodic temperature changes were incorporated. A "realistic" bottom-hole state of stress was generated and a micro-annuli generation was simulated by the tensile debonding of the cement-formation interface. Simone [5] and Liu [6] developed an analytical solution to assess the stresses during the drilling, construction and production phases. The reason for the difference between these models and the conventional model was theoretically analyzed.

In this paper, for in-situ development wells of shale oil, the thermal expansion coefficient of cement stone was measured to clarify the changing law of cement stone under high temperature conditions. In order to clarify the impacts of the thermal expansion coefficient on cement sheath, a stage finite element model of the wellbore with fully coupled temperature and pressure was established to analyze the changes of the temperature and the stress of the cement sheath under high temperature conditions during shale oil production, which can provide technical support for the in-situ exploitation of shale oil to guarantee the cement sheath integrity.

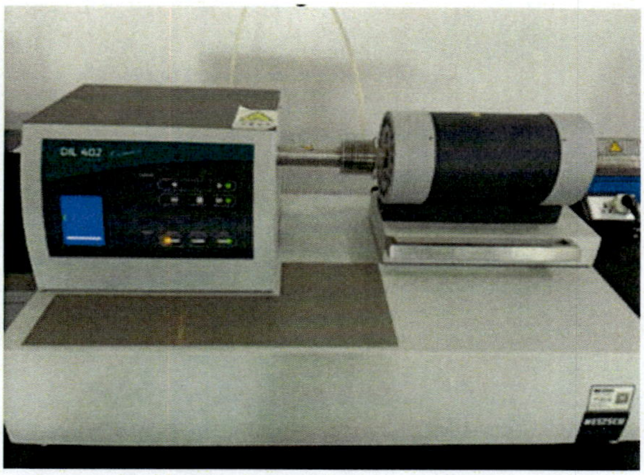

Fig. 1 German Netzsch DIL 402 classic thermal dilatometer

2 Cement Stone Thermal Expansion Coefficient Test

The differential thermal expansion method was used in the law, the thermal expansion coefficient of quartz material did not change with temperature, and the expansion of the measured sample was compared with the standard quartz sample to calculate the thermal expansion coefficient of the measured sample. This method required that the test piece should have a specific size that was mainly suitable for materials with a large difference in thermal expansion coefficient from quartz. The test temperature range was wide, and the results were intuitive and accurate. The temperature was controlled by a slow and constant rate of temperature increase (usually 5 °C/min), and a push rod dilatometer equipped with an Al_2O_3 carrier was used to measure the length of the sample material with temperature and the length of the carrier. In this paper, the German Netzsch DIL 402 classic thermal dilatometer was used to measure the thermal expansion coefficient of cement stone, as shown in Fig. 1.

3 Finite Element Model Establishment

3.1 Stage Finite Element Model

When predecessors performed finite element modeling, they ignored the initial state of the formation and directly fixed the casing, cement sheath and formation together, and then applied corresponding stress boundary conditions, which simplified the modeling process to a certain extent, but the far-field displacement calculated by the model failed to meet the actual far-field boundary conditions. At present, only a few scholars have noticed the limitations of this modeling process. For this reason, they are considering the initial stress state of the formation during the modeling

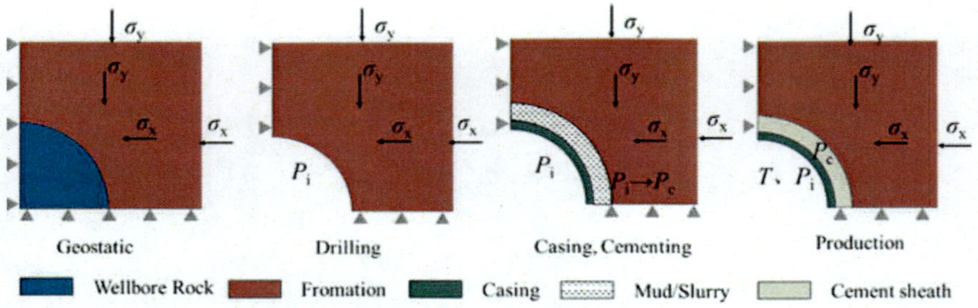

Fig. 2 The stage finite element modeling procedures

process, including the drilling, casing setting, cementing, and production process. This method is called stage finite element modeling method, as shown in Fig. 2.

The stage finite element modeling procedures can be divided into four parts:

Step 1: Geostatic. This step is mainly to simulate the initial state of the formation before drilling. The initial stress includes the maximum and minimum horizontal stresses and the pressure of the overburden. For simplicity, the model is simplified to a plane strain problem. The borehole column coordinate system was adopted, while the x and y directions were two mutually perpendicular directions on the borehole cross section. In this coordinate system, the geo-stress can be decomposed into the borehole coordinate system, and the coordinates, σ_x and σ_y, in the plane strain state can be obtained. These stresses were set as the boundaries in the model. At the same time, the same geo-stresses and temperature were added in the wellbore rock and the formation rock in the form of a pre-stress field to simulate the initial state of the wellbore.

Step 2: Drilling process. The rock that originally stays at the place of the wellbore was drilled and stress redistributed near the wellbore. During the simulation procedure, the rock elements in this part were killed by the keyword *Model change, and then the mud pressure of P_i was directly applied at the inner wall of the borehole, which balanced the wellbore stress state.

Step 3: Casing and cementing. After the casing was run to the bottom of the well, the inner and outer walls of the casing and the inner wall of the formation bore the drilling mud pressure of P_i. During cement, the annular pressure became the cement slurry pressure of P_c. At this time, the outer wall of the casing and the inner wall of the formation bore the pressure of P_c. In order to simplify the modeling process, the entire cement slurry hardening process was not considered. The cement was added to the model with a pre-stress field of P_c.

Step 4: Production. During this process, the high-temperature fluid would flow to the wellhead at a certain production rate, and the temperature field of the wellbore assembly would change greatly in a short time. Therefore, the high-temperature boundary condition T can be added to the inner wall of the casing, aiming to simulate the heating process of the assembly by the production fluid with high temperature.

Under the certain production rate and temperature conditions, the heat transfer coefficient of the production fluid and the inner wall of the casing were calculated.

According to the shape of the cement sheath and the well construction process, the finite element model was established, and the temperature and pressure of the production fluid were determined. The temperature and pressure boundary conditions of the inner wall of the casing were added into the model. From this, the influence of different factors on the cement sheath stress during the production process can be explored.

3.2 Basic Assumption

In order to simplify the analysis process, some assumptions need to be made. Firstly, the casing, the cement sheath and the surrounding rock of the well will be consolidated as a whole after the solidification of cement slurry. Furthermore, the contact interfaces will be well cemented without slippage between the casing and cement sheath or cement sheath and formation. Secondly, the casing, cement sheath and the surrounding rock are homogeneous and isotropic materials. Thirdly, based on the theory of elastic mechanics, stress concentration will occur near the circular hole in the infinite plate. When the boundary size exceeds 5–6 times of the wellbore, the effects of stress concentration will be small. Therefore, setting the formation boundary size appropriately can simulate the actual wellbore stress state with sufficient precision.

3.3 Parameter Setting

According to the design of different types of wellbore structure in the field pilot test, the wellbore structure of production well are displayed in Fig. 3. Based on the actual field conditions of the production well, a two-dimensional wellbore transient state temperature stress field model was established. The casing diameter is the production casing size: The drill bit diameter is 215.9 mm, the casing outer diameter is 177.8 mm, the wall thickness of casing is 9.19 mm, and the stratum size is set as 20,000 mm to minimize the simulation errors.

In the model, the thermal conductivity, specific heat capacity, density, thermal expansion coefficient, elastic modulus, Poisson's ratio of casing, cement sheath, and formation rock are illustrated in Table 1. Among them, the thermal conductivity, specific heat capacity, density, thermal expansion coefficient, elastic modulus and Poisson's ratio of cement stone and formation rock are determined by the experiment, and the casing parameters can be obtained from relevant literatures. The temperature at the inner casing is set as 400 °C, the yield stress of cement sheath is 30 MPa, and the internal friction angle is 17°. The mud pressure, slurry pressure and geo-stress in the formation are 15 MPa, 25 MPa and 15 MPa, respectively.

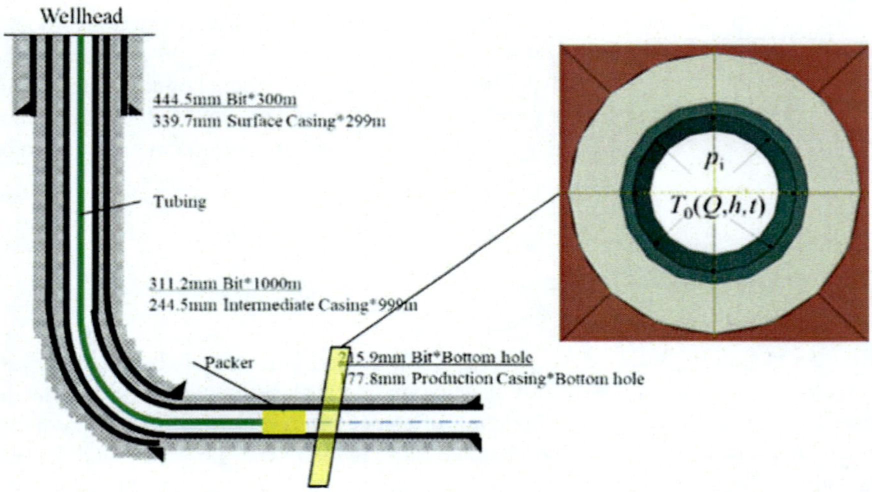

Fig. 3 Wellbore structure of shale oil production well

Table 1 Parameters setting for different parts in casing-cement sheath-formation model

Name	Casing	Cement sheath	Formation
Thermal conductivity, W/(m·°C)	50	0.94	4.32
Specific heat capacity, J/(kg·°C)	460	295	550
Density, g/cm^3	7.85	1.9	2.6
Thermal expansion coefficient, °C × 10^{-6}	15	25 ~ 20	1.5
Elastic modulus, GPa	210	6	23
Poisson's ratio	0.3	0.15	0.25
Convection heat transfer coefficient, W/(m^2·°C)	1000	/	/

4 Results and Analysis

4.1 Expansion Coefficient of Cement Stone

For the cement stone cured under different curing conditions, the cement sample with the size of 6 × 6 × 25 mm was prepared, as shown in Fig. 4a. The thermal expansion coefficient was measured with a thermal dilatometer. According to the measurement results, the thermal expansion coefficient of different samples was drawn, as shown in Fig. 4b.

It can be seen from Fig. 4b that the linear expansion coefficient of the cement sample cured for 2d is larger than 15 × 10^{-6}/°C for the temperature ranging from 30 °C to 100 °C. When the temperature exceeds 100 °C, it decreases, and then becomes zero at the temperature of about 250 °C, which means that the sample volume returns

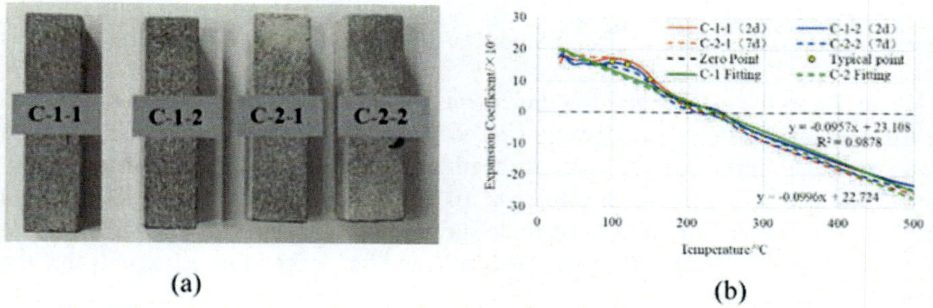

Fig. 4 Cement sample and thermal expansion coefficient results, **a** Cement sample, **b** Expansion coefficient

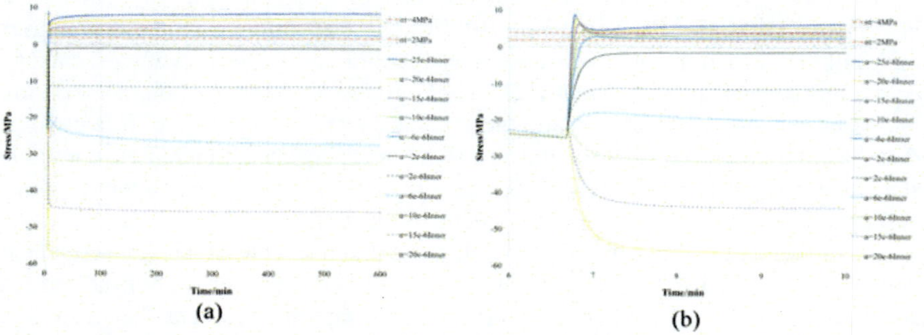

Fig. 5 The circumferential stress at the inner wall of cement sheath for different expansion coefficient of cement stone: **a** Time from 0 to 600 min, **b** Time from 6 to 10 min

to its original size. With the temperature continues rising, the expansion coefficient continues to reduce linearly, which indicates that cement shrinks. At the temperature of 500 °C, the shrinkage rate is about 25×10^{-6}/°C. For the cement sample cured for 7d, it is consistent with the aforementioned change trend. At the same time, when the temperature exceeds 100 °C, the linear expansion coefficient of the cement cured for 7d is slightly smaller than that of the sample cured for 2d. It is generally believed that as the temperature increases during the hydration process of cement, the hydration rate increases. When the temperature exceeds 200 °C, the crystal water after cement hydration dissipates, and the hydration products will also change, thus resulting in continuous decrease for cement volume, which has great impacts on the cement sheath integrity during the production process with high temperature for shale oil wells.

According to the curve fitting, the relationship between the linear expansion coefficient α and the temperature T of the cement samples cured for 2d and 7d are $\alpha_{2d} = (0.0957\ T + 23.108) \times 10^{-6}$/°C, $\alpha_{7d} = (0.0996\ T + 22.724) \times 10^{-6}$/°C, respectively. Based on this formula, the linear expansion coefficient can be quantitatively described according to the ambient temperature conditions of the cement stone.

4.2 Cement Sheath Stress

According to the established finite element model, simulations on the influence of different cement stone thermal expansion coefficients on the stress state of the internal and external interface of the cement sheath were carried out. The expansion coefficient was set from 20×10^{-6}/°C to 25×10^{-6}/°C in the model. The results are listed in Figs. 5 and 6. When the cement stone shrinks, the solid line is the circumferential stress on the inner wall of the cement sheath, and the dashed line represents that the cement stone expands.

It can be seen from Fig. 5 that when the expansion coefficient of cement stone is less than 10×10^{-6}/°C or greater than 10×10^{-6}/°C, the circumferential stress reaches its peak within a few seconds, rapidly decreases to a stable state within 1 min, and then remains almost without any change. When the expansion coefficient is between 10×10^{-6}/°C and 10×10^{-6}/°C, the circumferential stress increases rapidly in about 1 min, and then remains stable. For cement stone with a shrinkage coefficient greater than 15×10^{-6}/°C, the circumferential stress of the cement ring increases slowly within 60 min and then remains stable. From Fig. 6, it is obvious that the circumferential stress on the outer wall of casing sheath increases to a stable state in about 180 min for the expansive cement. It increases a little in a few seconds, and then decreases to a stable state in about 180 min.

For the cement with the tensile strength of 4 MPa, when the expansion coefficient is less than 15×10^{-6}/°C, the circumferential stress on the inner wall of the cement sheath exceeds the tensile strength, which causes the inner cement sheath failure. Similarly, when the expansion coefficient is less than 20×10^{-6}/°C, the outer cement sheath fails. Therefore, it can be obtained that if the cement stone shrinks with the shrinkage rate larger than 15×10^{-6}/°C, the cement sheath is prone to be failure.

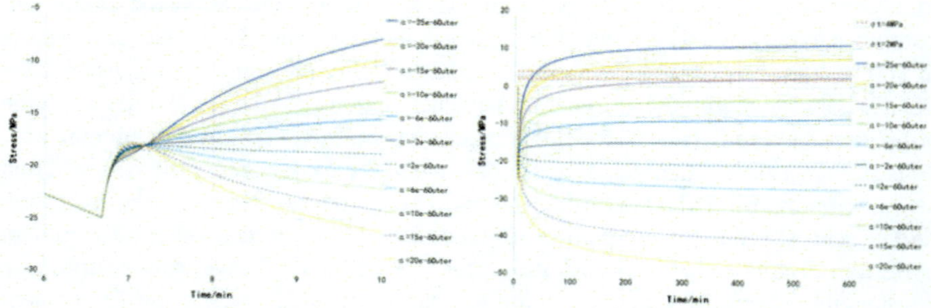

(a) (b)

Fig. 6 The circumferential stress at the outer wall of cement sheath for different expansion coefficient of cement stone: **a** Time from 0 to 600 min, **b** Time from 6 to 10 min

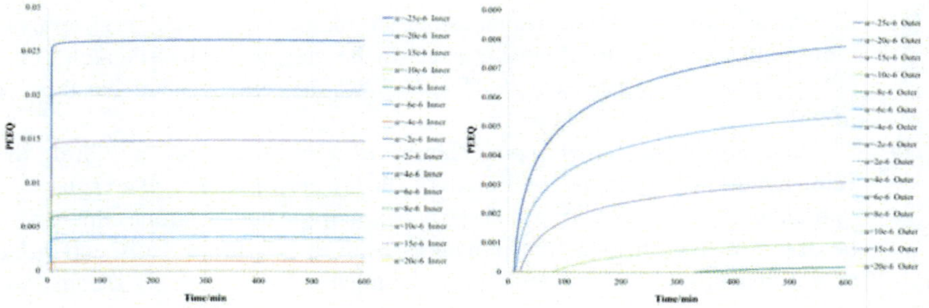

(a) (b)

Fig. 7 The PEEQ at the inner and outer wall of cement sheath for different expansion coefficient of cement stone: **a** Time from 0 to 600 min; **b** Time from 6 to 10 min

4.3 Cement Sheath Equivalent Plastic Strain

The equivalent plastic strain PEEQ on the internal and external interfaces of the cement sheath for different thermal expansion coefficients are shown in Fig. 7.

It is obvious that when the expansion coefficient is less than 4×10^{-6}/°C, the equivalent plastic strain PEEQ appears on the inner wall of the cement sheath, increases rapidly in about 1 min, and then slowly increases to the maximum. When the expansion coefficient is less than 8×10^{-6}/°C, the equivalent plastic strain appears on the outer wall of the cement sheath, and slowly increases to the maximum value.

In summary, if the cement stone shrinks and reaches a certain level, the high temperature production fluid will cause the cement sheath to fail during the heating process of the cement sheath.

5 Conclusion

For the production wells involved in in-situ extraction of shale oil, the temperature of the wellbore during the production process is relatively high, which has great impacts on the cement sheath integrity. Through the measurement of the thermal expansion coefficient of cement stone under the condition of high temperature, a stage finite element model of the wellbore assembly was established based on the experimental data, and the stress of the cement sheath under the high temperature conditions was analyzed as well. Researches on the sealing integrity of the formation/cement sheath/casing combination in the horizontal section of the production well were carried out. Some conclusions are as follows:

(1) The thermal expansion coefficient of cement stone decreases linearly with the increase of temperature, which indicates that the autogenous shrinkage of cement stone is more obvious under high temperature conditions.

(2) The analysis proved that the cement stone shrinkage occurs under high temperature conditions, which has great impacts on the stress state of the cement stone. When the shrinkage rate exceeds 15×10^{-6}/°C, the cement sheath has the risk of failure.
(3) The strength of the cement stone should be considered when designing the cement slurry property. In addition to the decay performance, when subjected to high temperature conditions, the expansion of the cement stone should also be designed to ensure that the cement stone does not shrink. Only under this condition can it provide protection for the integrity of the cement sheath.

Acknowledgements This work was supported by Scientific Research and Technology Development Project of China National Petroleum Corporation (Grant No. 2021DJ4403, 2020F-49, 2020B-4019, 2021DJ5203).

References

1. Pattillo, P. D, Kristiansen, T. G.: Analysis of Horizontal Casing Integrity in the Valhall Field: SPE/ISRM Rock Mechanics Conference[C], 20–23 October. Society of Petroleum Engineers, Irving, Texas, USA, 200210
2. Gray, K.E., Podnos, E., Becker, E.: Finite-element studies of near-wellbore region during cementing operations: part I[J]. SPE Drill Complet. **24**(01), 127–136 (2009)
3. Mackay, F., Fontoura, S.A.B.: The Description of a Process for Numerical Simulations in the Casing Cementing of Petroleum Salt Wells—Part I: from drilling to cementing: 48th U.S. Rock Mechanics/Geomechanics Symposium[Z],1–4 June. American Rock Mechanics Association, Minneapolis, Minnesota, 20149
4. Zhang, W., Eckert, A., Liu, X.: Numerical Simulation of Micro-annuli Generation by Thermal Cycling: the 51st U.S. Rock Mechanics/Geomechanics Symposium[C]. American Rock Mechanics Association, San Francisco, California (2017)
5. Simone, M.D., Pereira, F.L.G., Roehl, D.M.: Analytical methodology for wellbore integrity assessment considering casing-cement-formation interaction[J]. Int. J. Rock Mech. Min. Sci. **94**, 112–122 (2017)
6. Liu, W., Yu, B., Deng, J.: Analytical method for evaluating stress field in casing-cement-formation system of oil/gas wells[J]. Appl. Math. Mech. **38**(9), 1273–1294 (2017)

Research on Risk Evaluation of Featured Town Project Based on PPP Mode

Yang Song

Abstract According to the characteristics of characteristic town projects in the PPP model, Risks in the process of the project are identified, and a risk evaluation index system for characteristic town projects based on the PPP model is established. This paper adopts the construction and application of the entropy fuzzy matter-element evaluation model, and strives to qualitatively analyze and quantify, making the risk assessment of characteristic town projects in the PPP model more effective. Finally, an actual engineering project is taken as an example to introduce the application of this method.

Keywords PPP model · Characteristic town · Risk evaluation · Entropy fuzzy matter-element evaluation

With the gradual development of the rural revitalization strategy during the 14th Five-Year Plan period and the continuous improvement and development of the PPP model, the emergence of characteristic towns is an effective way to promote the integrated development of urban and rural areas, and is also an important starting point for solving the "three rural" issues. The development of characteristic town projects based on the PPP model has begun to spread throughout the country. This provides a new mode of exploration for the continuous consolidation of poverty alleviation, and there are also certain risks. Therefore, research on the risk management of characteristic town projects under the PPP model has emerged. Safety is the benefit. Controlling risks can effectively reduce costs and improve project operation efficiency.

Y. Song (✉)
Chongqing Three Gorges Vocational College, Chongqing, China
e-mail: dust4s@tom.com

H. Dai (ed.), *Computational and Experimental Simulations in Engineering*, Mechanisms and Machine Science 119,
https://doi.org/10.1007/978-3-031-02097-1_28

1 PPP Project Risk Evaluation Index System

This article comprehensively identifies project risks through literature research, case studies, and Delphi questionnaire survey. Then through Li's meta-classification methods based on the three different levels of risks of UK PPP projects: macro risks, meso risks and micro risks. Questionnaire surveys were used to send emails to the risk sources, the opinions of 12 experts were collected, and all project-related materials, documents and related information were summarized, and finally 25 project risk factors of characteristic small town projects that conform to the PPP model were identified:

Macro risks (F1): policy instability risk (R1), legal change risk (R2), government intervention risk (R3), inflation risk (R4), exchange rate risk (R5), interest rate risk (R6), financing risk (R7), climate risk (R8), geological risk (R9), environmental risk (R10), public acceptance risk (R11), force majeure risk (R12);

Meso risks (F2): planning and design risk (R13), construction cost risk (R14), construction period risk (R15), project quality and safety risk (R16), contract risk (R17), technical risk (R18), supply and demand risk (R19), operational risk (R20); Micro risks (F3): project lack of experience risk (R21), project entity credit risk (R22), unreasonable risk sharing (R23), unreasonable risk of defining rights and responsibilities (R24), organization coordination risk (R25) [1–4].

2 Construction and Application of Entropy Fuzzy Matter-Element Evaluation Model

Matter-element analysis theory can effectively solve incompatible problems in reality. Therefore, this paper introduces entropy theory when determining index weights, and establishes a project risk evaluation model based on fuzzy matter-element. Compared with previous scholars' evaluation of project risk. The adopted analytic hierarchy process, expert scoring and fuzzy comprehensive evaluation method, etc., It can effectively reduce the subjectivity in the determination of indicator weights and continuous project risk evaluation, and can objectively and comprehensively analyze and evaluate the risk of characteristic town projects of the PPP model. The results of calculations are intuitive and clear as well as provide a new idea for the research of characteristic town project risk management based on PPP mode.

Obtaining the weight of the index according to the entropy method can avoid the deviation caused by the inevitable subjective factors such as the analytic hierarchy process, and can reduce or eliminate the artificial influence as much as possible, so that the evaluation result is more consistent with the actual situation. The entropy method is used to determine the weight coefficient, and the specific calculation steps are as follows:

(1) Construction of compound fuzzy matter-element

The n-dimensional matter elements of m objects are combined to form the n-dimensional composite fuzzy matter element R_{mn} of m evaluation indicators.

$$R_{mn} = \begin{bmatrix} & M_1 & M_2 & \ldots & M_m \\ C_1 & X_{11} & X_{21} & \ldots & X_{m1} \\ C_2 & X_{12} & X_{22} & \ldots & X_{m2} \\ \vdots & \vdots & \vdots & \vdots & \vdots \\ C_n & X_{1n} & X_{2n} & \ldots & X_{mn} \end{bmatrix} \quad (1)$$

(2) Compound fuzzy matter-element with optimal membership degree, standard and sum difference square

According to the meaning of preferred degree of membership, the following formula can be used for calculation. Where Maxxij and Minxij indicate the maximum and minimum values of the index:

$$u_{ij} = \frac{x_{ij} - \min x_{ij}}{\max x_{ij} - \min x_{ij}} \quad (2)$$

$$u_{ij} = \frac{\max x_{ij} - x_{ij}}{\max x_{ij} - \min x_{ij}} \quad (3)$$

Uij is the degree of preferential membership. When the selected evaluation index Xij is marked as a positive index (the larger the better), the first formula is used to calculate; otherwise, the second formula is used to calculate, so the fuzzy matter element of the preferential membership degree can be constructed $\hat{R}_{mn}$. According to the maximum or minimum value of each index in $\hat{R}_{mn}$, a standard fuzzy matter element R0n can be constructed. Finally, the standard fuzzy matter element R0n and the compound fuzzy matter element $\hat{R}_{mn}$ can be used to form the square of the difference and the difference square compound fuzzy matter element Rη.

(3) Entropy method to determine the weight coefficient of the evaluation index

According to the definition of entropy, the entropy of m evaluation objects and n evaluation indicators is:

$$E(f_j) = -\frac{\sum_{j=1}^{m} f_{ij} \ln f_{ij}}{\ln m} \ (i = 1, 2..., n;\ j = 1, 2, ..., m) \quad (4)$$

$f_{ij} = \frac{u_{ij}}{\sum_{j=1}^{m} u_{ij}}$. Calculate the weight θ and entropy weight θ_j of the evaluation index:

$$\theta = (\theta_j)1 \times n \ ; \ \theta_j = \frac{1 - E(f_j)}{(n - \sum_{j=1}^{n} E(f_j)} \tag{5}$$

And satisfied $\sum_{j=1}^{n} \theta_j = 1$.

Through the above theories and mathematical models, the weight of each risk of the project is evaluated, and corresponding measures and methods are proposed to control the occurrence of risks, so as to ensure the realization of project goals.

3 Empirical Analysis

Taking a PPP characteristic town project as an example, the 25 identified risks will be scored by a questionnaire survey. Because the characteristic town project of the PPP model is a new model development project in recent years, I directed a total of 15 questionnaires to experts in the industry and related fields, and 12 of them were returned, with a recovery rate of 80%. The questionnaire has a total of 25 questions. Considering the three dimensions of macro-risk, meso-risk, and micro-risk, 12 experts are selected for evaluation. The experts score all the identified risk characteristics, and assign each grade standard as 5, 4, 3, 2, and 1, respectively. According to the expert score collection and processing, the composite fuzzy matter element is established, and the difference square fuzzy matter element is obtained through the standard fuzzy matter element according to step (1) and (2) (Table 1).

According to the scoring situation of 12 experts, SPSS software is used to measure the reliability of their scoring situation, and the results are as follows: Cronbach's Alpha = 0.753, Since the reliability coefficient is greater than 0.750, it shows that the measurement reliability of the questionnaire survey data has achieved a satisfactory degree of agreement.

Through the combination of the theory described in the previous article and the indicators to be evaluated in this article, an evaluation model can be constructed:

The first step is to establish a composite fuzzy matter element Rmn. T1 ~ T12 are used to represent the voting and scoring of 12 experts, and R1 ~ R25 respectively represent the 25 risk indicators in the risk evaluation of the characteristic town project under the PPP model.

The second step is to calculate the preferred degree of membership Uij. In view of the fact that all evaluation indicators in this paper are positive, the formula (2) is adopted in this paper to calculate the preferred degree of membership.

The third step is to construct the fuzzy matter-element $\hat{R}_{mn}$ of the preferential membership degree through the preferential membership degree. According to the maximum or minimum value of each index of $\hat{R}_{mn}$, a standard fuzzy matter element R0n can be constructed. Finally, the square of the difference between the standard fuzzy matter element R0n and the composite fuzzy matter element $\hat{R}_{mn}$ can be used

Table 1 List of the scoring of risk experts for characteristic town projects under the PPP model

	T1	T2	T3	T4	T5	T6
R1	2	4	3	3	4	2
R2	2	4	1	2	4	2
R3	4	5	2	2	3	4
R4	3	3	2	2	3	2
R5	2	2	1	2	3	2
R6	2	2	2	2	3	2
R7	3	3	3	2	4	4
R8	1	2	2	2	1	1
R9	1	2	3	2	1	2
R10	3	2	2	3	3	3
R11	5	2	2	5	3	2
R12	1	4	2	4	4	5
R13	3	3	2	1	3	2
R14	2	2	2	1	1	2
R15	3	2	2	1	1	2
R16	5	2	3	2	3	4
R17	3	3	2	5	1	4
R18	2	2	1	5	3	3
R19	3	1	3	5	3	3
R20	3	1	3	5	3	3
R21	3	2	2	5	3	3
R22	3	4	2	5	3	3
R23	3	4	2	3	1	3
R24	3	4	3	3	3	3
R25	2	1	2	3	3	3
	T7	T8	T9	T10	T11	T12
R1	5	2	5	4	2	4
R2	2	3	4	5	5	4
R3	3	4	4	3	3	4
R4	2	3	3	2	4	3
R5	1	2	2	2	3	1
R6	2	2	2	2	3	1
R7	4	3	3	3	5	5
R8	2	2	2	3	2	2
R9	2	3	2	3	3	2
R10	4	2	2	4	4	4

(continued)

Table 1 (continued)

R11	4	2	2	2	3	3
R12	2	4	2	3	1	2
R13	4	2	3	3	5	4
R14	3	2	2	2	2	2
R15	3	2	2	3	3	3
R16	3	1	2	5	4	5
R17	3	1	2	4	3	3
R18	4	2	2	4	5	5
R19	3	2	2	4	3	3
R20	3	2	2	3	3	4
R21	2	3	2	5	2	3
R22	4	3	5	4	5	5
R23	3	3	2	5	2	3
R24	3	3	3	5	4	4
R25	3	3	2	4	3	3

to form the difference squared composite fuzzy matter element $R\eta$, which is (Table 2).

The fourth step is to calculate the entropy value E(fj) and the weight θj respectively according to the following formula:

$$\begin{aligned} \text{E(fj)} = (&0.97730, 0.96490, 0.98671, 0.98912, 0.97669, 0.98842,\\ &0.98778, 0.98077, 0.97836, 0.98481, 0.97264, 0.95090, 0.96756,\\ &0.97989, 0.97151, 0.96001, 0.95890, 0.95642, 0.97144, 0.97144,\\ &0.97065, 0.98020, 0.96910, 0.98740, 0.97661). \end{aligned}$$

$$\begin{aligned} \theta \text{j} = (&0.03539, 0.05480, 0.02075, 0.01669, 0.03640, 0.01807,\\ &0.01908, 0.03002, 0.03379, 0.02372, 0.04272, 0.07665, 0.05065,\\ &0.03139, 0.04448, 0.06245, 0.06416, 0.06805, 0.04459, 0.04459,\\ &0.04582, 0.03091, 0.04831, 0.01970, 0.03652). \end{aligned}$$

Judging from the weight distribution of the above 25 evaluation indicators, the top fifteen elements in order are: R12, R18, R17, R16, R2, R13, R23, R21, R19, R20, R15, R11, R25, R5, R1. Among them, 5 indicators are macro risk indicators, 7 are meso risk indicators, and 3 are micro risk indicators. It shows that the degree of influence of the risk factors of the project is mainly concentrated on the intermediate risk; the weights of the last ten indicators are R9, R14, R22, R8, R10, R3, R24, R7, R6, R4 in order. There are 7 macro risks, 1 meso risk, and 2 micro risks. It shows that the macro risk of the project has a relatively small impact on the project, because

Table 2 Composite fuzzy matter-element of characteristic town project risk difference squared under PPP mode

	T1	T2	T3	T4	T5	T6
R1	0.36	0.04	0.16	0.16	0.04	0.36
R2	0.36	0.04	0.64	0.36	0.04	0.36
R3	0.04	0	0.36	0.36	0.16	0.04
R4	0.16	0.16	0.36	0.36	0.16	0.36
R5	0.36	0.36	0.64	0.36	0.16	0.36
R6	0.36	0.36	0.36	0.36	0.16	0.36
R7	0.16	0.16	0.16	0.36	0.04	0.04
R8	0.64	0.36	0.36	0.36	0.64	0.64
R9	0.64	0.36	0.16	0.36	0.64	0.36
R10	0.16	0.36	0.36	0.16	0.16	0.16
R11	0	0.36	0.36	0	0.16	0.36
R12	0.64	0.04	0.36	0.04	0.04	0
R13	0.16	0.16	0.36	0.64	0.16	0.36
R14	0.36	0.36	0.36	0.64	0.64	0.36
R15	0.16	0.36	0.36	0.64	0.64	0.36
R16	0	0.36	0.16	0.36	0.16	0.04
R17	0.16	0.16	0.36	0	0.64	0.04
R18	0.36	0.36	0.64	0	0.16	0.16
R19	0.16	0.64	0.16	0	0.16	0.16
R20	0.16	0.64	0.16	0	0.16	0.16
R21	0.16	0.36	0.36	0	0.16	0.16
R22	0.16	0.04	0.36	0	0.16	0.16
R23	0.16	0.04	0.36	0.16	0.64	0.16
R24	0.16	0.04	0.16	0.16	0.16	0.16
R25	0.36	0.64	0.36	0.16	0.16	0.16
	T7	T8	T9	T10	T11	T12
R1	0	0.36	0	0.04	0.36	0.04
R2	0.36	0.16	0.04	0	0	0.04
R3	0.16	0.04	0.04	0.16	0.16	0.04
R4	0.36	0.16	0.16	0.36	0.04	0.16
R5	0.64	0.36	0.36	0.36	0.16	0.64
R6	0.36	0.36	0.36	0.36	0.16	0.64
R7	0.04	0.16	0.16	0.16	0	0
R8	0.36	0.36	0.36	0.16	0.36	0.36
R9	0.36	0.16	0.36	0.16	0.16	0.36

(continued)

Table 2 (continued)

R10	0.04	0.36	0.36	0.04	0.04	0.04
R11	0.04	0.36	0.36	0.36	0.16	0.16
R12	0.36	0.04	0.36	0.16	0.64	0.36
R13	0.04	0.36	0.16	0.16	0	0.04
R14	0.16	0.36	0.36	0.36	0.36	0.36
R15	0.16	0.36	0.36	0.16	0.16	0.16
R16	0.16	0.64	0.36	0	0.04	0
R17	0.16	0.64	0.36	0.04	0.16	0.16
R18	0.04	0.36	0.36	0.04	0	0
R19	0.36	0.36	0.36	0.04	0.16	0.16
R20	0.16	0.36	0.36	0.04	0.16	0.16
R21	0.36	0.16	0.36	0	0.36	0.16
R22	0.04	0.16	0	0.04	0	0
R23	0.16	0.16	0.36	0	0.36	0.16
R24	0.16	0.16	0.16	0	0.04	0.04
R25	0.16	0.16	0.36	0.04	0.16	0.16

the characteristic town project of the PPP model is a state-supported project, and its development in the macro environment is more in line with the interests of all parties. Certain control measures can be taken for the risk factors corresponding to the risk weight.

Adopt a proactive risk response strategy, adopt dynamic control methods, and use dynamic control procedures to divide risk control into three stages, including preparation, dynamic tracking and control, and target adjustment. Combine precautionary measures, monitoring and control during the incident, and exemption measures after the incident.

4 Concluding Remarks

This paper uses the entropy fuzzy matter-element evaluation model to evaluate the risks of characteristic town projects in the PPP model. With a combination of qualitative and quantitative analysis, the risk identification and evaluation of PPP projects are carried out scientifically. The department provides a more scientific risk assessment tool. Make full use of the "two hands" of the government and the market to pinpoint the risk points of the industry and promote the characteristic towns to embark on a "track of rational development." However, this article still has shortcomings in the establishment of risk evaluation index system, which needs to be further improved.

References

1. Li, B.: Risk management of construction. Ph.D. thesis, Public private partnership projects (2003)
2. Qi, M., Wang, M.: Research on the evaluation of regional sustainable innovation ability based on fuzzy matter element—taking jiangsu province as an Example[J]. J. Chang. Univ. (Soc. Sci. Ed.) **15**(03), 33–36 (2014)
3. Chen, L.: On the theory and practice innovation of characteristic town construction[J] **2017**(10), 150–152 (2017)
4. Fei, P.: PPP project risk management research [D]. Anhui Jianzhu University, Hefei (2016)

A Study on the Early Warning Index System of Road Traffic Risks Under Extreme Weather Conditions

Yuepeng Cui and Zijian Liu

Abstract Extreme weather frequently occurs in today's society; it is easy to produce natural disasters like snow, water, or ice accumulation in the development of urban construction, which directly affects the regular road traffic operation. In recent years, extreme weather conditions by monitoring data analysis indicate that the use of case history, statistics, or methods like fault number analysis enables an accurate identification of extreme weather and addresses hidden troubles in traffic. This paper establishes the corresponding index system of risk prediction and puts forward effective countermeasures to guarantee urban residents regular travel.

Keywords Extreme weather · Risk warning · Cloud model classification · Fuzzy comprehensive evaluation

1 Introduction

With the increase of extreme weather events in urban construction and development, urban traffic problems have become more complex. Therefore, it is essential to accurately identify hidden traffic dangers under extreme weather conditions and propose risk warning indicators on road traffic based on accumulated experience. Regarding the current identification of road traffic hazards, it follows specific scientific principles. To accurately grasp the hidden dangers of road traffic under extreme weather conditions, it is vital to comprehensively identify the hidden dangers from various aspects such as road conditions and management work to propose effective treatment schemes [1]. Principles on operability must also be considered. The identification and management of hidden dangers should determine the appropriate identification mode based on the existing characteristics of the road traffic industry, risk warning index, and construction system, thus playing an active role in practice. There should also be an emphasis toward continuously improving the identification scheme and early warning index systems in practice [2]. Road traffic hazards should

Y. Cui · Z. Liu (✉)
Department of Transportation, Fujian University of Technology, Fuzhou 350118, China
e-mail: lzj3082@hotmail.com

H. Dai (ed.), *Computational and Experimental Simulations in Engineering*, Mechanisms and Machine Science 119,
https://doi.org/10.1007/978-3-031-02097-1_29

further abide by systematic principles. Under extreme weather conditions, traffic hidden dangers directly correlate with overall system operation, as a problem may be directly or indirectly related to multiple factors. Thus, a systematic and comprehensive anatomy should be conducted during an identification study to avoid over-formalizing warning indicators [3]. Finally, principles on comprehensiveness must be given significance. Road traffic risks and hidden dangers result from the comprehensive influence of various internal system factors [4]. Hence, during identification, various factors should be integrated and studied, emphasizing on the fundamental identification and analysis to improve traffic safety.

With the continuous improvement of urban construction and development in China, road traffic safety problems have become increasingly significant. Transportation research has always been critical to building a perfect early-warning index system for road traffic risks considering extreme weather conditions. Dobromirov et al. put forward a traffic safety evaluation of the St. Petersburg ring road by means of road traffic accident data accumulated by practical development [5]. Batrakova et al. clarified the influence of road conditions on road traffic safety in their practical research [6]. Li et al. applied an artificial neural network model to comprehensively evaluate various road safety factors [7]. Ma et al. study of evaluation trends and methods in regional road traffic safety implemented a system-based research; the outcomes indicated that the evaluation research on road traffic safety focuses on influence factors, and not on extreme weather conditions put forward by a systematic safety evaluation index system [8]. Based on the understanding of the existing social and economic development trends in various parts of China, this paper determines that the regional characteristics of road safety are highly apparent. With this phenomenon, this paper proposes a systematic study of risk warning index systems on road traffic under extreme weather conditions using a cloud model based on fuzzy comprehensive evaluation method.

2 Method

2.1 System Construction

There are several influence factors of road traffic safety. This paper primarily begins with extreme weather conditions and selects a comprehensive and straightforward evaluation index to construct an index system. According to existing safety cases, this system can show the area of road traffic risks and a hidden danger index to clear possible road traffic accident factors during operation. For instance, the various influence factors of extreme weather conditions in the evaluation index of traffic hidden dangers involve several types of safety accidents, including sandstorms, adverse weather, and road icing, which not only impact drivers' psychological factors but also increase the probability of traffic accidents during road transportation. Figure 1 illustrates the specific indicator system diagram.

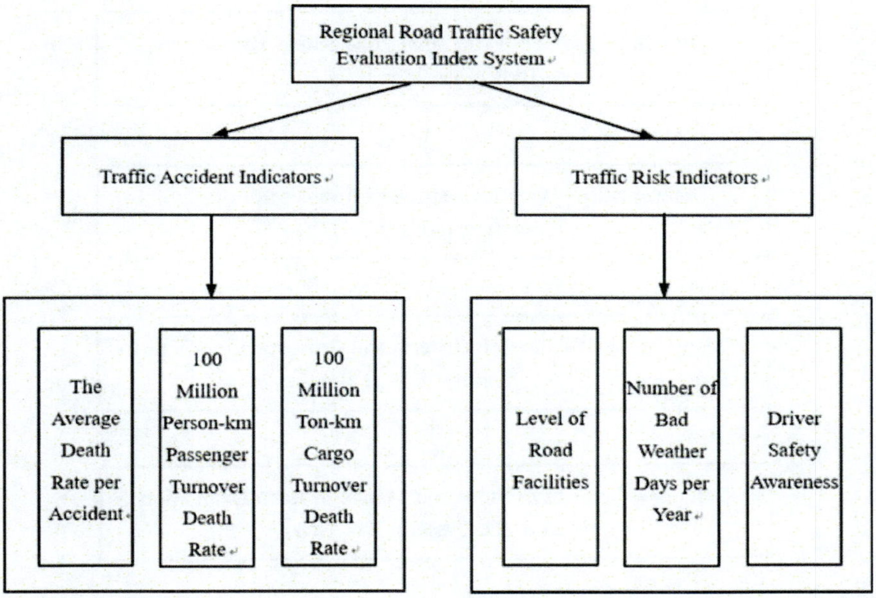

Fig. 1 A structure diagram of the index system of traffic safety evaluation on regional roads

2.2 *Cloud Model: Fuzzy Comprehensive Evaluation Method*

Fuzzy comprehensive evaluation method is an analytic hierarchy process (ahp) and fuzzy mathematics method. The core of its comprehensive utilization can grasp the evaluation index and its weight and membership degree based on the corresponding evaluation matrix. The matrix and the weight of evaluation index vector fuzzy operators and normalized processing can clear the results of fuzzy comprehensive evaluation eventually. The evaluation objects selected by this evaluation method has unique evaluation results. The set of the evaluated objects does not impact the final evaluation result, while the winning target is selected by sorting the results from good to bad based on the obtained results. Figure 2 depicts the operation process involving the fuzzy comprehensive evaluation method to evaluate road traffic safety in the region, conforming to the condition i, j = 1, 2, 3… N [9].

Cloud models consider ambiguous transitions between qualitative concepts and quantitative values. Suppose that X represents any set and A represents the fuzzy set of X, and there is a random value with a stable tendency relative to all elements of set X. In this case, X is the membership degree of A. X belongs to the base variable condition, and its elements are ordered. If the elements in set X are not simply ordered, a specific rule F should be used to map X to an ordered set X' with a unique correspondence. Hence, X' can be regarded as the essential variable, and the distribution of membership degree on X' is called a membership cloud. The numerical characteristics of the membership cloud involve expected Ex, entropy En, and super

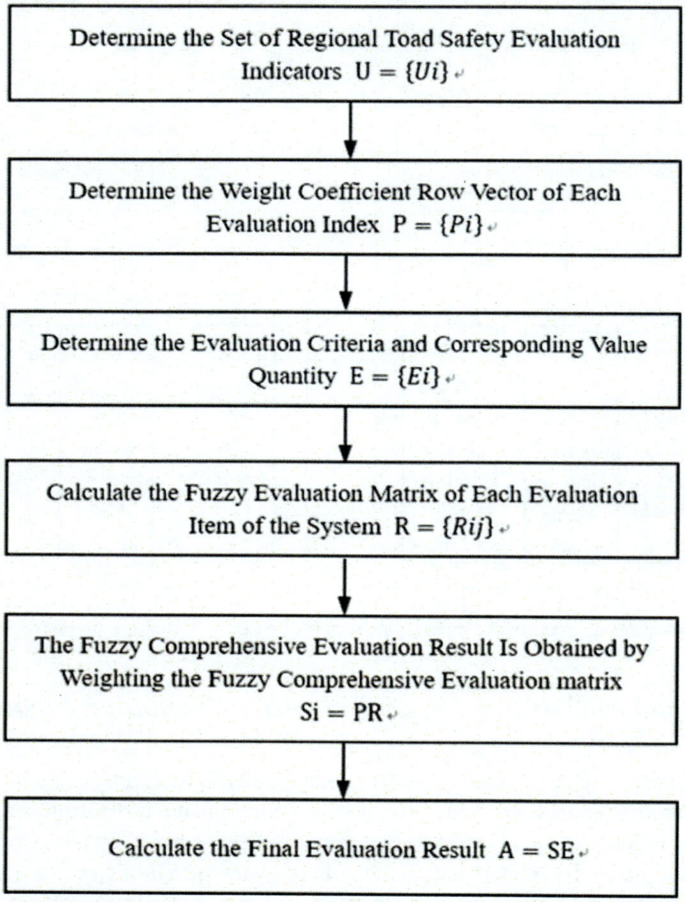

Fig. 2 Operation flowchart of safety evaluation using the fuzzy comprehensive evaluation method

entropy He. Ex is expected to represent typical sample data, while En is expected to denote the random probability of cloud droplets. With the gradual increase in entropy, fundamental fuzziness and randomness also increase. Super entropy He signifies the dispersion of cloud droplets. With the increase of this value, the dispersion of cloud droplets become larger. Figure 3 illustrates the operation flow of the cloud model, where X represents the clarity of the qualitative concept Y, and N is the number of cloud droplets generated. The specific calculation formula (1) is described as follows to present the relevant contents of qualitative and quantitative transformation [10]:

$$Y = e\frac{-(x - E_x)^2}{2(E_n')^2} \tag{1}$$

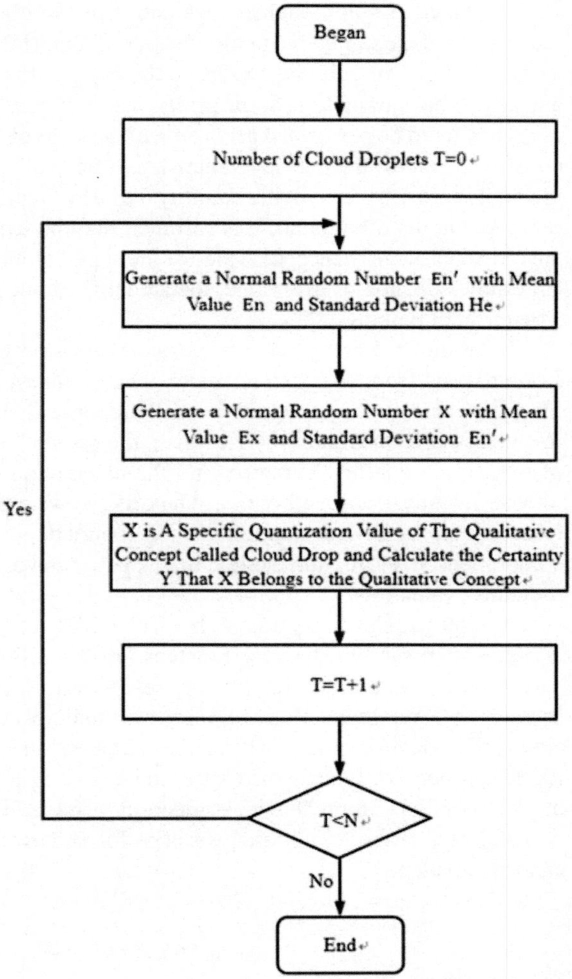

Fig. 3 Operation flowchart of cloud model algorithm

The cloud model-fuzzy comprehensive evaluation method is an evaluation method that integrates the fuzzy comprehensive evaluation method and a cloud model. Specifically, it uses the cloud model to replace the membership function and calculate the corresponding weight coefficient and evaluation matrix. Based on this method, the weight coefficient matrix P and the evaluation matrix R are calculated by reverse cloud generator, represented by cloud parameters. Eventually, the final comprehensive evaluation result is acquired by combining the fuzzy synthesis operator. The corresponding calculation formula (2) as follows:

$$B = P \cdot R = (E_x E_n H_e) \tag{2}$$

The objective, comprehensive evaluation results obtained by this method can accurately study their subjective randomness, ranking, stability, and other characteristics. They are sorted in order according to the expected parameters, entropy parameters, and ultrasonic parameters. Some principles should be observed. On the one hand, the ranking should be performed according to the sizes of the expected values. Assuming consistent values, the entropy value should be evaluated. As the entropy value gets lower, it is proved that a higher stability of the evaluation results entails a better actual ranking. On the other hand, it is assumed that the expected parameter and quotient are only consistent; hence, the value of the super quotient must be evaluated. With the continuous decline of this value, the actual randomness becomes smaller, enabling a better final ranking.

The traditional fuzzy comprehensive evaluation method performs only conversion analysis from fuzzy to accurate. The final result is only a single fixed value for the evaluation target, which is challenging to depict the fuzziness of the fundamental research and analysis. However, the research and analysis on the generation of the weight coefficient matrix and the calculation of the evaluation matrix have intense fuzziness and ambiguity, which is likewise the content that the traditional fuzzy comprehensive evaluation method cannot depict. Thus, the cloud model-fuzzy comprehensive evaluation method in this paper can effectively solve the problems in previous evaluations.

The cloud model-fuzzy comprehensive evaluation method evaluates and analyzes regional road safety. The specific steps are the following. First, the specific evaluation index system and its weight are defined. Second, the evaluation set is put forward in a reference element, and the standard cloud model forms a reference cloud of evaluation grade. Third, the membership cloud model of the evaluation index is proposed. Fourth, the expert score is integrated into the membership cloud model to obtain comprehensive evaluation results. Fifth, the comprehensive evaluation of digital characteristic analysis is performed to obtain the final cloud evaluation analysis results.

3 Analysis Results

Applying the proposed assessment scheme of a comprehensive regional evaluation of road traffic safety operations according to personal knowledge and the statistical yearbook 2016, and calculating various kinds of evaluation index data, the weight coefficient, and index evaluation level, the nine experts on the indicators of the evaluation grades were analyzed. Table 1 presents the results.

Instead of selecting based on cloud model using the membership function of a fuzzy comprehensive evaluation, the points are assigned into a membership cloud model, using Normrnd functions to constitute a normal distribution random number and clear expected initial data cloud parameters. Based on the flowchart of the cloud model algorithm for an integrated evaluation of the areas. Table 2 provides the final results.

Table 1 Analysis of expert scoring results

The evaluation index	The weight coefficient of evaluation index	The number of experts/people scoring				
		0.9	0.7	0.5	0.3	0.1
Average death rate per accident	0.15	0	3	5	1	0
100 million person-km passenger turnover mortality rate	0.10	2	5	2	0	0
100 million ton-km cargo turnover mortality rate	0.10	0	3	6	0	0
Level of road facilities	0.20	0	4	3	2	0
Number of bad weather days per year	0.20	1	4	2	2	0
Driver safety awareness	0.25	1	2	5	1	0

Table 2 Expected initial values of cloud parameters

The evaluation index	Weighted cloud parameter	Evaluate cloud parameters
Average death rate per accident	(0.17 0.06 0.003)	(0.11 0.12 0.002)
100 million person-km passenger turnover mortality rate	(0.17 0.06 0.003)	(0.14 0.15 0.004)
100 million ton-km cargo turnover mortality rate	(0.17 0.06 0.003)	(0.11 0.14 0.003)
Level of road facilities	(0.17 0.06 0.003)	(0.11 0.12 0.006)
Number of bad weather days per year	(0.17 0.06 0.003)	(0.12 0.10 0.001)
Driver safety awareness	(0.17 0.06 0.003)	(0.11 0.10 0.007)

Based on the formula analysis of the final comprehensive evaluation results calculated by fuzzy synthesis algorithm, the eventual results of the fuzzy comprehensive evaluation of the cloud model on road traffic safety in the examined region are:

$$B_1 = P \cdot R = \left(0.119\ 0.044\ 0.001\right)$$

Meanwhile, according to the primary data mastered by statistical analysis, the comprehensive safety evaluation results of a region can be determined, as shown in Table 3, containing the ranking rules of the three qualities. Finally, according to the data, the local road traffic safety situations are sorted from good to bad, which can master the scope of the urban areas with safety risks and proposed effective prevention countermeasures for common extreme weather conditions to ensure drivers' safety.

Table 3 Final results of road traffic safety evaluation in the region

Comprehensive evaluation results		
(0.119	0.044	0.001)
(0.231	0.142	0.011)
(0.148	0.115	0.006)
(0.224	0.042	0.004)
(0.125	0.016	0.017)
(0.212	0.011	0.002)
(0.212	0.022	0.013)
(0.241	0.013	0.002)
(0.082	0.014	0.001)
(0.163	0.085	0.010)
(0.163	0.085	0.004)

4 Conclusion

To sum up, according to the above research and analysis to establish an evaluation index system on regional road traffic safety, and by using a cloud model and the fuzzy comprehensive judgment method to conduct the road running situation of a regional comprehensive evaluation, an accurate sorting system is determined given the existing security risks, putting forward effective countermeasures in the comprehensive assessment. For future construction and development, amid increasing extreme weather conditions and their adverse impact on urban transportation, construction, and development, management personnel to clear extreme weather conditions may cause hidden traffic troubles after scale and quantity, alongside reasonable use of recognition with analysis. Different extreme weather conditions put forward various effective protection countermeasures, which is the focus of current research. Based on scientific management countermeasures, risk warnings for different types of extreme weather are helpful to effectively control road traffic risks caused by a complex interaction between extreme weather and traffic and avoid unnecessary casualties and property. Thus, in urban construction reformation, it is necessary to strengthen the training of professionals and to conduct in-depth research on road traffic risk warnings under extreme weather conditions from all perspectives based on the accumulated experience of practical explorations, to ensure the safety and stability of urban road transportation.

Acknowledgements This material is based upon work supported by Natural Science Foundation of Fujian Province (Grant No.2020J05194). Any opinions, findings, and conclusions or recommendations expressed in this material are those of the authors and do not necessarily reflect the views of the sponsor.

References

1. Sui, L., Chen, Z., Ni, S., Wang, L., Li, W.: Research on the methods of identifying traffic hidden dangerous points and guarantee countermeasures under extreme weather conditions. Highway Transp. Sci. Technol (Applied Technology Edition) **11**(09):232–234+269 (2015) (In Chinese)
2. Xie, Q., Gao, J., Wang, F., Ma, W., Yuan, T.: Research on regional power grid security early warning strategy considering the influence of extreme meteorological conditions. Electr. Appl. **37**(23), 39–45 (2018) (In Chinese)
3. Li, C., Zhang, G., Tang, J.: Research on the early warning and management system of meteorological disasters in expressway traffic. Road Traffic Saf. **03**, 16–19 (2008) (In Chinese)
4. Wang, X., Liu, D., Chen, Q., Zhao, J.: Study on the early warning index system of urban road traffic safety. Highway Automob. Transp. **02**, 48–51 (2010) (In Chinese)
5. Dobromirov, V., Evtiukov, S., Duncheva, E., et al.: Methodology and Results of the traffic safety evaluation on the Saint Petersburg Ring Road. Transp. Res. Procedia **20**, 151–158 (2017)
6. Batrakova, A., Gredasova, O.: Influence of road conditions on traffic safety. Procedia Eng. **134**, 196–204 (2016)
7. Li, X., Tian, P., Jiang, G.: Artificial neural network method for comprehensive evaluation of road traffic safety. J. Southwest Jiaotong Univ. **04**, 496–500 (2006)
8. Ma, S.: Theories and Methods of Regional Road Traffic Safety Evaluation. Beijing Jiaotong University (2012)
9. Li, J.: Application of power grid risk early warning system based on meteorological information. Guangxi Electr. Power **36**(05), 25–27 (2013) (In Chinese)
10. Hempel, U., Auge, J., Schütze, M., et al.: Sensor-actuator-based network for an early-warning system in extreme weather conditions. IFAC Proceed. **43**(23), 7–12 (2010)

Research on the Digital Image Processing Method Based on Parallel Computing

Zhen Kong

Abstract In the continuous innovation of science and technology, most scientific problems need to deal with more data information, and the actual processing speed and quality directly determine the problem-solving process. Although the current single core processing technology got rapid development both at home and abroad, but still can't meet the demand of scientific problem, and the theory of parallel computing technology effectively solve the above problems, in the practical use of Chinese super cool platform can achieve the processing capacity of one hundred million times per second, and with the innovation of practical technology is still in constant breakthroughs. Therefore, based on the understanding of digital image processing methods and research trends, according to the concepts related to parallel computing and unified computing equipment architecture, this paper deeply discusses the core image processing technology methods and experimental results, thus proving the application value of parallel computing technology.

Keywords Parallel computing · Unified computing equipment architecture · Digital · Image processing · Segmentation

1 Introduction

In order to optimize the processing efficiency of relevant data and speed up the research process of scientific problems, the method of multi-node simultaneous computation has been proposed in the development of scientific research. From the perspective of overall development, the processing mode with parallel computing as the core can improve the computational efficiency of the problem on the one hand. It takes a week for a single core processor to deal with an application problem requiring too much computing data, which will inevitably affect the overall solution process and hinder the efficiency of actual problem solving. Therefore, in order to improve the solving speed of scientific problems, the research results of parallel computing

Z. Kong (✉)
China Information Consulting & Designing Institute Co., Ltd., Nanjing, China
e-mail: 18686004285@163.com

H. Dai (ed.), *Computational and Experimental Simulations in Engineering*, Mechanisms and Machine Science 119,
https://doi.org/10.1007/978-3-031-02097-1_30

should be used to divide into several small problems that do not affect each other in the multi-node simultaneous collaborative parallel computing platform. Therefore, parallel computing can comprehensively improve the solving speed of problems. On the other hand, the computational scale of the problem can be enlarged. Processing large-scale data simulation computing process on a single core processor is bound to be limited by system resources, and only a small part of computing operations can be completed in a short time, so it is difficult to meet the needs of large-scale data processing in time or space, thus hindering the research results of scientific problems. The application of parallel computing technology can provide help for computing requirements. With the continuous improvement of social economy and scientific and technological level, the research on parallel computing technology is more and more in-depth at home and abroad. A variety of computing models have been proposed and applied accurately in the field of scientific life. For example, CUDA proposed by Nvidia in the early twenty-first century, as a universal parallel computing programming model with CPU stream processor as the core, not only optimizes data processing level, but also ensures scientific and effective actual computing. As a new content to meet the requirements of image processing, GPU in the traditional sense is mainly used for processing rendering or mapping, but has not fully demonstrated its advantages in concurrent processing. After CUDA is promoted, it not only improves the graphics processing level of previous Gpus, but also improves the programming efficiency of graphics and images, and provides a high-quality platform for the application of practical parallel computing technology. Nowadays, with the continuous optimization of CUDA technology, researchers have accumulated a lot of experience in practical exploration, and believe that it will definitely play a positive role in the development of heterogeneous computing model in future research and application [1–3].

2 Methods

2.1 Parallel Computing

Simple speaking, parallel computing is only in contains multiple computing nodes or computing core processing system, the application will be too big a task in a certain strategy is divided into multiple independent and contact smaller subsystems, and then assigned to the contained or processor core computing nodes, prompting them to complete the task in the cooperation. This can use the best way to improve the efficiency of problem calculation, and expand the actual problem calculation scale. In order to ensure the parallel computation model in this paper can be effectively used, the build processing system should satisfy the following conditions: first of all have two or more processors, and have the same position in the internal system, this helps to submit task processor in high speed connection, finally carried out large-scale computing tasks. Secondly, it is necessary to ensure the parallelism of tasks, only in this way can the application advantages of parallel computing technology be fully demonstrated. Finally, the parallel programming environment should be optimized to facilitate the organization and management of parallel algorithms, so as to efficiently process parallel tasks. At the same time, the parallel computing environment is developed and utilized on Windows platform, and the common programming models are Open MP, MPI, PVM and so on [4, 5].

2.2 *Cuda*

As a practical development, the new architecture of universal parallel computing can ensure that GPU has high-quality hardware facilities and software environment, enhance GPU's performance in mapping and graphics rendering, and guide developers to break through the limitations of graphics API and directly use GPU for computing unit management.

From a practical point of view, this architecture program structure consists of two parts, one is the HOST side code, the other is the DEVICE side code. The former needs to complete the task on the CPU, mainly for programmatic initial work, such as task division, storage display data, etc. The latter needs to perform tasks on the GRAPHICS processor, and finally complete the computational work efficiently in parallel. In the compilation process, CUDA compiler developed and designed by NVIDIA is scientifically divided into two parts. The code on the host side is compiled using the standard C+ language, and compiled and analyzed according to the C compiler installed on the host. Finally, the code is serialized on the CPU. The device side code needs to use the extensible C+ language to write Kernel functions, and use NVCC compiler to compile, and finally execute in THE SP stream of GPU.

2.3 *CUDA-BASED Image Segmentation*

Image processing technology is divided into two types according to the difference of information garden, one is the processing of analog image, the other is the processing of exponential word image. The former is the signal processing of analog images, which involves electronic processing and optical processing in two ways, such as remote sensing images and cameras are processed using the optical principle of glass lenses, and TELEVISION signals belong to electronic processing. The latter is to use the computer to process the image information with the two-dimensional array as the core, the array unit is pixel point. This processing technology has the advantages of precision and flexibility in operation, and can effectively deal with overly complex nonlinear images. So far, more high-quality processing algorithms have been developed.

This paper mainly analyzes the IMAGE segmentation technology based on CUDA. Simply speaking, the image segmentation technology is to divide the information contained in the digital image into multiple independent sub-regions according to a certain strategy, and extract the target with meaning in these contents. This technique is part of the important process of analyzing and identifying image information. Prewitt operator as the use of the first-order differential image edge detection algorithm, need to compute analysis in the field of pixel gray numerical difference between adjacent pixels, and use it as a vertical gradient, around the nearby pixel gray value differential can be regarded as a horizontal gradient, then calculate the gradient in the direction of two distance, the image edge [6–8].

3 Result Analysis

Because Prewitt operator needs to carry out convolution operation on the field and template of pixels, the actual calculation amount is large, so the output result of pixels does not affect the subsequent algorithm execution, and it does not need to involve the output result of already calculated pixels, so it is very suitable for the parallel algorithm requirements of CUDA. The following is an in-depth study of the Prewitt operator parallel algorithm with different processing strategies, so as to compare and analyze the application performance of CUDA parallel computing in digital image processing.

First, process the image line by line. This kind of algorithm needs to complete the initialization of the data on the host first, and then transfer the data image information to the host for storage. Apply for two Spaces in the device memory with the same capacity as the image data, one of which is used to store the data copy and the other is used to store the output results. Finally, the Kernel function on the device side is used to transfer the processing process to the device side. The corresponding pseudo-codes of C language are shown in Table 1 below [9, 10]:

Table 1 C language pseudocode

The host loads the input image data into the host memory
Unsigned Char*d-pImg Data;
Unsigned Char*d-pImg Data Out;
Cutil Safe Call(CudaMalloc((Void**)&d-pImg Data,nLineByte*nLmgHeight));
Cutil Safe Call(CudaMalloc((Void**)&d-pImg Data Out,nLineByte*nLmgHeight));
Cutil Safe Call(CudaMemcpy(d-pImg Data,pImgData,nLineByte*nLmgHeight));
CudaMemcpy Host To Device));
①Dim3 DimBlock = 256
②Dim3 DimGrid = (((nImgHeight-2) + DimBlock.X-1)/dimBlock.X);
Kemel <<< dimGrid,dimBlock >>> (nLmgWidth,nImgHeight,nPixelByte,nLineByte,
D-pImgData,d-pImgDataOut);

Table 2 Calculate the average result of the analysis execution time after 10 calculations

The resolution of the	Serial algorithm (MS)	Parallel algorithm (MS)	Speed up than
320*240	7.8	8.80	0.89
640*480	20.1	22.47	0.89
800*600	30.8	35.70	0.86
1024*768	49.2	67.47	0.73
3200*2400	492.5	975.99	0.50
6400*4800	1932.3	4355.24	0.44

During the experimental analysis, the average value of analysis execution time was calculated after repeated execution of Prewitt operator for 10 times for 24 bitmap images with different resolutions, and the results can be obtained as shown in Table 2.

It can be seen from this study that the processing effects of parallel algorithm and serial algorithm are basically the same, but when processing images with the same resolution, the execution time of parallel algorithm is longer, and the acceleration ratio is less than 1. With the continuous increase of image resolution, this gap will be bigger and bigger. It is proved that the parallel algorithm selected at this time has no acceleration effect.

First, optimize the processing by row. This kind of algorithm optimization requires processing pixels per thread, and the host side allocates ready-to-use images based on the number of lines contained in the image. In other words, the host side code does not change. The corresponding Kernel functions are shown in Table 3 below:

Combined with the analysis of the results shown in Table 4 below, it can be seen that the shared memory optimization program used in this study is effective. The acceleration ratio of images with different resolutions can reach up to 2.68, which is three times higher than the original performance and achieves the desired optimization effect.

Table 3 Kernel function representation diagram

_Global_Void Kemel(Int nImg Width,Int nImgHeight,Int nPixelByte,Int nLineByte,Unsigned
Char*d_pImgData,Unsigned Char*d_pImgDataOut){
_Shared_Unsigned Char SharePixels[256][3];
Int i,k,nRow;
nRow = BlockIdx.x*(BlockDim.x-2) + ThreadIdx.x;
If(nRow < nImgHeight){
For(i = 1;i < nImgWidth-1;i ++){
For(k = 0;k < nPixelByte;k ++){
SharePixels[ThreadIdx.x][0] = *(d-pImgData + nRow*nLineByte + (i-1)*nPixelByte + k);

Table 4 Comparison results

The resolution of the	Serial algorithm (MS)	Parallel algorithm (MS)	Speed up than
320*240	7.8	3.30	2.36
640*480	20.1	7.51	2.86
800*600	30.8	12.12	2.54
1024*768	49.2	22.18	2.22
3200*2400	492.5	250.66	1.96
6400*4800	1932.3	1003.02	1.93

Third, process the image by column. In C language programming, the sequence will be in accordance with the order of the navigation preferred store, and the digital image data is stored in the internal system according to the array manner, with a line of pixels is continuity of data calculation and analysis of press line processing scheme, multi-threaded concurrent execution acquired data is different, can't meet the demand of combined access. Even in shared memory, all threads read three rows at a time, but this approach is inefficient. Therefore, it is necessary to use the merge access global memory technology to allocate threads according to the number of image columns, and all threads are responsible for the data processing of a row of pixels. The final result is shown in Table 5.

Combined with the analysis of the above table, it can be seen that the performance of the case-by-row processing algorithm is higher than that of the line-by-row processing algorithm. For example, the acceleration ratio obtained by 320 × 240 small resolution image increases by 10 times, and as the actual resolution continues to increase, the acceleration ratio of the actual algorithm will also increase, and finally achieve the desired acceleration effect. At the same time, it is further proved that the parallel algorithm has unique advantages in processing large quantities of parallelized data. At this point, the code on the host side is shown in Table 6 below:

Table 5 All threads are responsible for the data processing result of a row of pixels

The resolution of the	Serial algorithm (MS)	Parallel algorithm (MS)	Speed up than
320*240	7.8	0.65	11.95
640*480	20.1	1.55	13.00
800*600	30.8	1.94	15.89
1024*768	49.2	2.34	21.07
3200*2400	492.5	17.63	27.94
6400*4800	1932.3	63.37	30.49

Table 6 Host-side code

Unsigned Char*d-pImg Data;
Unsigned Char*d-pImg Data Out;
Cutil Safe Call(CudaMalloc((Void**)&d-pImg Data,nLineByte*nLmgHeight));
Cutil Safe Call(CudaMalloc((Void**)&d-pImg Data Out,nLineByte*nLmgHeight));
Cutil Safe Call(CudaMemcpy(d-pImg Data,pImgData,nLineByte*nLmgHeight));
CudaMemcpy Host To Device));
Dim3 DimBlock = 256
Dim3 DimGrid = (((nImgWidth-2) + DimBlock.X-1)/dimBlock.X);
Kemel <<< dimGrid,dimBlock >>> (nLmgWidth,nImgHeight,nPixelByte,nLineByte,
D-pImgData,d-pImgDataOut);
Cutil Safe Call(CudaMemcpy(pImg Data Out,d-pImgData Out,nLineByte*nLmgHeight;

4 Conclusion

In summary, by understanding the basic classification of current image processing technologies in China, the edge detection algorithm and the basic principle of Prewitt operator, which are most frequently cited during image segmentation, are clarified. Then, the algorithm strategy and experimental results are compared and analyzed under CUDA universal parallel computing framework with GPU stream processor as the core. The final results show that Prewitt operator has a positive effect on edge detection of digital images, and the accelerometer can be obtained 10 times higher as long as the appropriate processing strategy is selected. At the same time, the array structure of digital image files can be fully understood to ensure the parallelism of large-scale data information contained in data processing, so parallel transformation and transplantation can be carried out in CUDA architecture, thus speeding up the efficiency of data calculation and analysis. In addition, it is necessary to strengthen the training of professionals during technology research and development, pay attention to grasp more application advantages of parallel computing technology in practical exploration, and then choose appropriate processing countermeasures according to the needs of digital image processing. Only in this way can the unique value of parallel computing be fully displayed in practical operation.

References

1. Tang, J.: Research on parallel algorithm for image processing in multiprocessor. Autom. Expo. **24**(006), 105–108 (2007)
2. Zhao, R., Liang, S.: Analysis of embedded image processing method based on parallel computing. Modern Property Manage. **10**, 80–81 (2012)
3. Zhang, R., Li, L.: Application of PVM based network parallel computing in remote sensing image processing. Comput. Simul. **20**(010), 55–56, 135 (2003)

4. Xiong, J., Liu, C.: Research on parallel image processing algorithm based on message passing interface. J. Chengdu Univ. Nat. Sci. **29**(002), 137–139 (2010)
5. Zhan, Z., Li, G., Zhang, X., et al.: Research on parallel image preprocessing based on CUDA. Mach. Electron. **7**, 64–67 (2014)
6. Liu, Q., Wang, Z.: Research progress of rock numerical simulation based on digital image processing. Chinese J. Rock Mech. Eng. **39**(S02), 11
7. Wei, X.: Research on fast algorithm of image processing based on expectation and variance extension. Sci. Technol. Wind **434**(30), 63–64 (2020)
8. Zhang, C., Yang, J.: Research on image processing algorithm based on GPU. J. Southwest Normal Univ. (Nat. Sci) **07**, 41–45 (2013)
9. Nasridinov, A., Lee, Y., Park, Y.H.: Decision tree construction on GPU: ubiquitous parallel computing approach. Computing **96**(5), 403–413 (2014)
10. Kobayashi, M., Toda, H., Kawai, Y., et al.: High-density three-dimensional mapping of internal strain by tracking microstructural features. Acta Materialia **56**(10), 2167–2181 (2008)

Author Index

H. Dai (ed.), *Computational and Experimental Simulations in Engineering*, Mechanisms and Machine Science 119,
https://doi.org/10.1007/978-3-031-02097-1